T0239923

HUMAN ECOLOGY

We face an environmental catastrophe of global proportions. The ecological rationality of modern society, and of science in particular, is in question. Science still responds to crises at the level of technocratic expertise, still treats society as an adoptive system.

By bringing together a number of integrative approaches to the human-environment problem, *Human Ecology* shapes a more radical, fundamental agenda for change. The book creates a framework for an anti-fragmentary discourse, for a 'new human ecology'. From the notion that the individual person is an agent mediating between society and environment, the individual contributors recognise that the environmental crisis is really a crisis of society – manifesting itself in an increasing fragmentation of lives in general and knowledge in particular. Arguing for environmentally sustainable lifestyles, the book envisages a new kind of consciousness and a new environment.

Dieter Steiner is Professor of Quantitative Geography and Human Ecology at the Swiss Federal Institute of Technology in Zürich. **Markus Nauser** is a Scientific Collaborator with Ecosens Ltd, an environmental management consultancy near Zürich.

HUMAN ECOLOGY

Fragments of anti-fragmentary
views of the world

Edited by
Dieter Steiner and
Markus Nauser

LONDON AND NEW YORK

First published 1993
by Routledge
2 Park Square, Milton Park, Abingdon, Oxon, OX14 4RN

Simultaneously published in the USA and Canada
by Routledge Inc.
711 Third Avenue, New York, NY 10017

Routledge is an imprint of the Taylor & Francis Group, an informa business

First issued in paperback 2016

Transferred to Digital Printing 2006

© 1993 Dieter Steiner and Markus Nauser

Typeset in 10/12pt Garamond by
Ponting–Green Publishing Services, Chesham, Bucks

British Library Cataloguing in Publication Data
A catalogue record for this book is available from
the British Library

Library of Congress Cataloging in Publication Data
Human ecology : fragments of anti-fragmentary
views of the world / edited by Dieter Steiner
and Markus Nauser.
p.   cm.
Includes bibliographical references and index.
ISBN 0–415–06777–4
1. Human ecology–Philosophy.   I. Steiner, Dieter.   II.
Nauser, Markus, 1957– .
GF21.H84   1993
304.2–dc20   92-33284
CIP

Publisher's Note
The publisher has gone to great lengths to ensure the
quality of this reprint but points out that some
imperfections in the original may be apparent

ISBN13: 978-0-415-06777-5 (hbk)
ISBN13: 978-1-138-00933-2 (pbk)

CONTENTS

v

CONTENTS

Part II   The implicit and the explicit

CONTENTS

CONTENTS

FIGURES

FIGURES

TABLES

CONTRIBUTORS

Gerhard Bahrenberg
Born in 1943, he has a PhD in geography from the University of Münster, Germany. Originally interested in physical geography, he has moved since to work in methodological issues in geography, quantitative methods and spatial modelling. His more recent research interest focuses on interrelations of society and environment. Formerly a scientist at the University of Duisburg, he has been Professor of Social and Economic Geography at the University of Bremen since 1975. He is a co-author of *Statistische Methoden in der Geographie* (2 vols, Stuttgart: Teubner 1990/92) and a co-editor of several volumes, including *Theorie und Quantitative Methodik in der Geographie* (Bremen: University of Bremen, 1984), *Recent Developments in Spatial Data Analysis* (Aldershot: Gower, 1984), *Zur Methodologie und Methodik in der Regionalforschung* (Osnabrück: University of Osnabrück, 1985), and *Geographie des Menschen. Dietrich Bartels zum Gedenken* (Bremen: University of Bremen, 1987).

Stephen Boyden
He holds a PhD in biology. He is a former researcher in immunology at the University of Cambridge in England, the Rockefeller Institute in New York, the Pasteur Institute in Paris, the Tuberculosis Immunization Research Centre in København and the John Curtin School at the Australian National University. Since 1965 he has been engaged in research and writing in human ecology and biohistory at the Australian National University, Canberra, where he held a position as a Professorial Fellow in Human Ecology at the Centre for Resource and Environmental Studies until his official retirement at the end of 1990. He was Director of the MAB Hong Kong Human Ecology Program (1972–77) and a UNESCO Consultant to the MAB Program (1974–89). Since 1988 he has been leader of the Fundamental Questions Program of the Centre for Resource and Environmental Studies. He has published several books, among them *Western Civilization in Biological Perspective: Patterns in Biohistory* (Oxford: Oxford University Press, 1987) and, as a co-author, *The Ecology of a City and its People: the Case of Hong*

Kong (Canberra: Australian National University Press, 1981) and *Our Biosphere Under Threat: Ecological Realities and Australia's Opportunities* (Melbourne: Oxford University Press, 1990).

Claudia Carello

She has a PhD in Experimental Psychology from the University of Connecticut, Storrs, Connecticut, USA. Her theoretical and experimental work is conducted within the framework of ecological psychology and concerns the general problem of characterizing the information that specifies environmental layout and events for a particular perceiver. Her research publications deal with the haptic perceptual system, visual guidance of coordinated movement, and dynamical constraints on pictorial information. With Claire Michaels she co-authored the book *Direct Perception* (New York: Prentice Hall, 1981). Since 1987 she has been Director of the Center for the Ecological Study of Perception and Action at the University of Connecticut.

Preparation of the manuscript reproduced in this volume was supported by National Science Foundation grant BNS 87–20144.

Gregor Dürrenberger

He holds a PhD in geography from the Swiss Federal Institute of Technology (ETH), Zürich. He is a researcher in the Human Ecology Group, Department of Geography, ETH, and lectures in the Environmental Sciences study program of this institution. In 1988 he was a visiting scientist at the Centre for Urban and Regional Development Studies (CURDS), University of Newcastle upon Tyne, UK. His main interests focus on the human dimensions of global change, specifically on climatic risks and socio-economic change, and on urban development in the light of the new information technology. He is author of *Menschliche Territorien. Geographische Aspekte der biologischen und kulturellen Evolution* (Zürich: Verlag der Fachvereine, 1989), a co-author of *Telearbeit – Von der Fiktion zur Innovation* (Zürich: Verlag der Fachvereine, 1987) and main author of *Das Dilemma der modernen Stadt. Theoretische Überlegungen zur Stadtentwicklung am Beispiel Zürichs* (Berlin: Springer, 1991).

Marek Dutkowski

Born in 1954, he has an MA degree and a PhD in geography from the University of Gdańsk, Poland. Since 1978 he has worked as Assistant and Assistant Professor in the Department of Economic Geography of the same university. His research and publications deal with regional policy and planning, and environmental policy and conflicts in Poland. From October 1988 to March 1990 he was a Research Fellow of the Alexander von Humboldt Foundation at the University of Bremen, Germany.

Huib Ernste

He was born in 1957 in the Netherlands. After having obtained his Master's degree in social and economic geography from the State University of Groningen in the Netherlands, he has worked as a researcher and lecturer at the Geography Department of the Swiss Federal Institute of Technology (ETH) in Zürich. He holds a PhD from this institution, based on a thesis on office location and new information technologies. Since 1987 his research has focused on flexible specialization, environmental consciousness and action and on the geography of trust relations. He teaches several courses on research methods. He is a co-author of *Das Dilemma der modernen Stadt. Theoretische Uberlegungen zur Stadtentwicklung am Beispiel Zürichs* (Berlin: Springer, 1991) and of *Umwelt zur Sprache bringen. Uber das umweltverant-wortliche Handeln, die Wahrnehmung der Waldsterbensdiskussion und den Umgang mit Unsicherheit in der Schweiz* (forthcoming). He is a co-editor of *Information Society and Spatial Structure* (London: Belhaven, 1989) and of *Regional Development and Contemporary Industrial Response: Extending Flexible Specialisation* (London: Belhaven, 1992).

Peter Gould

Born in 1932, is a geographer with doctorates from Northwestern University, Evanston, Illinois, USA and Strasbourg. He has maintained a long interest in spatio-temporal processes, starting with the diffusion of agricultural subsidies in Wisconsin, USA, the impact of transportation in Ghana, the spread of membership in a progressive farmers' club on Kilimanjaro, the diffusion of cotton cooperatives in Mwanza District, the development of modernization in Tanzania, the amplification and impact of radioactive fallout through physical, biological and human structures (see his book *Fire in the Rain – the Democratic Consequences of Chernobyl*, Cambridge, Polity Press, 1990), and the geographic diffusion of the AIDS epidemic modelled with spatial adaptive filtering and neural computing (see his book *The Slow Plague – A Geography of the AIDS Pandemic*, Oxford: Blackwell, in press). He is now tracing the geographical spread and intellectual impact of the 1755 Lisboa earthquake throughout Europe in the middle of the Enlightenment, which will form his fourteenth book. He gets bored easily.

Markus Huppenbauer

Born in 1958, he has studied philosophy and theology at the University of Zürich, Switzerland. He has been working on Luhmann's sociology of religion and more recently on the present interpretation of mythological thinking of the Antiquity. The latter was also a topic of his PhD work. Since 1986 he has been a scientific collaborator and, more recently, director of the *Evangelische Studiengemeinschaft an den Zürcher Hochschulen* in Zürich.

Ingela Josefson

She has a PhD in linguistics. Until 1979 she was a researcher at the University of Göteborg, Sweden. Since then she has been at the Swedish Center for

Working Life in Stockholm where she is engaged in research related to epistemological questions in connection with 'skill' and 'knowledge'. She is particularly concerned about the current trends, arising from the development of computers and expert systems, to change the professional training of hospital nurses to more theoretical knowledge. She is author of the books *From Apprentice to Master* (in Swedish; Lund: Studentlitteratur, 1988) and *Forms of Knowledge* (in Swedish; Stockholm: Carlssons förlag, 1991), editor of *Language and Experience* (in Swedish; Stockholm: Carlssons förlag, 1985), and a co-editor of *Skill and Artificial Intelligence* (Berlin: Springer, 1988).

Alfred Lang

Born in 1935, he is Professor of Psychology at the University of Bern, Switzerland. He received his academic education in Switzerland and Canada. His interests have focused on personality and infant development, including problems of perception and action. In the early seventies he started to concentrate on person–environment relations, particulary the psychology of residential activity. He has published extensively on such topics as infant perception, time perception, absolute pitch, the field theory of Kurt Lewin, human–computer interfaces, psychological ecology, residential activity and the psychology of things. His major publications include 'Uber zwei Teilsysteme der Persönlichkeit: Beitrag zur psychologischen Theorie und Diagnostik' (Berne: Huber 1964). 'Ist Psychodiagnostik verantwortbar? Wissenschaftler und Praktiker diskutieren Anspruch, Möglichkeiten und Grenzen psychologischer Erfassungsmitten' (eds U. Pulver, A. Lang and F.W. Schmid, Bern: Huber, 1978). Presently he is involved in a larger project to be documented in a monograph provisionally entitled 'Semiotic Ecology: An Anthropologic Framework'.

Roderick J. Lawrence

He studied at the University of Adelaide, South Australia and at St John's College, Cambridge, England. He holds a Master's and a Doctoral degree. His various appointments include work at the Ecole Polytechnique Fédérale de Lausanne and the School of Architecture at the University of Geneva, both in Switzerland, the School of Social Sciences at the Flinders University of South Australia and at the University of Adelaide. He has been a consultant to the Committee for Housing, Building and Planning of the Economic Commission for Europe. He is currently appointed to the Centre for Human Ecology and Environmental Sciences at the University of Geneva. He has widely published on the reciprocal relations between architectural and behavioural parameters in housing, building and land-use planning from cross-cultural, societal and psychological perspectives that address historical processes. Among his publications are the two books *Le seuil franchi: logement populaire et vie quotidienne en Suisse romande, 1860–1960* (Genève: Georg

Editeur, 1986) and *Housing, Dwellings and Homes: Design Theory, Research and Practice* (Chichester: John Wiley, 1987).

Hans-Joachim Mosler

He first studied zoology with a thesis on ethology and later psychology at the University of Zürich with a thesis on social psychology. In 1990 he obtained his PhD in psychology with research on the self-organization of environmentally adequate action, with special interest in the influence that trust has on the way resources are used. He is presently a scientific collaborator in the Social Psychology Section of the Department of Psychology at the University of Zürich.

Markus Nauser

Born in 1957, he studied geography, economics and media science at the University of Zürich, Switzerland. His Master's thesis was concerned with the interplay of environmental problem awareness and individual behaviour in theory and community practice. After graduating in 1987 he joined the Human Ecology Group of the Geography Department at the Swiss Federal Institute of Technology (ETH) in Zürich. At the same time he was lecturer for environmental psychology and sociology in the Postgraduate Program on Environmental Sciences of the University of Zürich. His academic interest focuses on the perception, communication and interpretation of environmental problems in the private, economic, and political context. Since 1992 he has been a scientific collaborator with Ecosens Ltd, an environmental management consultants' company near Zürich.

Thanks are due to Hans Kastenholz and Wolfgang Zierhofer for helpful comments on an earlier draft of the manuscript reproduced in this volume.

Gonzague Pillet

He studied economics at the University of Fribourg, Switzerland, where he later received a *venia legendi* to become a university lecturer ('Privatdozent'). Meanwhile he started to develop his interests in environmental economics. He has been a Research Associate in the Department of Environmental Engineering at the University of Florida (USA) and an economist at the Centre for Human Ecology and Environmental Sciences of the University of Geneva. Presently he is working at the Paul Scherrer Institute (PSI), a research institution annexed to the Swiss Federal Institute of Technology in Zürich. He is also President of Ecosys Ltd, a consulting company in Geneva. He is author of *Comptes économiques de l'environnement* (Bern: Swiss Federal Statistical Office, 1992) and of the textbook *Manuel d'économie écologique* (Genève: Georg Editeur, in preparation), co-author with H.T. Odum of E^3 – *énergie, écologie, économie* (Genève: Georg Editeur, 1987) and co-editor with T. Murota of *Environmental Economics – The Analysis of a Major Interface* (Genève: Leimgruber, 1987).

Thanks are due to Ola Söderström, Lausanne, Carlo Jaeger, Zürich, Eric F. Berthoud, Neuchâtel, as well as an anonymous referee for commenting on an earlier version of the manuscript reproduced in this volume, and to the Centre for Human Ecology and Environmental Sciences, University of Geneva, for financial support.

Dagmar Reichert

She was born in 1957 in Wien, where she also studied geography, physical education and philosophy and worked as a high school teacher, radio-journalist and university teacher. Later she worked and studied in Canada (University of Toronto), Sweden (Nordplan Stockholm), Switzerland (ETH Zürich) and England (Cambridge University). Since 1989 she has been a researcher and lecturer at the Department of Geography of the ETH in Zürich. Together with Wolfgang Zierhofer she has written the book *Umwelt zur Sprache bringen* (to be published) and is currently finishing her PhD dissertation entitled 'Über Zirkularität und Selbstreferenz in den Human-wissenschaften'.

Ola Söderström

Born in 1959, he has an MA and a PhD in geography from the University of Lausanne in Switzerland, where he is at present a researcher on urban geography and social theory. His work deals with the social production of the built environment with an emphasis on urbanism considered as a field for the social study of science. He is co-editor with L. Mondada and F. Panese of *Paysage et crise de la lisibilité* (Lausanne: Presses centrales, 1992).

Dieter Steiner

Born in 1932, he studied geography at the University of Zürich, Switzerland. He is a former expert in remote sensing and quantitative methods and in this capacity held academic positions at the University of Chicago, USA, the University of Zürich and the University of Waterloo in Canada, and served on the Commission on Data Sensing and Processing (formerly Commission on Air Photo Interpretation) of the International Geographical Union. Since 1975 he has been a Professor at the Swiss Federal Institute of Technology (ETH) in Zürich where in 1985 he established a Human Ecology Group which has become a focus of social science-oriented research and teaching in the area of human–environment problems. He is co-editor with M. Ostheider of *Theorie und Quantitative Methodik in der Geographie* (Zürich: Department of Geography, ETH, 1981), with H. Elsasser of *Räumliche Verflech-tungen in der Wirtschaft* (Zürich: Department of Geography, ETH, 1984), and with B. Wisner of *Human Ecology and Geography* (Zürich: Department of Geography, ETH, 1986), and he is a co-author of *Das Dilemma der modernen Stadt. Theoretische Überlegungen zur Stadtentwicklung am Bei-spiel Zürichs* (Berlin: Springer, 1991).

Peter Weichhart

Born in 1947 in Vienna, he studied geography, German philology and philosophy at the Faculty of Arts of the University of Salzburg, Austria. He has held various academic appointments at the University of Salzburg and the University of München, Germany, and was a visiting scientist and lecturer at the International Institute for Environment and Society in Berlin and at the Swiss Federal Institute of Technology, Zürich. Presently he is head of the Department of Human Ecology at the Institute of Geography at the University of Salzburg. His research interests include the methodology and philosophy of geography, behavioural and action–theoretical geography, housing and migration, urban geography, place identity, the theory of human–environment transactions and human ecology. He has published the following books: *Geographie im Umbruch. Ein methodologischer Beitrag zur Neukonzeption der komplexen Geographie* (Wien: Deuticke, 1975), *Wohnsitzpräferenzen im Raum Salzburg* (Salzburg: Institute of Geography, 1987), *Verbrauchermärkte in Salzburg* (Salzburg: Institute of Geography, 1988), and *Raumbezogene Identität. Bausteine zu einer Theorie räumlich-sozialer Kognition und Identifikation* (Stuttgart: F. Steiner, 1990).

Benno Werlen

Born in 1952, he has a PhD in geography from the University of Fribourg in Switzerland. He held various academic appointments at the Universities of Kiel (Germany), Fribourg and Zürich. As a visiting scholar at the University of Cambridge, England (Faculty of Social and Political Science and Department of Geography) he worked on a project for a new research concept for regional studies. At present he is lecturer in methodology, social and urban geography at the University of Zürich and the Swiss Federal Institute of Technology. His interests are focused on epistemology and action theory and their application to the society–space and society–environment relationships, different types of identities and their interrelations. He has widely published on these topics and is the author of *Gesellschaft, Handlung und Raum* (Stuttgart: Franz Steiner, 1987); title of the English version: *Society, Action and Space. An Alternative Human Geography* (London: Routledge, 1992).

Thanks are due to Dagmar Reichert for her critical and constructive reading of the first draft of the manuscript reproduced in this volume, to Ulrike Erichsen for substantial comments on a later version, and to Pip Hurd for the translation.

PREFACE

In 1986 the Swiss Association of Geographers decided to establish a working group to be concerned with 'Theory and Integrative Approaches'. The idea had emerged from the feeling that the existence of an ecological crisis was sufficient reason to re-examine possibilities of creating a bridge between natural science-oriented physical geography and social science-oriented human geography. That this should be necessary is rather ironical, as geography originally set out from the background of a philosophy which stressed the importance of a holistic view of regions and landscapes. Yet the modernization of the discipline has led, in parallel to the general development of science, to an increasing internal segmentation and divergence. At the same time, however, this has also meant the establishment of external links to neighbouring disciplines. Such links have a fairly extensive history in physical geography (for example, contacts between geomorphology and geology, hydrogeography and hydrology) but are less common in human geography. Only more recently have we witnessed the interest of certain branches of human geography in their outwardly corresponding sciences, for example, of economic geography in economics and of social geography in sociology. Even so, the outward connections remain rather one-sided; geography tends to import theories and methods from the neighbouring disciplines, whereas the latter are largely unconcerned with developments in geography.

In this situation it soon became apparent to the Working Group that the problem is not confined to the discipline of geography, but is of a more general nature. In fact, it can be restated in terms of a human ecology that attempts to look at human lifespaces and at the interplay occurring in those spaces between human beings and the biophysical environment. This has a bearing on many different disciplines. As it happened, a Human Ecology Group had been set up in 1985 within the Department of Geography at the Swiss Federal Institute of Technology (ETH) in Zürich under the direction of Dieter Steiner, co-editor of this volume. As he also held the chair of the Working Group on Theory and Integrative Approaches it was natural that the work of this group developed into a kind of human ecology project. This undertaking had an interdisciplinary flavour that reached beyond the traditional confines of geography, while at the same time recalling some basic tenets of the old integrative notions of that discipline. Precisely because of the latter concern it also became apparent that integration cannot be brought about by science alone; any attempt to treat the human–environment theme in a holistic way invariably becomes entangled in questions of a philosophical nature and issues directly related to life–world realities; hence our interest in the possible change of dominant worldview or cosmology.

To stimulate a discussion within wider circles, the Working Group organized a workshop in December 1987. Statements presented to the workshop were subsequently published under the title *Jenseits der mechanistischen Kosmologie – Neue Horizonte für die Geographie?* ('Beyond the Mechanistic Cosmology – New Horizons for Geography?', edited by D. Steiner, C. Jaeger and P. Walther, Zürich: Department of Geography, ETH). From here we proceeded to the preparation of an international conference that we called 'Person – Society – Environment'. It was held on 24–26 May 1989, at Appenberg, Canton of Bern, Switzerland. Invited speeches were given by five people from five different disciplines: Stephen Boyden (biology), Claudia Carello (psychology), Ingela Josefson (linguistics), Gonzague Pillet (economics), and Peter Gould (geography), all of which are reproduced in this volume. The remainder of the contributions contained in this book are based on papers that were presented in abridged form at the conference and reworked later. Among the authors of this latter collection are a philosopher and theologian (Markus Huppenbauer), a psychologist and system theorist (Hans-Joachim Mosler), an architect and social scientist (Roderick Lawrence), and a psychologist (Alfred Lang). The others (Peter Weichhart, Dagmar Reichert, Gerhard Bahrenberg, Marek Dutkowski, Benno Werlen, Gregor Dürrenberger, Huib Ernste, Ola Söderström, Markus Nauser, and Dieter Steiner) are all geographers. The discipline of sociology was represented, if not in person then in spirit, by the theory of structuration of Anthony Giddens which, for reasons discussed in several papers of this volume, became a central focus of attention from which an attempt was made to start our networking enterprise. Several of the authors made an effort to read some of the collected papers, and gave cross-references to or made critical statements about them in their own contribution, thus adding to our general aim of tying loose ends together.

Although this volume, as evidenced by the list just given, leans heavily in the direction of geography, it is not intended to be a treatise of this discipline. Rather it is, as suggested by the title, meant to be a book about human ecology. It is an attempt at stimulating the development of a human ecology in the form of 'anti-fragmentary views of the world'. Such views are urgently needed if we ever hope to overcome the ecological crisis while preserving humankind and Planet Earth somewhat intact. Yet it is not likely that any kind of 'grand theory' can be developed. Instead the task amounts to trying to establish a network between existing disciplinary theories and methods, a procedure that might be called 'constructive eclecticism'. It goes without saying that this has its dangers. Indeed, the project seems to develop into a tightrope walk at times. Nevertheless, it is one that should be undertaken. And the danger may be lessened by the fact, as also suggested by the book's title, that there cannot be one single view. The bits and pieces presumably can be fitted to each other in several ways. The goal of a general human ecology, then, is not to find *the* one right answer to our problems, but to set a stage on which some kind of creative interdisciplinary and trans-scientific communication can unfold, whereby the stage itself can be questioned and reworked again.

The editors are grateful to the members of the Working Group on Theory and Integrative Approaches of the Swiss Association of Geographers as well as to the members of the Human Ecology Group at the Department of Geography of the ETH in Zürich for their support in organizing the Appenberg conference and providing materials and ideas on which we could draw, in alphabetical order: Lisbeth Bieri, Gregor Dürrenberger, Huib Ernste, Franco Furger, Carlo Jaeger, Markus Huppenbauer, Hans Kastenholz, Räto Kindschi, Verena Meier, Ruedi Nägeli-Oertle, Dagmar Reichert, Armin Reller, Erna Rüedi, Alec Schaerer, Ola Söderström, Pierre Walther, Jakob Weiss, Benno Werlen, Wolfgang Zierhofer. Some of them, of course, appear as authors in this volume. Among the people listed, however, Carlo Jaeger deserves special mention for providing his knowledge and drive which substantially furthered the development of the human ecology idea in general and the conception of the present volume in particular.

Thanks are also due to Susan Braun for reformulating clumsy or faulty passages in the papers written by authors whose mother tongue was not English, and to Brigitte Göldi for her assistance in the text formatting process. Finally, we should like to express our gratitude to Tristan Palmer who, as our Routledge contact, patiently bore with us in the tedious process of putting this volume into shape.

Dieter Steiner and Markus Nauser
Zürich, June 1992

1

GENERAL INTRODUCTION

The story of human ecology is part tradition, part response to crisis. Its origins can be traced back to developments over several decades of scientific interest in human–environment relationships which, for the most part, occurred separately in a number of disciplines. A brief overview of this background is given in the following first section of this introduction. An intensified pursuit of the topic set in with the advent of the planetary ecological crisis. This new kind of human ecology strives for integration in the form of interdisciplinary linkages and trans-scientific connections. A number of remarks regarding this development can be found in the second section of this introduction.

It is the aim of this volume to contribute to this more recent development. Conceptual and theoretical considerations related to the establishment of an integrated human ecology are discussed in the four contributions of Part I. Part II contains five papers that deal in various ways with the different levels of consciousness that humans (as agents whose actions may be relevant to the environment) possess, and this leads to questions related to epistemology, methodology and morality. Part III is composed of three articles that discuss aspects of structuration which we see as a central theme of a general human ecology. Structuration refers to the fact that there is a duality with regard to human agents acting as members of human societies: they act within social as well as spatial–environmental structures, but, by so doing, also help to re-establish those structures. Finally, the five contributions of Part IV in various ways address the importance of the regional dimension. A regional differentiation of modern society rather than a global, functional one is expected to constitute an important step towards the goal of ecological sustainability.

Each part of the volume begins with an introduction. The aim here is not simply to summarize what is to come, but rather to pick up some important themes relevant to the part in question and to enrich the discussion with further thoughts in such a way that threads are drawn together which otherwise might remain loose.

1

1 TRADITION: INTRADISCIPLINARY ROOTS OF HUMAN ECOLOGY

The following brief summary sketches the development of concepts of human ecology that can be called traditional. The interested reader can find more detailed information in a paper by Young (1974) and in a reader edited by the same author containing a collection of seminal papers by representatives of the human ecology orientations of the past (Young 1983). For further remarks about the development of human ecology see also Jaeger (1991, Introduction) and Lawrence (this volume).

1.1 Biology

Ecology is known as a branch of biology. The term 'ecology' was coined by Ernst Haeckel in 1866 to describe the science dealing with the relationships of organisms to the surrounding outer world (*Aussenwelt*). Since then this understanding of ecology has basically remained the same. A definition in a recent textbook reads: 'Ecology, if one goes by the usual definition, is the study of the relationship between organisms and their environment' (Smith 1986: 3). However, the author adds a note of caution: 'that definition can be faulted unless you consider the words *relationship* and *environment* in their fullest meaning. Environment includes not only the physical but also the biological conditions under which an organism lives; and relationships involve interactions with the physical world as well as interrelationships with members of other species and individuals of the same species.' As pointed out by Young (1974: 4), to the extent that ecology is founded on the notion of organisms adapting to a selecting environment, 'the formal beginnings of the ecological approach . . . lie in evolutionary biology, in the work of Darwin'.

The usual understanding of 'environment' is one that is specific to the type of organism concerned: it refers to those features of the surrounding world which affect or are affected by the organism in question. Nevertheless, the use of the term is a source of frequent confusion, in particular in the context of human ecology. Do we, by 'environment', mean natural components in our surroundings as opposed to human-made features and modifications? Or do we include the latter? Clearly a comprehensive human ecology must refer to the environment 'as is'. However, the recent development of human ecology has grown out of a concern for the dwindling degree of naturalness of this environment. A different kind of distinction is made by Knötig (1972). He contrasts what he calls a 'Haeckelian environment' with an 'Uexküllian environment'. The former refers to the common notion held in ecology about an environment being given and an organism being largely a passive respondent to variables describing this environment, such that physical, chemical, biological or psychological relations result. The latter refers to notions developed by Uexküll (1928). He sees an organism rather as an active being which selects and modifies its surroundings such that, in fact, it creates its own environment according to its needs. To the extent that this process is based on the use of an 'inner model' we have here the beginning of a concept that becomes particularly useful in a context involving humans as organisms.

To start with, however, one can obviously be interested in humans as purely biological organisms and their interactions with the environments, disregarding any potential mental faculties, and then call this interest the subject of human ecology. For some people this is indeed what human ecology is all about: a concern with questions related to the biophysical foundations of human existence, on the one hand, and to the influence of humans on the biophysical environment, on the other. More recent examples of approaches along this line include Ehrlich *et al.* (1973) and Freye (1986). Typically these books comprise topics such as population dynamics, carrying capacity of the land, food and energy production and consumption, nutritional problems, pollution affecting ecosystem functioning, and health hazards under natural conditions (e.g. parasites) as well as resulting from anthropogenic toxic waste. The concern is with individuals and populations as aggregates of individuals and their immediate relationships to the environment in terms of biophysical variables. The fact that ultimately the quality of these relationships is heavily influenced by the kind of society a population forms and the kind of culture it carries is recognized but not considered any further.

1.2 Sociology

An outstanding feature of human beings is precisely, however, that they are very eminently social animals. They form societies with arrangements that tend to influence the purely biophysical relationships of humans to their environment. Early in this century, starting with Park in 1916, the sociologists of the so-called Chicago School became interested in this question. As they were concerned primarily with cities, they were dealing with built-up and socially defined rather than natural environments (for this reason the Chicago human ecology is also referred to as 'social ecology'). They held the opinion that there are underlying biological motives to human actions such as a drive for competition, but that these drives become restrained by superimposed social rules. The resulting processes, however, in the view of these Chicago sociologists, could still be described by a biological language: the migration of populations through a city and the change of neighbourhoods could be understood in analogy to the succession of plant societies. Variants of this sociological style of human ecology have survived to this very day. A system–theoretic version is presented by Hawley (1986): human populations as collectives are ecosystems in the sense that they are capable of adapting themselves to an environment.[1] An overview of the development of social ecology can be found in Rojo (1991). An extensive treatment of the subject is

3

also provided by Friedrichs (1977). Of a quite different nature, representing an opposite point of view, are recent contributions that criticize the discipline for disregarding relationships between social and nonsocial facts, in particular human environmental impact, and hence demand an ecological perspective on human societies which stresses questions such as carrying capacity and sustainability (see, for example, Catton 1983).

1.3 Cultural anthropology

The question of the relationship between nature and culture in human societies has been the focus of a branch of cultural anthropology, known as cultural ecology, a term coined by Steward (1955). Since the latter part of the previous century there has been an alternation between opinions favouring an environmental determinism and notions postulating the dominance of culture over nature in a way which Orlove (1980: 236) describes as 'a number of swings on intellectual pendulums'. The more recent development is marked by positions that are more intermediate with respect to this question. On the other hand a new controversy has developed concerning the 'sociological question': is it appropriate to treat human societies in terms of system theoretic approaches or should one rather use action-oriented models? A comprehensive introduction to cultural ecology is provided by Bargatzki (1986), an overview of its history and its significance to a development of a general human ecology by Steiner (1992). For the latter the findings of cultural ecology are of interest despite or perhaps precisely because of the fact that classical cultural anthropology investigates pre-industrial societies almost exclusively. We cannot avoid noting that these societies to a large extent have shown a marvellous adaptation to different kinds of environments with equilibrating relationships (see, for example, Campbell 1983) and hence the question arises as to how this situation has been guided by the cultural values of those societies.

1.4 Geography

There is a long-standing claim by geography to be a scientific discipline dealing especially with the theme of human–environment relationships. Geography indeed is an unusual discipline because, with the internal distinction between physical and human geography, it is anchored in the natural as well as in the social and cultural sciences. We can note a number of parallels with the development of cultural ecology mentioned above. Classical geography, in a similar way, went through periods alternating between environmental and cultural determinism, the difference being that the focus was on the environment in terms of landscapes and regional patterns rather than on society and culture. It must be pointed out, however, that, while early German geographers such as Alexander von Humboldt, Carl Ritter and

Friedrich Ratzel still had a truly integrated view of their subject, the later discussion about the interplay between environment and humans was relegated to a large extent to the realm of the philosophy of geography. The actual scientific work was marked by an increasingly strong division of labour between the physical and the human geographers. Geography thus has not really made good its promise and has succumbed also to the general trend of increasing scientific specialization. This has happened despite early attempts to create a geographical human ecology. In fact, Barrows in 1923 maintained that geography *is* human ecology, but his kind of thinking did not bear much fruit. Typically, in a paper with the same title fifty years later, Chorley (1973) argues that the spreading ecological crisis is not a reason for trying to bring physical and human geography together, but, on the contrary, it is a proof that their union is impossible.[2] At the same time, however, we also witness the attempts of a number of distressed geographers to swim against the current. Examples are Hewitt and Hare (1973) who maintain that geography should become a 'modern human ecology',[3] and Weichhart (1975) who tries to establish foundations of an 'ecogeography'.

1.5 Psychology

Compared with the other disciplines mentioned above the variety of ecological approaches is in psychology perhaps greatest, and so is probably the resulting confusion. These approaches are variously called 'ecological psychology', 'environmental psychology', and 'psychological ecology'. Also, the history of an ecological orientation is shorter. Young (1974) earmarks the work by Barker starting in the fifties as pioneering (it is summarized in his book published in 1968). Decisive is 'the movement out of the laboratory with its controlled experiments on human behavior and into the field for direct observation of behavior under uncontrolled field conditions' (Young 1974: 23). However, one can see this innovation as a consequence of the work by Lewin (1936), whose student Barker was. Lewin developed a concept of lifespace as a field of activating and deactivating forces in which the individual person represents a point. Although his notions and terminology are highly physicalistic it is interesting to note that he saw this field as purely psychologically defined. In contrast, Barker stressed the material, biophysical reality of the environment as a 'behavior setting', a setting within which human behaviour can unfold. A more recent development includes Gibson's brand of ecological psychology which sees an animal or a human being and its environment as merely two aspects of one and the same irreducible system (Gibson 1979). The contribution by Carello in this volume is representative of this kind of thinking. A comprehensive compendium of the various approaches is provided in the handbook by Stokols and Altman (1987), and a useful overview by Miller (1986).

2 RESPONSE TO CRISIS: HUMAN ECOLOGY AS A GENERAL PERSPECTIVE

2.1 Integration through a common denominator?

A common feature of the developments mentioned above is that they represent endeavours within particular disciplines and are grounded in the general scientific motive of knowledge accumulation. Human ecology thus has quite a variety of backgrounds and different people mean different things by it. This is still largely true and presumably to some extent unavoidable. There cannot be any single view on such a highly complex topic as the relationship between humans and the environment.[4] Ample evidence for the existing diversity of backgrounds and opinions can be found in a number of recent conference volumes such as those from four conferences of the Society for Human Ecology (SHE)[5] and those edited by Glaeser (1989) and by Kilchenmann and Schwarz (1991).

Nevertheless, more recent developments are based on at least one common denominator: a growing awareness of the ecological crisis. We seem to have here a case of science responding to a public concern. Ever since the publication of Rachel Carson's *Silent Spring* (Carson 1962) an environmental movement growing in size and scope has made itself felt, even within scientific circles. First of all, this has affected the further development of the intradisciplinary human ecologies mentioned above. For example, the main concern of classical cultural ecology has been human–environment relationships in pre-industrial societies. More recently some authors such as Rappaport (1979) have extended their thinking to a critical analysis of modern society and possible reasons for the ecological disorder it creates. Within geography there is a mounting feeling that any further argumentation about the scientific incompatibility of its physical and its human branch is rather cynical. It was precisely for this reason that the working group mentioned in the Preface was set up by the Swiss Association of Geographers, an undertaking which led to the Appenberg conference and finally to the present book. Also, of course, there are new, environmentally oriented developments within scientific disciplines which previously have disregarded to a great extent the existence of a biophysical environment. A case in point is economics with its new environmental or ecological branches.

It was an early understanding of the members of the geographical working party just mentioned that substantial contributions to the ecological question can, if at all, eventually be expected only from some kind of interdisciplinary human ecology. Indeed this understanding simply reflects an already widespread recognition of a need for integration, one that is reflected in a growing number of scientific organizations (the foremost example at present being the previously mentioned US-based Society for Human Ecology)[6] and meetings (such as the International Conference on Human Ecology held in June 1991 in Göteborg, Sweden).[7] A persisting problem of these organizations and conferences is, however, that people get together who are specialists in some field but who, although they are likely to be interested in the same or similar issues, may have difficulties in finding a common language. An obvious conclusion is that we need a comprehensive kind of human ecology, one capable of establishing bridges between the various disciplines concerned with the topic.

This kind of thinking seems to take for granted that an ecological crisis really exists. Of course, it is legitimate to ask: 'Is there an ecological crisis at all?' This is exactly the question asked by Dryzek. As he points out, 'the empirical evidence cannot conceivably be decisive' (Dryzek 1987: 16). However, 'human encroachment on natural systems is occurring and expanding at unprecedented levels and rates. One need only look at deforestation, the build-up of carbon dioxide in the atmosphere, desertification, and current rates of energy use' (Dryzek 1987: 22). On the one hand, it is obvious that what we care to call 'crisis' now is a result of social perception. Some people even seem to believe that the ecological crisis is something that is made up by the media. On the other hand, the one good thing that science can do, if nothing else, is to point to the distinct possibility that humankind, if it continues on its present course, can destroy the ecological conditions on this planet to such an extent that life for future generations of human beings, let alone for scores of other species, will become impossible. Even if the signs are not clear now, it is nevertheless irresponsible to continue until they are clear, as the situation may have developed beyond the point of possible return by then. To say that this is irresponsible is to assume that the possession of a capability to wipe out much of life on Earth, including humankind itself, raises a very eminent and fundamental moral issue. There seems, at least, to be widespread agreement on this point.

2.2 Another round of Enlightenment

The growing recognition of ecological problems and of the fact that humans are causing them may be regarded as a necessary continuation of the *rationalization process* that set in with the Enlightenment (see, for example, Bahrenberg 1987, Ulrich 1987). The shedding of the dogmas of the church made a development possible that initially led to the autonomy of individuals, but eventually turned into the autonomy of societal subsystems, which seem to function regardless of whether the individuals like it or not, and hence *result in the 'subordination of individual variety'* (Hewitt and Hare 1973: 35). Science, technology and economy are the paramount examples of such subsystems, each operating according to its own special logic. Today our task would thus seem to be to challenge the dogmatic aspects associated with these logics and to sever the bonds established by them. A general human ecology should help this process by being a critical power, critical of previous beliefs,

7

such as that science invariably leads to truth,[8] that technology solves any practical problems, and that economic growth furthers the well-being of humans at all times. What Giddens (1986: 2) claims for sociology, namely that it 'necessarily has a subversive quality' and that 'it is exactly because sociology deals with problems of such pressing interest to us all (or should do so), problems which are the objects of major controversies and conflicts in society itself, that it has this character', should be true in an even larger context for human ecology, which tries to address problems at the interface of society and environment.

At one time the Enlightenment set a process of societal differentiation in motion which finally overstepped its mark. The further enlightenment thus must be a process of reintegration. What this may mean can be better understood in the context of the evolutionary perspective discussed in Introduction to Part I, Section 4. Suffice it to point out here the presumably most important aspect, that is, the need for an embedding of the economy in the wider context of ecology, which is quite in line with the notion expressed by Hösle (1991) in his philosophical treatment of the ecological crisis, namely of a shift from a predominantly economic to a predominantly ecological paradigm.

2.3 An apology for omissions

Two further remarks concerning the thematical scope of the present volume are in order. First, most of what is said generally refers directly to Western civilization and its development over the ages. This is not an 'ethnocentric' oversight, but a consciously made restriction. Of course, a total picture of the global nature of the ecological crisis can be obtained only by including the non-Western parts of the world. The restriction, however, can be justified on grounds that are similar to those underlying the recognition that, in a wider context, anthropocentrism is unavoidable if not necessary. If we agree that modern economic society with its trend to globalization is largely responsible for all of the ecological disasters that we find on Earth today, then perhaps we should delve first into the reasons for the degree of destructiveness that our society has attained. This is, of course, not to belittle all of the efforts made by scores of human ecology oriented researchers working in Third World countries. There are tremendous problems there that certainly cannot wait (see, for example, Steiner and Wisner 1986). And to the extent that we are responsible for these problems it is part of our duty to help. It is simply our feeling that we might be in a much better position to give fruitful assistance if we have gained an insight into the ills of our own society first, as this might enable us to avoid making the same mistakes there as we have made here.[9] This, therefore, is our stance with regard to the development of a general human ecology and hence to the scope of the present book.

A second omission is less easily defended. It refers to the relationship between females and males in our society as expressed by positions and roles or, simply said, to the question of patriarchy and its historical background. With the emergence of political societies after the neolithic revolution[10] males started to develop (more or less consciously, and with the assistance of the women themselves) structures of domination over women. This eventually became the model for mastery over people in general, in particular in the form of their enslavement (Lerner 1986). Although women today have, in Western countries, a legal status that puts them officially on an equal footing with men, structures of male domination persist in more subtle forms and to various degrees. As the male principle has been associated with the mind and the female principle with nature all through the history of Western philosophy (Lloyd 1984), one can readily surmise that the domination of men over women has much to do with the destructiveness that Western society has attained towards nature. The gender issue thus would seem to be a very central topic for a general human ecology. However, except for some remarks and footnotes here and there we have not managed to address it in this present volume for the simple reason that when we embarked on our project of human ecology we were not very well aware of the significance of this issue. We are now, but as it stands we will have to postpone its discussion to further work and later publications. It is, we think, an excellent example of how the consciousness of people becoming involved in human ecology can change. Human ecology is work in progress, and it may stay that way forever.

3 NOTES

1 Hawley's approach to human ecology is heavily criticized by Berry (1988: 138) for being deterministic (environment determines culture), mechanistic and material-istic: 'Humankind dances, but an external puppet-master pulls the strings; the dance can affect the environment, but reason cannot dictate the environmental changes. Adaptation is all.'

2 Today, in the light of the present environmental situation, some of Chorley's statements may sound rather cynical. An example: 'If the proponents of geography ... wish to continue to reflect the relationships between society and nature they cannot afford to adopt models which ignore the glaring probability that this relationship is one which exists between an increasingly numerous, increasingly powerful and progressive, if capricious, master and a large, increasingly vulnerable and spitefully conservative serf' (Chorley 1973: 157).

3 Their concluding argument is: 'the similarities between modern ecology and geography are considerable. Though its theory remains poorly articulated, to say that geography is primarily "human ecology" still remains a better description of the range of geographical work, than emphasizing the more narrow ecological subset of man's spatial organization' (Hewitt and Hare 1973: 36).

4 Cf. Introduction to Part I, Section 1.

5 See Borden (1986), Borden and Jacobs (1988), Pratt and Young (1990), Sontag et al. (1991).

6 Besides the conference volumes mentioned earlier (cf. Note 5) SHE has also published a directory of human ecologists (Borden and Jacobs 1989).

9

7 See Hansson and Jungen (1992).

8 The notion that science is very much a historical undertaking and that its results are in essence social constructs has, ever since Kuhn (1962), been widely recognized. Less commonly agreed upon is a statement to the effect that the *progress provided by science* and its technological consequences has become dubious. However, this second point is exactly related to the fact that new branches of science develop and old disciplines reorient themselves because of the advent of the ecological crisis, a crisis for which science itself has to bear partial responsibility. Let us illustrate this as follows. Suppose we wish to document the development of science by looking at the courses listed in university calendars year after year. During the 'Golden Age' of physics in the 1920s and 30s, for example, on detecting the new topic of 'quantum physics', we may very well have classified this as real progress. Today we find scores of new courses on environmental topics. Is this now progress or is it not? In the end, we may recognize that the question of whether or not science provides progress cannot really be asked meaningfully without a notion of what 'progress' is supposed to mean. Is it related to something that furthers our education or rather to something that establishes growing structures of domination?

9 A devastating documentation of mistakes in development and technical aid made in the past is provided by Hagen (1988).

10 Cf. Steiner, this volume, Section 3.2.

4 REFERENCES

Bahrenberg, G. (1987) 'Unsinn und Sinn des Regionalismus in der Geographie', *Geographische Zeitschrift* 75 (3): 149–60.

Bargatzki, T. (1986) *Einführung in die Kulturökologie*, Berlin: Dietrich Reimer.

Barker, R.G. (1968) *Ecological Psychology: Concepts and Methods for Studying the Environment of Human Behavior*, Stanford, CA: Stanford University Press.

Barrows, H.H. (1923) 'Geography as human ecology', *Annals of the Association of American Geographers* 13 (1): 1–14.

Berry, B.J.L. (1988) 'Review of A.H. Hawley: Human Ecology', *Contemporary Sociology* 17 (2): 137–9.

Borden, R.J. (ed.) (1986) *Human Ecology: a Gathering of Perspectives. Selected Papers from the 1st International Conference, 1985, University of Maryland, College Park, MD*, College Park, MD: Society for Human Ecology, University of Maryland.

Borden, R.J. and Jacobs, J. (eds) (1988) *Human Ecology: Research and Applications. Selected Papers from the 2nd International Conference, 1986, College of the Atlantic, College Park, MD: Society for Human Ecology, University of Maryland.*

Borden, R.J. and Jacobs, J. (eds) (1989) International Directory of Human Ecologists, Bar Harbor, ME: Society for Human Ecology, College of the Atlantic.

Campbell, B. (1983) *Human Ecology*, London: Heinemann.

Carson, R. (1962) *Silent Spring*, Boston, MA: Houghton Mifflin.

Catton, W.R., jun. (1983) 'Need for a new paradigm', *Sociological Perspectives* 26 (1): 3–15.

Chorley, R.J. (1973) 'Geography as human ecology', in R.J. Chorley (ed.) *Directions in Geography*, London: Methuen.

Dryzek, J.S. (1987) *Rational Ecology. Environment and Political Economy*, Oxford: Blackwell.

Ehrlich, P.R., Ehrlich, A.H. and Holdren, J.P. (1973) *Human Ecology*, San Francisco and London: W.H. Freeman.

Freye, H.-A. (1986) *Einführung in die Humanökologie für Mediziner und Biologen,* Heidelberg and Wiesbaden: Quelle & Meyer (UTB 1402).

Friedrichs, J. (1977) *Stadtanalyse. Soziale und räumliche Organisation der Gesellschaft,* Reinbek b. Hamburg: Rowohlt.

Gibson, J.J. (1979) *The Ecological Approach to Visual Perception,* Boston, MA: Houghton Mifflin.

Giddens, A. (1986) *Sociology. A Brief but Critical Introduction,* London: Macmillan.

Glaeser, B. (ed.) (1989) *Humanökologie. Grundlagen präventiver Umweltpolitik,* Opladen: Westdeutscher Verlag.

Haeckel, E. (1866) *Generelle Morphologie der Organismen, Bd. 2: Allgemeine Entwickelungsgeschichte der Organismen,* Berlin: Reimer.

Hagen, T. (1988) *Wege und Irrwege der Entwicklungshilfe. Das Experimentieren an der Dritten Welt.* Zürich: Verlag Neue Zürcher Zeitung.

Hansson, L.O. and Jungen, B. (1992) 'Human responsibility and global change, proceedings from the International Conference in Göteborg, 9–14 June, 1991', *Humanekologiska Skrifter* 12, Göteborg: Section of Human Ecology, University of Göteborg.

Hawley, A.H. (1986) *Human Ecology: A Theoretical Essay,* Chicago, Ill.: University of Chicago Press.

Hewitt, K. and Hare, F.K. (1973) 'Man and environment. Conceptual frameworks', *Commission on College Geography Resource Paper* 20, Washington, DC: Association of American Geographers.

Hösle, V. (1991) *Philosophie der ökologischen Krise. Moskauer Vorträge,* München: C.H. Beck.

Jaeger, C. (1991) 'The puzzle of human ecology. An essay on environmental problems and cultural evolution', habilitation thesis, Zürich: Department of Environmental Sciences, ETH (publication forthcoming).

Kilchenmann, A. and Schwarz, C. (eds) (1991) *Perspektiven der Humanökologie,* Berlin: Springer.

Knötig, H. (1972) 'Bemerkungen zum Begriff "Humanökologie"' *Humanökologische Blätter* 1(2/3): 3–140.

Kuhn, T.S. (1962) *The Structure of Scientific Revolutions,* Chicago, Ill.: University of Chicago Press.

Lerner, G. (1986) *The Creation of Patriarchy,* New York and Oxford: Oxford University Press.

Lewin, K. (1936) *Principles of Topological Psychology,* New York: McGraw.

Lloyd, G. (1984) *The Man of Reason. 'Male' and 'Female' in Western Philosophy,* London: Methuen.

Miller, R. (1986) *Einführung in die ökologische Psychologie,* Opladen: Leske & Budrich.

Orlove, B.S. (1980) 'Ecological anthropology', *Annual Review of Anthropology* 9:235–73.

Park, R.E. (1916) 'The city: suggestions for the investigation of human behavior in the urban environment', *American Journal of Sociology* 20: 577–612.

Pratt, J. and Young, G.L. (eds) (1990) *Human Ecology: Steps to the Future Proceedings of the 3rd International Conference, 1988, Golden Gate National Recreation Area, CA,* Sonoma, CA: Society for Human Ecology and Institute for Human Ecology.

Rappaport, R.A. (1979) *Ecology, Meaning, and Religion,* Berkeley, CA: North Atlantic Books.

Rojo, T. (1991) 'Sociological contributions to human ecology: A European perspective', in M.S. Sontag, S.D. Wright and G.L. Young (eds) *Human Ecology: Strategies for the Future,* Fort Collins, CO: Society for Human Ecology and National Ecology Research Center.

Smith, R.L. (1986) *Elements of Ecology*, New York: Harper & Row.

Sontag, M.S., Wright, S.D. and Young, G.L. (eds) (1991) *Human Ecology: Strategies for the Future. Selected Papers from the 4th International Conference, 1990, Michigan State University, East Lansing, MI,* Fort Collins, CO: Society for Human Ecology and National Ecology Research Center.

Steiner, D. (1992) 'Auf dem Weg zu einer allgemeinen Humanökologie: Der kulturökologische Beitrag', in B. Glaeser and P. Teherani-Krönner (eds): *Humanökologie und Kulturökologie. Grundlagen–Ansätze–Praxis,* Opladen: Westdeutscher Verlag.

Steiner, D. and Wisner, B. (eds) (1986) 'Humanökologie und Geographie – Human ecology and geography', *Zürcher Geographische Schriften* 28, Zürich: Department of Geography, ETH.

Steward, J.H. (1955) *Theory of Culture Change,* Urbana, Ill.: University of Illinois Press.

Stokols, D. and Altman, I. (eds) (1987) *Handbook of Environmental Psychology,* 2 vols, New York: Wiley.

Uexküll, J.V. (1928) *Theoretische Biologie,* Berlin: Springer.

Ulrich, P. (1987) *Transformation der ökonomischen Vernunft. Fortschrittsperspektiven der modernen Industriegesellschaft,* Bern and Stuttgart: Haupt.

Weichhart, P., (1975) *Geographie im Umbruch. Ein methodologischer Beitrag zur Neukonzeption der komplexen Geographie,* Wien: F. Deuticke.

Young, G.L. (1974) 'Human ecology as an interdisciplinary concept: a critical inquiry', in A. Macfadyen (ed.) *Advances in Ecological Research,* 8: 1–105, London and New York: Academic Press.

Young, G.L. (ed.) (1983) *Origins of Human Ecology,* Stroudsberg, PA: Hutchinson Ross.

Part I
HUMAN ECOLOGY

2

INTRODUCTION TO PART I

The first three contributions in Part I of this volume are explicit attempts at establishing components of an integrative framework, while the last paper looks at this endeavour in a critical fashion. There are certain commonalities between the papers. For example, variants of the human ecological triangle, person–society–environment, play a constitutive metaphorical role. But it should be clear that to establish an integrative framework cannot mean to develop an overarching theory, because this would mean to try the impossible. If this is so, then the question obviously arises of how still to achieve some degree of interdisciplinarity in practice. The futility of the search for a 'grand theory' and the ensuing interdisciplinarity question is the topic of Section 1 of this Introduction.

On the other hand, in Part I there are pertinent shifts in emphasis from one contribution to the other. In Sections 2–6 of this Introduction, we wish to draw the reader's attention to some aspects which, in our understanding, are particularly important. First, we refer to the fact that the possibility of human life on Earth has its biophysical limits. Second, we discuss the separation of socio-culturally created realities from the supporting biophysical reality. This may in part explain the environmentally destructive consequences of human action that are growing in degree and scope throughout cultural evolution. However, not only may human agency unfold within the 'false' logic of a social system, but the actions may also have unintended consequences. Third, we want to remind ourselves that actions of human individuals are the intermediary between socio-cultural premises and environmental conditions. There are no direct causal influences of a society on its environment or vice versa. Fourth, we stress the importance of an evolutionary perspective which can connect present problems to earlier socio-cultural developments. And finally, we maintain that a human ecology restricted to pure science is no genuine human ecology at all.

1 NO GRAND THEORY, BUT TRANSDISCIPLINARITY

As will be argued below and in Section 3, a general human ecology should be centred in the social sciences and the humanities rather than in the natural sciences. This aggravates the task of integration, as segmentation and disarray on the side of the former seem to be much more pronounced than on the side of the latter. A vivid illustration of the multitude of approaches common in social theory is provided by Giddens and Turner (1990). In the introduction they write: the 'lack of consensus . . . may be endemic to the nature of social science. At the very least, whether there can be a unified framework of social theory or even agreement over its basic preoccupations is itself a contested issue' (Giddens and Turner 1990: 1). It may be, therefore, that a degree of pluralism is unavoidable, given the fact that the human sciences have as their object of investigation complex wholes in the form of individual human beings or societies. One may surmise that such wholes cannot be described adequately by any single view and that their essence can be comprehended, if at all, only by a series of different complementary views.[1] If this is true, then there can be no unified grand theory for the human sciences, at least not in a strict scientific sense, let alone for a general human ecology. As Geertz (1983: 4) points out, 'though those with what they take to be one big idea are still among us, calls for "a general theory" of just about anything social sound increasingly hollow, and claims to have one megalomanic'.[2] Von Weizsäcker (1977: 17) suggests why, in anthropology, a unification has been impossible: 'A dominating paradigm of a comprehensive scientific anthropology has not existed so far. There have been scattered single disciplines with anthropological concerns. As soon as we ask for an all-embracing anthropology we are involved in philosophical discussion.'[3]

The lack of an overall theory, or the impossibility of finding one, should not, however, serve as an excuse for not striving towards some kind of 'organized pluralism' allowing for a fruitful interdisciplinary discourse. Wallerstein (1990), in connection with his concept of 'world-systems analysis', makes a plea to tear down the fences between such disciplines as anthropology, economics, political science and sociology because the presumed boundary criteria 'either are no longer true in practice or, if sustained, are barriers to further knowledge rather than stimuli to its creation' (Wallerstein 1990: 312). In reality, however, if one looks at the whole spectrum of human sciences and their relations to each other, one can easily get the impression that these sciences are in a state of increasing divergence, if not confusion. In particular, connections between psychology and sociology are badly developed. As Jaeger (1991:190–1) puts it: 'the human sciences will suffer, unless both disciplines develop the modesty to acknowledge that they cannot formulate a satisfying general theory in isolation from one another'. This is a situation to which we cannot be indifferent as long as we claim that the problem of the ecological crisis must be approached from the human side of the scenery and can only be solved, if at all, from this side. It is not, therefore, that pluralism is good and a unified theory is bad, but the desired route should strike a balance between the chaos of a babel on the one hand and the sterility of just one common language on the other.

This, of course, raises the question of how to realize interdisciplinarity in practical terms. As reported by Messerli (1988), a distinction of three levels has been made in the UNESCO 'Man and the Biosphere' (MAB) program. At the lowest level we have a mere 'pluridisciplinarity' in which various disciplines entertain channels of communication without having, however, a common goal. The middle level is taken up by interdisciplinarity proper: here the participating disciplines coordinate their endeavours in the direction of a common goal (that is, an answer to a common problem). In addition, one of the disciplines plays a leading role in the process of integration. Finally, the highest level is given to a concept of 'transdisciplinarity' in which this process runs through a hierarchy of goals such that there is a larger number of primary goals, a smaller number of intermediary goals and finally one ultimate goal. Accordingly, there may be more than one leading discipline.

A very basic problem common to all three of these models is that the flow of information is hampered by the disciplinary specificity of the languages used. True, the problem is alleviated on levels 2 and 3 through the function of coordination taken over by one or more disciplines. In reality, however, the success of this approach depends on an intensive, very personal involvement of one or more individuals representing those leading disciplines. In this sense we plead for a transdisciplinarity which does not end with a commonly defined problem, but is first of all based on the efforts of integrating persons. This is the stance taken in both the contributions by Boyden and Steiner. It is an attempt at solving the language problem by using persons as the 'medium of integration' in a very specific sense.[4] A really integrated science requires the existence of integrated scientists. The aim of this book is precisely to provide educational material for people willing to become such scientists in the realm of human ecology.

2 THE BIOPHYSICAL ENCASING

That living beings affect their environment is a basic fact of evolution; it is a necessary principle of life. Interactions of organisms with the environment have a bearing on the resource configuration: existing resources can be diminished or augmented, or new resources can be created. 'By the mere process of living, organisms change the very conditions upon which they depend for subsistence' (Freese 1988: 71).[5] One might add that organisms, by becoming components of food chains, create primary living conditions for other organisms and thus contribute to the development of whole ecological networks.

For a long time during the history of humankind humans in their dealings with the environment have not deviated drastically from this principle. True, they have cleared rather extensive tracts of forest to gain land for agricultural purposes, but this in fact more often than not led to the establishment of new and more varied habitats for other species. It is not surprising, therefore, that much of what today runs under the title of 'nature conservation' is actually a conservation of some aspects of traditional cultural landscapes. Indeed, it is only with the advent of modernity that humans have started to change their environment – which meanwhile has become the whole planet – to an extent that is increasingly ecologically destructive. Worse, they continue to act destructively even though they have become aware of the risks created by modern civilization. As the previously mentioned circular relationship between organisms and their environment also applies to humans, it becomes clear now that the changes induced by them, initially intended to better their living conditions, have started to backfire. With urban sprawl continuing, forests disappearing, poisons spreading and unusable waste materials piling up, not only will other species die out at an increasing rate but also the survival of humankind itself becomes threatened.

This will be true notwithstanding the unusual mental faculties that humans possess. They cannot avoid being still also biological organisms with a body whose proper functioning depends on healthy living conditions.[6] There is for our existence, so to speak, a biophysical encasing from which we cannot escape. Any kind of integrative human ecology, before speculating about cultural solutions to the problem, must therefore recognize the limitations set by biophysical actualities. It may be appropriate, therefore, to begin the present volume with a contribution written by a biologist, Stephen Boyden. Taking as his premise the fact that the human 'aptitude for culture' is a product of biological evolution, he demonstrates how this very product becomes a source of conflicts with basic bioecological principles. With his notion of 'ecodeviation' he surmises that there is a growing discrepancy between the statics of our genetically defined heritage and the dynamics of a rapidly changing environment, giving rise to maladaptations. As an example he mentions the emergence of 'diseases of civilization'[7] over the ages. He also points to the human tendency for 'technoaddiction', resulting in an increasing dependence of humans on their own products and hence in 'techno-metabolism' which surpasses any kind of basic and real needs by far. To avoid a continuing violation of bioecological principles and disregard for bio-physical limitations humankind should observe a number of 'ecological-imperatives', thereby creating a basis for a sustainable society with a sustainable environment.

Within a general human ecology, a manner of thought in terms of biophysical limits should be applied above all to our present economic system. This is exactly what 'ecological economics' is trying to do by establishing rules for possible sustainable economies on a finite Earth with finite resources (see, for example, Daly and Cobb 1989). In other words, the economy is integrated into an ecological framework.[8] This is in stark contrast to the economic viewpoint which is still predominantly held with regard to the ecological crisis, namely that of 'environmental economics': integrate nature into the economic system by monetarizing environmental goods, thereby bringing them under the control of the market forces, and all will be well.[9]

3 'COGNIZED MODELS' OF THE ENVIRONMENT AND 'AUTONOMOUS PROCESSES'

Boyden thus rightly stresses the importance of the regard for the biophysical actualities that are involved in any human ecological situation. As farming, manufacturing, building, and so on are largely governed by cultural arrangements, however, he also points to the need for a more comprehensive view which comprises a concern for the interplay between biophysical and human variables. Here we can take the development as a guideline to what has taken place in cultural ecology: the shift from looking at society and culture as a kind of anonymous superorganism which tends to adapt itself to changing environmental conditions to a more differentiated perspective which recognizes that societies are composed of individual persons acting within given structures.[10]

Human beings become what they are as the result of a person–society coevolution, generation after generation. A human can become a responsible person only within a social context that comprises other humans. He or she carries responsibility to the collectivity and its rules of behaviour. The important point is that, in the course of cultural evolution (see Section 5 below), what is right or wrong becomes less and less environmentally, but more and more socio-culturally determined. Human societies dispose of language as a means of communication which can refer to the environment in an abstract way or disregard it altogether. Actions that make sense in a particular society and culture are those that are properly embedded within a web of rules, a web that is based on an underlying view of the world, a 'cognized model', as Rappaport (1979) calls it, which has gradually solidified out of the conversation taking place within that society. Or as Peter Weichhart puts it in his paper in this volume: the process of socialization, during which 'meaning becomes implanted as an individual configuration of values, ... represents one of the central means by which individuals are bound to and integrated into a specific social and cultural context'. What it eventually means is that human societies create their own realities. The question then arises as to what extent such realities are compatible or incompatible with the biophysical realities of the environment mentioned above.[11] The reality created by our modern society, as documented beyond any reasonable doubt by the amount of ecological destruction it creates, is clearly incompatible.

19

Expressed in a very simple way one could say that existing socio-cultural structures enable or necessitate a kind of agency and associated mental states of the members of the society that is environmentally destructive. This is not to say that this destruction is all intended. Some of it surely is, or it is taken as unavoidable. Much of what happens, however, is the result of unintended consequences, the greenhouse effect being a good example. In this context the notion of some historians is interesting, that is, that unintended consequences may in fact become driving forces of civilization in the form of 'autonomous processes'. Hoyningen-Huene (1983), with reference to Faber and Meier (1978), explains it thus: the process gets underway if certain initial conditions are met which induce certain motives in individuals. As a result the participants in the process carry out particular actions which have partly intended, partly unintended consequences. The decisive point now is that the unintended, not the intended consequences reproduce the initial conditions such that, cybernetically speaking, there is a positive feedback loop. A good example is the growth of motor traffic. In the beginning, when cars became available, some people got the idea (and had the necessary money) to drive one. Later, a growing rate of private car possession made the development of a settlement structure possible with increasing distances between residences and work places. This structure in turn strengthened the motivation of more and more people to own and drive a car.[12]

The recognition of the human source of environmental problems is stressed as a necessary starting point for human ecology in the contribution by Steiner. An understanding of the interplay between persons and society is seen as a key to the solution of the ecological crisis. Consequently, it is the human sciences that should be given priority over the natural and engineering sciences. This runs counter to the still widespread belief that our society can stay as it is, that is, that it can keep its own 'cognized model' (cf. Note 11), and that, in fact, the natural sciences and technology provide solutions to all problems.[13]

Another aspect of the fact that human societies are language communities, and one that is a source of hope, is the ability for self-reflection. Humans are not puppets on the strings of the social reality they encounter. They can question the validity of the dominant principles in their own society as well as that of their own actions. And they can see themselves as beings that live not just in an environment, but in a universe. Indeed one can say that humans are open not simply to their environment but to the world at large.[14] They can speculate about the birth of the universe and the possibility of its being ruled by a supernatural mind. The history of humankind can be seen as a succession of different kinds of worldviews, a topic taken up in the contribution by Steiner, and one that forms a crucial background to the question of the chances we might have to solve the ecological problem.

4 THE PERSON AS A MEDIUM OF INTEGRATION

At the beginning of the development of a general human ecology is a call for integration. By referring to the human ecological triangle one can pose the question of the suitable 'place of integration': person, society or environment? This question assumes that it should be feasible somehow to project the human ecological problem on to one of the corners of the triangle. Thus, to avoid the trap of a one-sided determinism, it would have to be possible for such a projection to afford an investigation of the *mutual* relationships between humans and the environment.

For classical geography the place of integration was the environment in the shape of regions or landscapes: here one could study the interplay between natural conditions in terms of climate, relief, soils, natural vegetation cover and the human use of the land in terms of agriculture, settlements, transportation routes, and so on. Phenomena visible in the landscape could be taken as spatial indicators for underlying socio-economic processes.[15] Eventually, however, there was a mounting dissatisfaction with this approach as the social processes and human agencies behind the alteration of the environment were not the object of investigation, only their results were mapped and interpreted. The practical limitations were, for example, felt in regional planning, which became very much a business of allocating areas to different uses after the fact.[16]

A projection on to the corner of 'society' presupposes the possibility of arguing that human ecological systems are ecological systems in which one species, namely humans, are clearly the dominant part, so that one should be able to solve the problem by asking how nature becomes incorporated into and appropriated by society and culture. As much of sociology has developed in a kind of 'splendid isolation' from environmental questions, there is no established disciplinary tradition here comparable to that of geography on the other side. Nevertheless, the natural environment or components thereof play a role in some of the existing theories of socio-cultural evolution. Relevant in this context is, for example, the work by Eder (1988), which also contains a historical overview of the respective thinking from Marx to Habermas and Luhmann. The latter's theory is described as an instance in which nature reduces from real environment to a factor of social self-creation.[17] At any rate, in comparison with geography we are confronted here with the inverse problem: how can we translate a socially cognized environment back to the real environment? In both cases relationships that would be important for building a bridge between the two corners of the triangle become excluded from consideration.

A more comprehensive view lets us recognize that the causal path between society and environment runs via the person as an acting individual; without human agency there would be no anthropogenic alteration or even destruction of the environment. If, as postulated by Weichhart in his paper, we look at persons in a sufficiently holistic way, it may be appropriate to use some kind of methodological individualism. In other words, we can argue for the possibility of projecting the problem on to the corner of 'person'. Indeed, by just looking at the mental states of human beings, we realize that part of their self and consciousness are products of social processes, other parts of past relations to the environment.[18] This does not mean that the psyche of a person is determined completely by external influences. Instead, a selective appropriation of elements takes place, so that each person carries with him or her an individual inner world. In a sense we could say that this inner world contains a 'small' human ecological triangle[19] and that a partial explanation for the ecological crisis can be found in its insufficient integration.

Nevertheless, we should not restrict ourselves to defining personal identities exclusively in terms of the accumulated effect of past relations to society and environment. A person is, as Weichhart calls it, 'the pivot of a double duality'. One of those dualities is the interplay between agency and social structures that we have been discussing above. The other is the interplay between agency and the environment. These dualities do not just express themselves as residues of the past, but as continuously living relationships in the sense that a social as well as an environmental identity become constantly reproduced and recreated by appropriate observable external relations, that is, human actions. Weichhart stresses the fact that to acknowledge the existence of dualities entails the necessity of choosing relational approaches for scientific investigation. With reference to the psychological literature he points to a worldview shift to what is called a 'transactional view'.[20] On the person–society side he identifies the structuration theory by Giddens (see Part III of this volume), on the person–environment side the ecopsychological approaches by Carello (this volume) and by Lang (this volume) as instances of such a view.[21] Going yet a step further one might say that the ecological crisis can be understood as a case of inappropriate or insufficient personal identities: they do not include those external relations as constitutive parts. To overcome the problem we need in the sense of Naess (1989) a larger Self (as opposed to a small self) within which those relations become internal. Presumably a similar point could be made with regard to social identity and social relations.

5 EVOLUTIONARY PERSPECTIVE

We have pinpointed the root of the environmental problem in the interplay between persons and society and the way this interplay can lead to a created reality which is more or less in violation of biophysical principles that would afford sustainability. It should thus be clear, that if we agree on the premise that something should be done to stop the wholesale planetary destruction taking place at present, we must look at the history of the development of socio-cultural realities and the possible resulting effects on the environment.

We must try to understand what has happened in the past, why it has happened, and what possible bearing it might have on our present predicament and on possible attempts to steer clear of the impending disaster.[22] In other words, we must investigate what messages we can get by looking at the process of cultural evolution.[23] Boyden does this by using, for his consideration of biophysical actualities, what he calls a 'biohistorical' approach. To him this is important 'not only to understand the past, but also to plan for the future'. He distinguishes four phases of socio-cultural development: the primeval, the early farming, the early urban, and the high energy phase. His interest is (as already indicated by the labelling of the phases) in the biophysical consequences of this sequence as, for example, the amount of energy use. By just looking at those statistics it becomes evident that we cannot go on like this.

Much weight to an evolutionary perspective is also given in the contribution by Dieter Steiner. According to the standpoint taken there the emphasis is on the human side, that is, on the development of socio-cultural structures and their possible role in fostering environmentally unsound practices. Cultural evolution is described as a process which unfolds at an increasing pace from the stage of archaic societies to that of political societies and culminating in today's economic society spanning the whole globe. If we note that the first two stages are separated by the so-called neolithic revolution, during which the practices of agriculture and urban living developed, and the last two by the industrial revolution, we recognize, by collapsing Boyden's second and third phase into one, the parallel with his classification. The evolutionary analysis permits a search for signs of a possible 'malevolution' and to discuss the question of what a future post-industrial or a post-economic society might look like.[24]

Beyond trying to reconstruct the sequence of events that led from the early days of hominids to the humans of the present in order to understand what happened and to differentiate the reversible from the irreversible, an evolutionary perspective at a more general level can be a valuable orientation, particularly when arguing about the priority of various system rationalities. Evolution can be understood as a series of recursive systems embedded within recursive systems.[25] As an outer system provides a background that is existential for the functioning of an inner system, it would seem to be obvious that the rationale inherent in an outer system must take precedence over the rationale associated with an inner system. But what is rational? For the present discussion we may adhere to the concept of 'functional rationality' as used by Dryzek (1987: 25). Rationality applies first of all to human social systems: 'To describe a human social structure as functionally rational means, first and foremost, that its organization is such as to consistently and effectively promote or produce some value.' But Dryzek (1987: 35) extends the concept to non-human ecosystems: 'Setting aside ... the question of human interest, an ecologically rational *natural* system is one whose low entropy is manifested in an ability to cope with stress and perturbation, so that such a structure can consistently and effectively provide itself with the good of life support.' Dryzek (1987: 58) then argues for the primacy of ecological rationality, because 'the preservation and enhancement of the material and ecological basis of society is necessary not only for the functioning of societal forms . . ., but also for action in pursuit of *any* value in the long term.' If viewed against the background of an evolutionary perspective this argument gains in lucidity. Moreover, it would be possible more generally to derive a hierarchy of rationalities, whereby an evolutionarily older rationality is more encompassing than a younger one, something which Dryzek does not do. Applied to the sequence of basic recursive systems appearing during cultural evolution it would mean that socio-cultural rationality[26] should be prior to political rationality, and political rationality in turn prior to economic rationality.

If we can develop a concept of priorities or hierarchy regarding rationalities, and if rational systems have, as described by Dryzek, the property of value production, then by the same token we might want to apply an evolutionary concept to an order of values. This is what Rolston (1985: 39) tries to do. His scheme does not refer explicitly to an evolutionary framework, but it definitely would make sense in such a context. Thereby all is not as simple as it sounds. Consider the following example: 'Murder is not justified to obtain wilderness, even supposing society preferred this.' Still, one can talk about the fact that we live today with an upside-down hierarchy of values, and this is discussed as an instance of 'moral inversion' in Steiner's contribution. Getting things in the right order again amounts to a reattaching of a posteriori value systems to the whole of nature. We may wish to describe such a reattachment by the Latin term *religio* to insinuate our belief that it will unavoidably lead to the question of a religious dimension, a topic taken up in the following section.

6 TRANS-SCIENTIFIC SELF-REFERENCE

We have noted the human ability for critical self-reflection above, and this, of course, also applies to a situation in which we start developing an inter-disciplinary human ecology. As pointed out by Weichhart in his contribution, the use of a metaphorical framework such as the human ecological triangle (person–society–environment) entails the danger of a conservative approach in which the triangle is analytically decomposed into its corners as units or subsystems. Such a distinction implies oppositions between them. It is for this reason that Weichhart's paper carries the subtitle 'from dualism to duality'. As we have noted above, he claims that we need an integrative relational approach in which person and environment on the one hand and person and society on the other appear simply as complementary aspects of one and the same larger unit of investigation. However, how do we actually achieve an integration by dealing with relations? This is a question asked by Markus Huppenbauer in his paper. He sees some usefulness in doing this by means of a physical or bioeconomic approach which relies upon the tracing of material or energy flows (compare with the method proposed by Pillet in this volume). Eventually, however, all of this is clearly reductionistic in that a complex human ecological situation is collapsed into one dimension. How can we do better and find what 'connects such different things as personal emotions, social symbols and processes of the biosphere?' (Huppenbauer). Do we need a new kind of science for this? Or is it unavoidable that we have to transcend science, regardless of its being old or new?

Huppenbauer also points to a perhaps even more difficult problem. Almost everybody who has recognized the present existence of a serious ecological threat to humankind and the whole biosphere reacts by saying that we should stop talking about this threat and start some corrective action. However, the 'primacy of practice' can be identified as a basic characteristic of the technological orientation of our modern economic society,[27] and one that has given rise to the environmental crisis in the first place. What should we do? Nothing? Indeed there are people, such as Chargaff (1989), who do not see any alternative to present science (that is, in the form of alternative science) except its complete abandonment. An opposite view (expressed, for example, by Primas 1992) maintains that science should and can be transformed into a more holistic undertaking in which the acquisition of knowledge becomes of interest for its own sake and not only when it can be applied. Also, one may argue that we still need science even in its present form to help us to repair some of the damage and to serve as an alarm clock (as an example, consider the detection of the ozone hole). However, being at one and the same time a source of destruction and a means to detect and possibly counteract that same destruction is a fundamental conflict which present science will not be able to bear for too long: a transformation is unavoidable.

A further point concerns the choice of an evolutionary perspective. This is itself an event taking place in the ongoing cultural evolution. What is right today may be wrong tomorrow. This is perfectly obvious. What it clearly means is that any kind of serious human ecology cannot be a business engaged in the search for some absolute and permanent truth. Rather it should be regarded as a delicately balanced, intellectually and emotionally motivated participation in a process in which humankind tries to find alternative ways of living such that it treads on this planet more lightly than hitherto. To this end, knowledge in the form of 'true' scientific facts may be gathered in large quantities, but they will remain useless if their selection, interpretation and use is not guided by superimposed ethical principles that refer to the fundamental question of how humans should relate to this world.[28]

It is in this sense that we cannot envisage a general human ecology which is based solely on scientific thinking, but only one which embraces trans-scientific components of a philosophical and religious nature. The need for such an opening is stressed by Steiner as well as by Huppenbauer. They see the importance of some form of religious binding, be it to provide a basis for a more participatory attitude of humans towards the rest of the world, or as a means to remind us of our limits. As Mynarek (1986) argues, the ecological aspect is a key element of any religion; all genuine religions are basically ecological religions, though this fact may not be within the consciousness of the people adhering to it.[29]

Finally, let us note that the step to philosophy, or religion for that matter, as something 'above' science eventually is at the same time a step to the real lifeworld 'below' science. This is so because only those philosophical and religious arguments can have any impact which deal with the real problems and experiences of real humans in this real world. It is the closure of what is called the 'human ecological circle' in the paper by Steiner. According to Mynarek (1986) the religious dimension, the search for sense and meaning in this world, is eventually a fundamental existential problem of any human being. For a time we may have suppressed the insight that the ecological aspects of our existence are part of this religious dimension, or we may have forgotten completely that there are ecological aspects to start with. We have begun to relearn this basic fact. But, as pointed out by Mynarek, knowledge is not equal to religious consciousness right away. The one can become the other, however, if our knowledge about being embedded in nature turns into an experience. In a closer, self-referential sense it also means that those people claiming of themselves to be human ecologists will be so convincingly only if it is evident not only in their scientific activities, but equally well in their everyday life, in the way they act, think, relate to other humans and treat the environment. Human ecology cannot be business as usual.

7 NOTES

1 The concept of complementarity derives from quantum physics, where it has a very precise meaning (Primas 1992). This is not so for the human sciences and, consequently, the term is employed here in a fairly loose fashion.

2 An example of such a 'big idea' is the concept of 'sociobiology' by Wilson (1975), of which Geertz (1983: 21) says: 'that curious combination of common sense and common nonsense'. The case of sociobiology leads one to surmise that, at least outside physics, grand theories can come about only in the wake of extreme forms of reductionism.

3 Translated from the German original by the editors. It is interesting to note that the same author elsewhere sees the 'unity of nature' in the fact that all world phenomena can be reduced to the level of physics and rebuilt so to speak from there by means of cybernetic approaches (von Weizsäcker 1972).

4 Cf. Section 4 in this Introduction.

5 Clearly this is a first step towards a 'structuration' of the biophysical environment, a topic taken up in Part III.

6 Compare with the list of 'life conditions conducive to health in *Homo sapiens*' in Boyden (1987: 79). The first item mentioned is 'clean air (i.e. "palaeolithic air" – not contaminated with hydrocarbons, sulphur oxides, lead, etc.)'!

7   A present example is AIDS, the topic of Gould's paper in this volume.

8   The problem of 'finite resources' must be looked at in material terms (raw material consumption as well as garbage production), in terms of the availability of land, and in terms of energy and entropy. The first aspect is discussed, for example, by Jaeger (1991: 193) with his concept of 'technomass', which 'refers to material entities whose existence and/or position must be explained as an effect of human action'. The question then is how much technomass our planet can bear. A promising example for the treatment of the second aspect is provided by Wackernagel (1991) with his concept of 'appropriated carrying capacity' (ACC) applied to human populations (rather than regions) and expressed in terms of land needed '(a) to provide all the resources the people consume, and (b) to absorb all the waste they discharge'. It provides a tool to assess the performance of an economy associated with a given human population in ecological terms, i.e. sustainability. Finally, the last aspect is discussed in terms of the 'entropy margin' by Jaeger (1991) and Dryzek (1987), for example. 'The severity of ecological problems can ... be captured in terms of the extent to which low entropy is being depleted' (Dryzek 1987: 22). The entropy discussion builds on the work of Georgescu-Roegen (1971) and is being followed up by the activities of a newly formed 'European Association for Bioeconomic Studies'.

9   For a thorough critique of conventional environmental economics and a discussion of its fallacy in a larger socio-cultural context, see Furger (1992).

10 In cultural anthropology, for example, this shift is marked by a transition from more system–theoretic, adaptionist models with certain affinities to structuralism to more actor-based models (see Orlove 1980).

11 Rappaport (1979) contrasts his notion of 'cognized models' with that of the 'operational model'. The latter looks at the interrelationship between a human society and its environment on the grounds of scientific ecology. By assuming that an operational description is closer to reality, such a model may be taken as a basis for comparison with the ecological rationality of a cognized model. However, we should not forget that even a scientific model is, in the end, a human construct and, hence, a cognized model, albeit of a special kind. Conversely, it may even be that a cognized model, even though according to scientific standards it is based on conceptual nonsense, is superior in terms of ecological sustainability, simply because in the context of archaic people it provides a basis for implicit regulation, while a scientific description in the context of a modern society does not.

12 De Rougement (1980) argues that the advent of the car has led to the most intensive transformation of Western society since Napoleon, and he calls it 'a story of madness'.

13 We note that there may be reasons other than those associated with a 'technofix' mentality for claiming a converse priority for the natural sciences. For example, Golley (1991: 52–3) argues that 'human ecology viewed from biological ecology is the broadest, most inclusive, wholistic perspective' while human ecology seen from the human sciences 'is a specialty, a subject with no clear home', because the 'anthropocentric disciplines' are 'fractured into a complex mosaic'. Yet this is exactly why we stress the need for an interdisciplinary human ecology.

14 This is a point made by philosophical anthropologists such as Max Scheler who, in his work 'Die Stellung des Menschen im Kosmos' (1928), says that human beings are in a certain sense 'environment-free', that is, they do not live in an environment, but they *have* world (after Störig 1985).

15 According to Hartke (1959) such indicators enable one to register actions and reactions in a way similar to the registration of images on a photographic plate.

16 However, this does not mean that the regional approach is not useful at all or that

it is completely dead. In fact, it re-emerges in a new form because of the practical significance of regions in a world which tends to globalization, a world which without them as buildings blocks would end in chaos. Several aspects of this topic will be the theme of the contributions in Part IV of this book.

17 For a brief discussion of the theories by Habermas and Luhmann, see Part IV.

18 Cf. the discussion of different levels of consciousness in Steiner (this volume).

19 I owe this metaphor to Dagmar Reichert.

20 Note the kinship to the notion of self-organizing recursive systems within an evolutionary worldview in the article by Steiner.

21 In another paper, Lang (1988: 93, Abstract) 'pleads for a renewal of the image of man in psychology in the sense of a Copernican Turn' and he continues to say that 'foremost a decentration of the psychological research object from the individual towards the person–environment–unit is proposed'.

22 What Lerner (1986) says about the importance of the historical approach in the narrower context of the search for an explanation of the emergence of patriarchy also applies to the wider context of human ecology in general. She points to the fact that we probably will never know exactly what happened and, therefore, must rely on speculations about what might have been possible, speculations which have an important function: 'to know what might have been possible opens us up to new interpretations. It allows us to speculate about what might be possible in the future, free of the confines of a limited and entirely outdated conceptual framework' (Lerner 1986: 38).

23 Our interest in the cultural evolution is therefore largely of an empirical nature. Also the statement made in the contribution by Steiner, namely that evolution as a whole can be seen as a sequence of recursive systems, is a *post factum* interpretation suggesting an observable repetition of patterns. The notion of an evolutionary perspective should not be confounded with the adherence to an evolutionist theory which postulates underlying structures and forces necessitating a certain kind of, usually progressive, development. Such theories are rightly criticized by Giddens (1984).

24 A comprehensive treatment of aspects of cultural evolution as a background for human ecology is provided by Jaeger (1991). What we call 'malevolution' he discusses as a case of 'common madness', as 'a view of mental illness which applies not only to relatively rare individual cases but also to whole populations. . . . Common madness . . . can be studied as the outcome of clusters of nonsensical beliefs' (Jaeger 1991: 168–9).

25 Cf. Figure 4.4 in the contribution by Steiner (this volume).

26 'Social rationality' should be understood here in a narrow sense, referring to social interactions as face-to-face encounters between humans who know and trust each other. As such it is the basic integrating force in archaic societies.

27 At the root of this orientation is our drive toward quantification and operationalization. As a result, the modern Western mind is, in its nature, technical (Zimmerli 1988).

28 Cf. the discussion of 'the moral dimension' in Introduction to Part II, Section 3.

29 This statement implies a critical attitude toward the present state of Christian religion. In the introduction to his book, Mynarek (1986: 9) says: 'it became obvious to me that many religious people, among them also quite a number of Christians, are simply not ready any more to live their religiosity purely vertically in the direct relationship "humans – God"; that, on the contrary, they cannot be religious any more without also including nature, the universe, the animals, the plants, the elements and landscapes into their now broader religious feeling' (translated from the German original by D.S.).

8 REFERENCES

Boyden, S. (1987) *Western Civilization in Biological Perspective. Patterns in Biohistory*, Oxford: Clarendon Press.

Chargaff, E. (1989) 'Gibt es Alternativen zu unseren gegenwärtigen Naturwissenschaften?' *Neue Rundschau* 100 (4): 5–15.

Daly, H.E. and Cobb, J. (1989) *For the Common Good: Redirecting the Economy towards Community, the Environment, and a Sustainable Future*, Boston, MA: Beacon Press.

Dryzek, J.S. (1987) *Rational Ecology. Environment and Political Economy*, Oxford: Blackwell.

Eder, K. (1988) *Die Vergesellschaftung der Natur. Studien zur sozialen Evolution der praktischen Vernunft*, Frankfurt a/M: Suhrkamp.

Faber, K.-G. and Meier, C. (eds) (1978) *Historische Prozesse*, München: Deutscher Taschenbuch-Verlag.

Freese, L. (1988) 'Evolution and sociogenesis. Part 1: Ecological origins', in E.J. Lawler and B. Markowsky (eds) *Advances in Group Processes 5*, Greenwich, CT.: JAI Press.

Furger, F. (1992) 'Ökologische Krise und Marktmechanismen – Umweltökonomie in evolutionärer Perspektive', PhD thesis, Zürich: Department of Geography, ETH (publication forthcoming).

Geertz, C. (1983) *Local Knowledge. Further Essays in Interpretative Anthropology*, New York: Basic Books.

Georgescu-Roegen, N. (1971) *The Entropy Law and the Economic Process*, Cambridge, MA: Harvard University Press.

Giddens, A. (1984) *The Constitution of Society. Outline of the Theory of Structuration*, Berkeley and Los Angeles: University of California Press.

Giddens, A. and Turner, J.H. (eds) (1990) *Social Theory Today*, Cambridge: Polity Press.

Golley, F.B. (1991) 'Reasoning from ecological knowledge to problem solving', in M.S. Sontag, S.D. Wright and G.L. Young (eds): *Human Ecology. Strategies for the Future*, Fort Collins, CO: Society for Human Ecology.

Hartke, W. (1959) 'Gedanken über die Bestimmung von Räumen gleichen sozialgeographischen Verhaltens, *Erdkunde* 13 (4): 426–36.

Hoyningen-Huene, P. (1983) 'Autonome historische Prozesse – kybernetisch betrachtet', *Geschichte und Gesellschaft* 9 (1): 119–23.

Jaeger, C. (1991) 'The puzzle of human ecology. An essay on environmental problems and cultural evolution', habilitation thesis, Zürich: Department of Environmental Sciences, ETH (publication forthcoming).

Lang, A. (1988) 'Die kopernikanische Wende steht in der Psychologie noch aus! Hinweise auf eine ökologische Entwicklungspsychologie', *Schweizerische Zeitschrift für Psychologie* 47 (2/3): 93–108.

Lerner, G. (1986) *The Creation of Patriarchy*, New York and Oxford: Oxford University Press.

Messerli, P. (1988) 'Erfahrungen aus einem inter-disziplinären Experiment: Das schweizerische MAB-Programm', in D. Steiner, C. Jaeger and P. Walther (eds) 'Jenseits der mechanistischen Kosmologie – Neue Horizonte für die Geographie?', *Berichte und Skripten* 36, Zürich: Department of Geography, ETH.

Mynarek, H. (1986) *Ökologische Religion. Ein neues Verständnis der Natur*, München: Goldmann.

Naess, A. (1989) *Ecology, Community and Life Style. Outline of an Ecosophy*, translated and revised by D. Rothenberg, Cambridge: Cambridge University Press.

29

Orlove, B.S. (1980) 'Ecological anthropology', *Annual Review of Anthroplogy* 9: 235–73.

Primas, H. (1992) 'Umdenken in der Naturwissenschaft', *Gaia* 1 (1): 5–15.

Rappaport, R.A. (1979) *Ecology, Meaning, and Religion,* Berkeley, CA: North Atlantic Books.

Rolston, H. (1985) 'Valuing wildlands', *Environmental Ethics* 7 (Spring): 23–48.

Rougement, D. de (1980) *Die Zukunft ist unsere Sache,* Ch. 7: 'Erste Geschichte des Wahnsinns: Das Auto', Stuttgart: Klett-Cotta.

Störig, H.J. (1985) *Weltgeschichte der Philosophie,* Zürich: Ex Libris.

Wackernagel, M. (1991) 'Assessing ecological sustainability: indicating the sustainability of a region by measuring its appropriated carrying capacity', manuscript, Vancouver, BC: School of Community and Regional Planning, University of British Columbia.

Wallerstein, I. (1990) 'World-systems analysis', in A. Giddens and J. Turner (eds) *Social Theory Today,* Cambridge: Polity Press.

Weizsäcker, C.F. von (1972) *Die Einheit der Natur,* München: Carl Hanser.

Weizsäcker, C.F. von (1977) *Der Garten des Menschlichen. Beiträge zur geschichtlichen Anthropologie,* München: Hanser.

Wilson, E.O. (1975) *Sociobiology. The New Synthesis,* Cambridge, MA and London: Harvard University Press.

Zimmerli, W.C. (1988) 'Technik als Natur des westlichen Geistes', *Information Philosophie* 16 (5): 5–17.

HUMAN ECOLOGY AND BIOHISTORY

Conceptual approaches to understanding human situations in the biosphere

Stephen Boyden

1 INTRODUCTION

Any discussion on the theme 'human ecology as an integrative science' immediately presents us with a problem, because 'human ecology' means quite different things to different people. Thus, when sociologists speak of human ecology, they often have in mind the 'Chicago School of Human Ecology' which came into being in the 1920s, and which was concerned essentially with the spatial distribution within cities of different social groups (Burgess 1925, Shevky and Williams 1949, Shevky and Bell 1955, Abu-Lughod 1969). To another group of workers, human ecology is concerned almost solely with the interaction between the human species and micro-organisms, and when used in this sense it would seem to be more or less synonymous with the epidemiology of infectious disease (Banks 1950). To other authors, human ecology can be synonymous with anything from home economics or the psychology of personal relationships to the study of patterns of energy flow in urban ecosystems.

It would not be sensible to enter here into a lengthy discussion and criticism of the various meanings of 'human ecology'. The subject has been well discussed by a number of authors over the years (Quinn 1950, Bates 1953, Shephard 1967, Craik 1972, Bruhn 1974, Sargent 1974, Young 1974, Young 1983). Because of this semantic problem, it is necessary at the outset for me to explain the sense in which I am using the term 'human ecology' in this paper. In fact, in order to avoid confusion with other forms of human ecology, my colleagues and I are now referring to our approach to human ecology as *biohistory* (although we referred to it simply as 'human ecology' for many years).

The biohistorical approach to the study of humans in their environments that I will be discussing in this paper is based on an inescapable aspect of reality, namely, that all human situations involve a highly significant interplay between biophysical and cultural processes. It is reasonable to argue, therefore, that if we wish to improve our understanding of human situations we should pay attention to this interplay in its own right. I will also be putting forth the view that knowledge of patterns of culture–nature interplay in human history can be helpful in our efforts not only to understand the past but also to plan for the future.

There are two points I should make clear at the outset. First, I am not intending to discuss the interrelationships and distinctions between the paradigms of different academic disciplines that are relevant to our understanding of the human condition, such as geography, sociology, biology, psychology and history, nor do I intend to compare them with the conceptual framework of biohistory. Nevertheless, the proper development of biohistory as a system of knowledge in its own right depends, of course, on information derived from these, and other, areas of learning.

The second point that needs stressing is that biohistory does not involve 'organic analogy'. That is, it does not try to apply biological principles and laws to societal or cultural systems. However, it does recognize that such systems contain biological parts, and that biological principles are of the greatest significance in our efforts to understand the behaviour of, and parts played by, these biological components of human situations.

I define biohistory, therefore, as an approach to the study of human situations which reflects the broad sequence of happenings in the history of the biosphere, from the beginning of life to the present day (Figure 3.1). Its starting point is the study of the history of life on earth, of the basic principles of evolution, ecology and physiology, and of the sensitivities of living organisms and of ecosystems. Next it considers the evolutionary background, biology and innate sensitivities of humans , and the emergence in evolution of the human aptitude for culture and its biological significance. It then considers abstract culture itself, and is especially concerned with the interactions between cultural processes and biophysical systems, such as ecosystems and human populations. Patterns of culture–nature interplay are an overriding theme in biohistory.

Biosphere ⟷ Humans ⟷ Culture

Figure 3.1 Conceptual framework (Version 1)

A slightly more elaborate version of the model is shown in Figure 3.2. This model will be used as a framework for much of the discussion in this paper.

In Figure 3.2 a number of additional clusters of variables are introduced:

(a) Artefacts: these are the human-made components of the biophysical environment or the biosphere, including the built-environment, roads, machines, clothing and works of art.

32

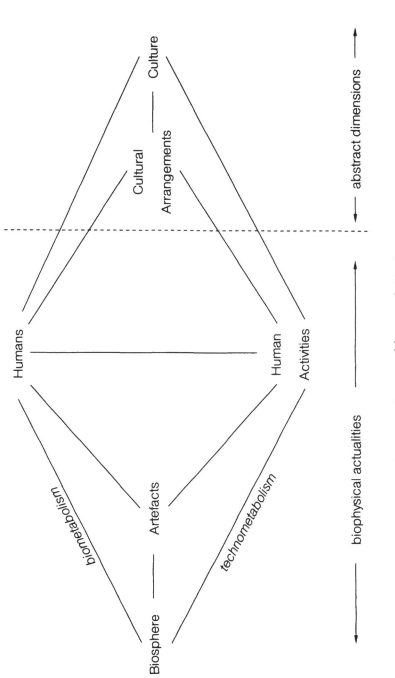

Figure 3.2   Conceptual framework (Version 2)

(b) Human activities: as an aspect of the human component of the total system, these cover all things that people do, including such societal activities as farming, manufacturing and transportation.

(c) Cultural arrangements: these are an aspect of abstract culture and include such factors as the economic system, social hierarchies and legislation.

(d) Technometabolism and biometabolism: technometabolism refers to those inputs and outputs of materials and energy into and out of society which are the consequence of technological processes. Biometabolism is the inputs of materials and energy (in air, food and water) into human organisms and outputs of organic wastes (including carbon dioxide). From the ecological standpoint, technometabolism is one of the most significant outcomes of the human aptitude for culture.

Biohistory is especially concerned with the impacts of societal activities on the biophysical variables of the biosphere, and on humans as biological beings. It also embraces the study of the adaptive processes, biological and cultural, that come into play when societal activities have impacts on biological systems which are disadvantageous, or which are perceived to be disadvantageous, for humankind. It pays attention especially to the processes of cultural adaptation that may be brought into action in response to culturally induced threats to human survival and well-being. Such cultural adaptive responses have been very important in human history; and whether or not humankind survives the next century will depend on the extent to which they are successful in the near future.

Biohistory is concerned not only with unravelling and describing significant interrelationships in human ecosystems between cultural and biophysical variables, but also with the identification of fundamental principles that help us to understand the nature of the constraints imposed on human society by virtue of its dependence on biological systems and processes. Some of these principles derive directly from the biological and social sciences and include, for example, principles relating to thermodynamics, biogeochemical cycles, soil ecology, natural selection, physiology, health and disease, alienation, anomie, and corporate behaviour.

Other principles derive from biohistory itself and specifically concern the interplay between biophysical and cultural processes. I will briefly mention a couple of examples to illustrate this point. The first, the principle of evodeviation, is, in my view, of great significance in biohistory. It is, in fact, a general biological principle, but in the context of the human species living under conditions of civilization it takes on special significance, and it is of relevance to our understanding of many aspects of culture–nature interplay in human history and at the present time. The principle arises from the fact that the processes of evolution produce populations with genetic characteristics such that the individuals perform optimally, in terms of survival and reproduction, in the ecological niches in which they are evolving. Thus, evolutionary forces select for optimum performance in the given environment. It follows, therefore, that if the conditions of life suddenly deviate from those of the natural habitat (to which the species has become adapted through natural selection), then it is likely that the individual animals will be less well suited, in either their physiological or their behavioural characteristics, to the changed conditions. Consequently some signs of maladjustment, physiological or behavioural, are likely to be evident. The adjective 'evodeviant' is used to describe life conditions which are different from those which prevailed in the natural habitat of the species. Disturbances in physiology or behaviour which result from evodeviations are referred to as examples of phylogenetic maladjustment.

This principle does not infer that every conceivable evodeviation in the conditions of life of a species will necessarily give rise to maladjustment. It does suggest, however, that any definite deviation in conditions from those that prevail in the natural habitat of an animal should be regarded as a potential cause of maladjustment until proven otherwise.

The principle of evodeviation applies as much to humans as it does to any other species. The phylogenetic characteristics of our species, including its universal health and survival needs, were determined by the selection pressures operating before and during the long hunter-gatherer phase of human existence. Cultural developments over the past 10,000 years have brought about many biologically significant changes in the conditions of life of the mass of humanity. As a consequence of these culturally induced changes in human life conditions, there have occurred countless examples of phylogenetic maladjustment, including numerous forms of nutritional, infectious and organic diseases. In fact, all 'diseases of civilization', ranging from scurvy, beri-beri, typhoid, cholera, smallpox, and influenza to coronary disease and most forms of cancer are examples of phylogenetic maladjustment (Boyden 1987).

Another biohistorical principle of considerable importance is the principle of technoaddiction. In human history it has frequently been the case that, when new techniques have been introduced into a society, they have not been really necessary for the satisfaction of the survival and health needs of the population. Sometimes they have been introduced simply for curiosity and sometimes because, in one way or another, they have benefited a particular individual or group within society. With the passing of time, however, societies reorganize themselves around the new techniques and their populations gradually become more and more dependent on them for the satisfaction of basic needs. Eventually a state of complete dependence is reached. Clearly, already by 7,000 BC, the population of Çatal Huyuk, in southern Anatolia, had become dependent on farming for its survival, despite the fact that a small part of the food supply still came from hunting. The dependence of the populations of high-energy societies on machines driven by fossil fuels is a more recent example.

This principle of technoaddiction is of importance in our attempts to understand the ecological and biosocial problems facing human society today, and in our efforts to plan for a better future.

2 THE CURRENT HUMAN SITUATION IN BIOHISTORICAL PERSPECTIVE

Brief comment is necessary here on the current global situation. Much has been written in very recent times about the ecological load now imposed by the human species on the biosphere (Ehrlich *et al.* 1977, Boyden 1987, World Commission on Environment and Development 1987, Goldsmith and Hild-yard 1988, Worldwatch Institute 1989). Here we can only touch on a few salient points.

In biohistorical terms, we can recognize four distinct ecological and biosocial phases in human history. These phases are as follows:

1 The primeval (hunter-gatherer) phase, which was by far the longest of the four phases.
2 The early farming phase, which began in some regions of the world 11–12,000 years ago.
3 The early urban phase, which began in some regions 5–9,000 years ago.
4 The high-energy phase, which phase began in some regions 150–200 years ago. It will be by far the shortest of the four phases.

There is not time to discuss here the distinguishing ecological and biosocial characteristics of these four phases. However, from the standpoint of the theme of this paper, the important fact is that the fourth, high-energy phase has a number of outstanding characteristics which are, in the long term, not ecologically sustainable. The most intensive examples of high-energy societies are those of western Europe, the USA, the USSR, Japan, Australia and parts of southern Africa. However, the impact of these societies and the value systems on which they are based is now felt throughout the world.

Since the time of the domestic transition, which marked the beginning of the early farming phase, the human population has increased about 1,000 fold. Even more significant ecologically is the impact that human society has had on the ecosystems of the biosphere during ecological Phase 4 that has resulted from the invention and use of machines powered by extrasomatic energy (e.g. fossil fuels). As a consequence of this development, the ecological impact of the human species on the biosphere (as expressed in terms of energy use) is now about 10,000 times greater than it was at the time of the domestic transition. Half of this increase has occurred in the past 25 years and three-quarters in the last 40 years.

It is certain that the biosphere, as a dynamic system capable of supporting the human species, will be unable to tolerate indefinitely this continuing intensification of technometabolism (that is, use of energy and resources and discharge of technological wastes by the human population). Phase 4 human society is not ecologically sustainable.[1]

Not all human populations are contributing equally to this extra ecological load on the biosphere. The developing Third World countries, although containing over three-quarters of the world's population, contribute only one-fifth of its technometabolism.

Associated with this overall intensification of technometabolism, there has been a massive increase in the per caput technometabolism in the Western countries. In North America, the average individual uses about 100 times as much energy as his or her hunter-gatherer ancestors, and about twenty times as much as the average individual in Shakespearian Britain. In the high-energy societies we are each consuming, for example, about half a tonne of iron each year, over 20 kg of stones and gravel per day, and giving off 20 tonnes of carbon dioxide a year (which is about 100 times the amount we give off through biological respiration).

It is not known how much longer the biosphere would be able to withstand the increasing ecological impacts being imposed on it by the high-energy societies, nor is it known what specific ecological changes represent the greatest threat to the system. However, there are already ominous signs of ecological disturbance at the global level, including the greenhouse effect, the hole in the ozone layer, the effects of acid rain, and widespread desertification and changes in the oceans.

That these signs should already be evident is not surprising, in view of the fact that the additional ecological load now imposed on the biosphere by human society is, in energy terms, equivalent to about 5 per cent of the total ecological load imposed by all other animals and plants put together. If the technometabolism of human society as a whole were to continue to intensify at the same rate as it has over the past twenty years, by the year 2100 human beings would be using as much energy, and consequently having as much impact on the biosphere, as all other existing forms of life. In fact, it is clear that the biosphere as a system capable of supporting humanity would not be able to survive this eventuality. The authors of the Brundtland Report are of the opinion that it would not be able to withstand an increase in technometabolism in the developing countries such that it reached the present level of the Western nations (that is five times the present global level of technometabolism; World Commission on Environment and Development 1987: 14).[2] Indeed, as already mentioned, it is clear that the biosphere will not be able to tolerate indefinitely even the present pattern of techno-metabolism.

The present situation can be summarized in terms of certainties and uncertainties. The certainties are:

1   There are limits to the absorptive capacity (of technological waste products and toxic substances), resilience, and adaptability of biological systems – a

principle which applies as much to the biosphere as a whole as it does to local ecosystems and to individual organisms.

2   The present pattern of increasing per caput consumption of resources, discharge of technological wastes and use of energy in human society is ecologically unsustainable. If it is not brought under control through deliberate societal action, it will come to an end either as a consequence of depletion of mineral resources or, more seriously, as the result of irreversible damage to the biosphere caused by technological waste products.

The uncertainties are:

1   How much longer the biosphere can tolerate the present pattern of industrial productivity.
2   Which particular culturally induced environmental changes represent the greatest threat to the integrity of the biosphere.

Basically, with respect to the future, three possibilities exist. First, the biosphere may, as a consequence of the ecological overload imposed on it by human society, collapse as a system capable of supporting humanity and civilization, bringing an end to the human species. Second, there may occur a major ecological catastrophe, but one that leaves some human survivors, who may then begin a new civilization – possibly based on wiser patterns of use of resources and energy.

The third possibility is that the dominant culture of the high-energy societies comes to embrace an appreciation of the monstrous ecological absurdity of the current situation, and sets about designing a new society that is capable of existing in harmony with the natural environment, satisfying the health needs both of the biosphere and of the human population. That is, the possibility still exists that humans might yet, through their aptitude for culture, overcome the current culturally induced threats to their survival, and move into a fifth ecological phase of existence that is ecologically sustainable and humanly desirable.

To conclude this section on the contemporary ecological and biosocial challenges facing humankind, I would like to refer to the fact that a biohistorical analysis of the present human situation leads inevitably to the recognition of a number of ecological imperatives that must be satisfied if civilization is to survive in the biosphere. Four of these imperatives can be stated as follows:

1   The size of the human population must be stable.[3]
2   The overall rate of resource and energy use and of technological waste production by society (that is the intensity of technometabolism) must be steady (or decreasing). This rate must be considerably lower than that characteristic of the high-energy societies at the present time.[4]
3   The organization of society and the economic system must be such that

human health, well-being and enjoyment of life do not depend on continually increasing per caput use of resources and energy and production of technological wastes.

4   The organization of society and the economic system must be such that high rates of employment[5] are not dependent on increasing consumption of the products of resource-intensive and energy-intensive industry or on activities that decrease the productivity and integrity of ecosystems.

3   THE APPLICATION AND ADVANTAGES OF BIOHISTORY

The usefulness of the biohistorical approach relates first and foremost to three inescapable and closely connected aspects of reality.

First, human beings are totally dependent, for their sustenance, health and enjoyment of life, on the underlying set of biological systems and processes which operate in the biosphere, in its ecosystems and in their own bodies. Dependence on these underlying processes is very basic. All of the products of culture – our institutions, ideas, knowledge, machines, computers, high technology, military strength, politics and economics – count for nothing if the societal system of which they are all a part does not satisfy the biologically determined health needs of the biosphere and of our bodies.

Second, every human situation, at the level of individuals, small groups or whole societies, involves continual interplay between biological and cultural elements, and the outcome of this interplay is often very important for human health and well-being or for the ecosystems on which we depend.

Third, human culture has influenced, and now increasingly influences, the biological processes on which we depend and of which we are a part. Some of these influences may be seen, from the anthropocentric point of view, as good and desirable, while others may be seen as bad and undesirable; and some of them threaten the survival of humanity.

The conceptual approach of biohistory has significant contributions to make, I suggest, to our understanding of human situations and problems by providing a realistic integrative framework for both research and education. It formed the basis of a study which my colleagues and I carried out on the ecology of Hong Kong (Boyden 1979, Boyden et al. 1981). This program was adopted by Unesco as the first pilot study in the section of the Man and the Biosphere Program dealing with human settlements (MAB Program, Project Area 11, Integrated ecological studies on human settlements). In this work, we examined changing ecological characteristics of the system as a whole (e.g. patterns of flow of energy and nutrients, production and disposal of wastes and pollutants, the built environment – linking them to societal activities), as well as the actual life conditions and patterns of health and disease in the human population. The study was aimed at improving our understanding of the dynamic interrelationships in urban ecosystems between, for example, patterns of use of energy and resources, the quality of the biophysical environment, and the quality of life of human beings.

4 THE LIMITATIONS OF BIOHISTORY

To repeat my main theme, I advocate an approach to the learning about human situations that is based on the study of the history and processes of life, of the biological background and characteristics of *Homo sapiens*, and of the interplay in history and at the present time between cultural systems on the one hand and biophysical systems on the other. I suggest that it makes good sense to bring together what we know about these aspects of reality as a single coherent system of knowledge – one of immense significance in our efforts to understand ourselves, the society of which we are a part, and our place in the natural order of things.

It is also important, however, to be aware of the limitations of biohistory. One possible difficulty comes immediately to mind. The systems with which it is concerned, containing biophysical, cultural, and human variables – some tangible and some intangible – are very complex. Are they, indeed, too complex to be properly comprehensible by the human mind, too complex to be susceptible to holistic analysis? In my view, this complexity does not present a major problem, because the conceptual framework of biohistory is, itself, quite simple, and the degree of detail in any biohistorical analysis of a human situation is clearly a matter of choice. Again I would emphasize the importance of identifying principles pertinent to our understanding of culture–nature interactions, and of recognizing in human situations, past and present, recurring patterns of interplay between human society and the biophysical environment. Even at a very general level of analysis, biohistory engenders a realistic and useful sense of perspective.

There is, however, a more important limitation of biohistory. Biohistory, as I have defined it, is concerned primarily with biophysical actualities, as distinct from cultural arrangements (see Figure 3.2). Nevertheless, of course, some of the biophysical actualities, such as practices in farming, mining, manufacturing, building and transportation (all of which have important impacts on the ecosystems of the biosphere and on humans), are largely determined by economic, institutional and other cultural arrangements. In Thailand, for example, highly significant changes have taken place over the past few decades in the distribution of the human population, the life experience of different sectors of the population, the quality of air and water and the bioproductivity of ecosystems; and these changes are the results, direct and indirect, of deliberate changes in the economic arrangements and societal organization initiated by governmental decision-makers. These cultural arrangements, in turn, have important links with culture itself, including, for example, the dominant value system, worldview and assumptions of modern Thai society.

While it is thus clear that culture and cultural arrangements are major determinants of societal activities, and consequently of ecological and bio-social conditions, it is also apparent that they involve processes of a kind that are quite different from those which take place at the level of biophysical actualities. Consequently, in attempting to understand the role of the economic system and other aspects of cultural arrangements in human ecosystems, or in considering societal options for the future, the biohistorian needs to collaborate with social scientists and others who are familiar with how the abstract cultural aspects of society operate.

Because of the critical importance of these interrelationships between biophysical actualities and cultural arrangements, there is an urgent need to devise much more effective means than exist in academia today for bridging the gap between students of these two different aspects of reality. We need to put more effort into deliberately addressing the interface between these two aspects of the total system, and between the two corresponding areas of academic endeavour.[6]

5 THE FUNDAMENTAL QUESTIONS PROGRAM

I would finally like briefly to mention a project of research and communication which is at present underway at the Australian National University and which we hope provides a framework for promoting more effective interaction between students of biohistorical actualities and students of culture and cultural arrangements. This project is known as the Fundamental Questions Program, and it is based on appreciation of the fact that the biosphere, as a system capable of supporting humankind, will not tolerate indefinitely the present pattern of resource and energy use and waste production characteristic of modern society. The program is designed to promote research and systematic discussion on the implications of this fact for human society in general, and for Australian society in particular.

The program can be regarded as consisting of three parts as follows (see also Figure 3.3).

Part I: Integrative ecological and biosocial assessment of the human situation in Australia

This part introduces the integrative conceptual framework of the program and summarizes ecological and biosocial trends in the human situation in historical perspective, first at the global level but concentrating especially on the Australian scene. It focuses mainly on actualities, as distinct from cultural arrangements, which are mainly the concern of Part II of the program. Thus, it deals with changes in the biophysical world and in human experience, and discusses the relationships between these changes and societal activities, such as farming, manufacturing, mining and transportation.

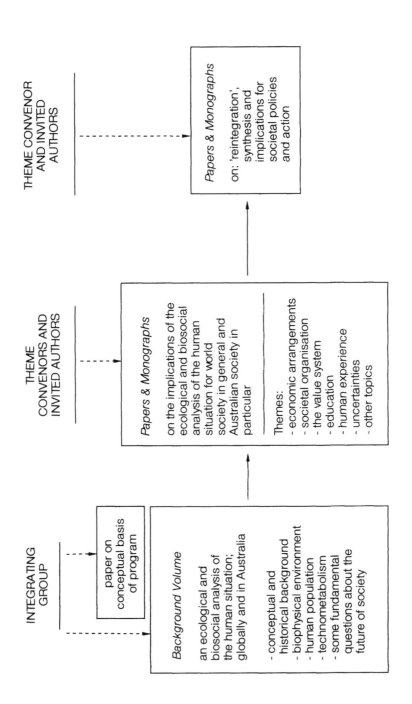

Figure 3.3   Design of the Fundamental Questions Program

Publications include a background volume entitled *Humans in Australia: ecological realities and future directions* and a series of papers analysing and describing aspects of the human situation in biohistorical perspective.

This part of the program raises important questions about the future of human society relating, for example, to the economic system, social organization, educational programs, and value systems.

Part II: Consideration of the societal implications of the ecological and biosocial assessment

Part II of the program is concerned largely with cultural arrangements. It consists of responses of social scientists and other interested individuals and groups to questions about the future of society arising out of the ecological and biosocial assessment discussed in Part I. Authors in Part II of the program will develop their ideas on possible societal scenarios for the future that would be ecologically sustainable and humanly desirable.

Initially, the following themes have been selected for special attention: ecological sustainability and the economic arrangements of society; ecological sustainability and social organization; education for, and in, an ecologically sustainable society; the quality of life in an ecologically sustainable society; societal values in an ecologically sustainable society; the built-environment and ecological sustainability; societal responses to ecological uncertainties; energy options for an ecologically sustainable society.

Part II of the program results in the publication of a series of *Fundamental Questions Papers*, as well as some *Monographs*, each focusing on a particular societal theme. Some of the monographs will be the outcome of workshops on the different themes.

Part III: Reintegration, synthesis and practical policies

This part of the program is the responsibility of a convener and a small team of individuals who are experienced in integration and transdisciplinary scholarship. Their task is: (1) to consider the responses of the contributors to Part II of the program in relation to the ecological and biosocial assessment of the situation developed in Part I; (2) to consider the relationships between the different inputs in the different themes of Part II of the program (e.g. economic arrangements, societal organization, built-environment, value system, education); (3) to analyse the contributions to Part II of the program in terms of agreement and disagreement, and consistencies and inconsistencies; (4) to consider the implications of the recommendations arising from Part II of the program for societal policies at the practical level.

The Fundamental Questions Program is in its early stages, and it is too soon to assess its success. However, the developments to the present time are indeed encouraging. It is hoped that the various publications arising out of the program will fill a serious gap in the socio-environmental literature and that they will ultimately make a significant contribution to policy formulation in Australia.

Incidentally, the design of the Fundamental Questions Program reflects a key principle to be borne in mind in the organizing of any integrative transdisciplinary program of research or education. For such a program to be successful it is essential that there be, at the centre of the program and on a full-time basis, a core of individuals whose main interest and enthusiasm is integrative scholarship. There is a great deal more to integrative, holistic, transdisciplinary studies than merely bringing together a number of experts from different academic disciplines to sit around a table and discuss a human situation, only to have them return to their areas of specialism after their meeting. Someone has to remain at the table, to consider in-depth and systematically the relationships between the different aspects of the overall situation on which the experts have pronounced. In the Fundamental Questions Program, an integrating group plays this role.

6 CONCLUSION

In this paper, I have described a conceptual approach to the study of human situations that I have called biohistory. I am suggesting that, while this approach is not new, it needs to be treated much more seriously as a coherent and meaningful system of knowledge that warrants considerable intellectual attention. This attention should be aimed at (a) improving and making more rigourous its conceptual base, (b) considering its implications for humankind, and (c) communicating its messages as widely as possible.

We also need to explore ways and means of encouraging useful discourse between students of biophysical actualities and students of culture and cultural arrangements. Reference has been made to a program of research and communication at present under way at the Australian National University which aims to achieve this.

Finally, I will conclude with a hypothesis; we refer to it as our 'naive hypothesis'. According to this hypothesis, if our culture were to embrace a biohistorical perspective as a central all-pervading theme, there would come about significant changes in the dominant understanding, assumptions, values and motives of our society. These changes would in turn lead to new policies and new societal arrangements of a kind which would be consistent with the long-term ecological sustainability of civilization and with the equitable satisfaction of the health needs of the human population.

7 NOTES

1 For the purpose of this paper, an ecologically sustainable human situation is defined as one in which the bioproductivity (production of organic matter through photosynthesis) of the ecosystems of the biosphere is indefinitely maintained and in which prevailing conditions satisfy the universal health needs of the human population.

2 It is noteworthy that this conclusion is completely inconsistent with some of the recommendations of this report calling for further intensification of industrial growth.

3 The actual size of this sustainable stable population will depend on its pattern of resource and energy use. The biosphere would not be able to support the present world population at a per caput intensity of technometabolism equal to that of the present-day high-energy societies.

4 For the purposes of this discussion, one-fifth of the present per caput intensity of technometabolism of North America will be taken initially as a reasonable societal objective.

5 The word 'employment' is used here in a broad sense to include all direct or indirect (e.g. wage-earning) subsistence activities that are associated with a sense of personal involvement and purpose.

6 The biohistorical approach does, in fact, have something to offer also to our understanding of societal organization, economic arrangements and abstract cultural variables. This theme is discussed elsewhere (Boyden 1987: chs 9 and 11, Boyden 1992).

8 REFERENCES

Abu-Lughod, J.O. (1969) 'Testing the theory of social area analysis: the ecology of Cairo, Egypt', *American Sociological Review* 34: 198–212.

Banks, A.L. (1950) *Man and His Environment*, Cambridge: Cambridge University Press.

Bates, M. (1953) 'Human ecology', in A.L. Kroeber (ed.) *Anthropology Today*, Chicago, Ill.: University of Chicago Press.

Boyden, S. (1979) *Integrative Ecological Approach to the Study of Human Settlements*, MAB Technical Note 12, Paris: Unesco.

Boyden, S. (1987) *Western Civilization in Biological Perspective: Patterns in Biohistory*, Oxford: Oxford University Press.

Boyden, S. (1992) *The Human Aptitude for Culture and Its Biological Consequences*, Perth: University of Western Australia (in press).

Boyden, S., Millar, S., Newcombe, K. and O'Neill, B. (1981) *The Ecology of a City and its People: The Case of Hong Kong*, Canberra, London and Miami: Australian National University Press.

Bruhn, J.G. (1974) 'Human ecology: a unifying science?' *Human Ecology* 2: 105–25.

Burgess, E.W. (1925) 'The growth of the city: an introduction to a research project', in R.E. Park, E.W. Burgess and R.D. McKenzie (eds) *The City*, Chicago, Ill.: University of Chicago Press.

Craik, K.H. (1972) 'An ecological perspective on environmental decision-making', *Human Ecology* 1: 69–80.

Ehrlich, P.R., Ehrlich, A.H., and Holdren, J.P. (1977) *Ecoscience: Population, Resources, Environment* (2nd edn), San Francisco, CA: W.H. Freeman.

Goldsmith, E. and Hildyard, N. (1988) *The Earth Report*, London: Mitchell Beazley.

Quinn, J.A. (1950) *Human Ecology*, New York: Prentice-Hall.

Sargent, F. (1974) *Human Ecology*, Amsterdam: North-Holland.

Shephard, P. (1967) 'Whatever happened to human ecology?', *Bioscience* 17: 891–4, 911.

Shevky, E. and Bell, W. (1955) *Social Areas Analysis: Theory, Illustrative Application, and Computational Procedures*, Stanford, CA: Stanford University Press.

Shevky, E., and Williams, M. (1949) *The Social Areas of Los Angeles: Analysis and Typology*, Berkeley, CA: University of California Press.

World Commission on Environment and Development (1987) *Our Common Future* (the Brundtland Report), Oxford and New York: Oxford University Press.

Worldwatch Institute (1989) *State of the World: A Worldwatch Institute Report on Progress toward a Sustainable Society*, 1989 edn, New York: W.W. Norton.

Young, G.L. (1974) 'Human ecology as an interdisciplinary concept: a critical inquiry', in A. MacFadyen (ed.) *Advances in Ecological Research*, 8, London and New York: Academic Press.

Young, G.L. (ed.) (1983) *Origins of Human Ecology*, Stroudsberg, PA: Hutchinson Ross.

4

HUMAN ECOLOGY AS TRANSDISCIPLINARY SCIENCE, AND SCIENCE AS PART OF HUMAN ECOLOGY

Dieter Steiner

> The world is not at all imperfect;
> imperfect are our language, our knowledge,
> and our consciousness.[1]

In this chapter I shall develop some ideas concerning a framework for what I call a general human ecology. By 'general' I mean a human ecology which transcends the traditional endeavours to deal with the human–environment relationship within the confines of single disciplines in two ways: its character is transdisciplinary and it contains trans-scientific components. At the same time it is being developed against the background of a newly emerging evolutionary cosmology. The following presentation clearly conforms to a line of personal argumentation. However, as it has been influenced by many different sources, I hope it to be general enough to serve as a contribution to the establishment of a context within which a structured dialogue between scientists, philosophers and practitioners interested in human ecology may unfold.

1 THE ECOLOGICAL CRISIS AND THE INSUFFICIENCY OF CONVENTIONAL ANSWERS

For a while the development of modern Western society, driven by the trinity of science, technology and economy, seemed to pave the way to progress in the form of a growing quality of human life. However, what once started out as a respectable attempt at revealing the secrets of nature in order to put them to work in an orderly fashion for the benefit of humankind, has developed into an accelerating maelstrom driven by a mutually reinforcing self-dynamic of the components of this trinity. There does not seem to be any time left to pursue things carefully and cautiously nor perhaps any willingness to do so. The largely unintended consequences of this race beyond any reasonable limits make themselves felt as an ecological crisis which in its scope and intensity is unprecedented and endangers our long-term survival.

As we know, the gradual recognition of the environmental problems has led to the discussion of possible remedies and the introduction of actual measures. Institutionalized environment-related rules and practices have been or are being developed in fields such as planning, environmental law and environmental economics. Science and technology are called upon to develop and provide practical solutions for pollution control and alternative resources. And there is much talk about the need for environmental education. It is also true that the discussion about environmental risks has intensified with the recent realization that the climatic changes expected from the greenhouse effect may become a major factor in the threat to our survival. With all of this, however, environmental destruction has by no means come to a halt. It continues, and perhaps its pace has even increased. As one critic remarked rather sarcastically, referring to the success of planning in Switzerland: 'Landscape destruction is not prevented, only channelled into an orderly process. It continues in coordinated fashion' (Thélin 1987)[2].

One may argue, of course, that the lack of success in curtailing the environmental impact of human action is due to the fact that the measures taken so far are not sufficient and/or thorough enough. Also, one may point out that usually a concerted concept of a sensible combination of different measures to be taken simultaneously is lacking. There is certainly some truth to this and, consequently, I am not suggesting that such measures should not be taken at all, but I would claim that, even if bundled into an integrated combination, they cannot be the final answer. Precisely such a belief, however, is common for most advocates of environmental science, planning, law, economics, and education. The implicit assumption is that the same kind of thinking which has led to the present troubles in the first place can also assume the role of the troubleshooter. In other words, the notion is widespread that we should be able to continue, business as usual. As an example, consider what Bennett and Chorley (1978: 16) have to say in view of the threatening collapse of the world-wide ecosystem: 'it is apparent that, far from encouraging man to relax his overall environmental grasps, such tendencies may result in proposals for increased planning and control. ... thus ... "God's design", having been replaced by "Nature's design", may be in turn supplanted by "Man's design".'

The reliance on the availability of control and prediction can be recognized as the child of a mechanistic worldview which has been the mainstay of the development of modern science (and about which more will be said in Section 2.2). My contention is that, if we really want to retain any hope for getting out of our present predicament, we need a fundamental change of scenery. This means that mechanistic thinking must be overcome. The notion of control should become replaced by a notion of creative participation in the adventure of evolution on this planet. It means that we should do less, and do everything more cautiously, and this can happen only in a new kind of society. In other words, we should remind ourselves time and again that the so-called environmental crisis is not really a crisis of the environment, but a crisis of we human beings.

2 CONCEPTUAL FOUNDATIONS OF A GENERAL HUMAN ECOLOGY

Starting from this premise, namely that the crisis we are dealing with is a human crisis, I propose here a way to look at the human–environment problem which I call 'general human ecology'. In the following sections I wish to emphasize three aspects of such a human ecology:

1   As human beings cannot be dealt with satisfactorily by traditional science alone, we need a change of science in such a way that it allows conscious connections to philosophy and the lifeworld of people, that is, it becomes embedded in a science-transcending background perspective on human affairs and on the world at large.
2   Such a background perspective is fundamentally shaped by the dominant worldview (or cosmology)[3] which up to now has been of the mechanistic type. It must change itself to one that is capable of fostering a respectful attitude of humans towards nature.
3   Within the scientific realm we should consider that (a) priority in environment-related research should be given to the social sciences and the humanities, and not to the natural sciences and engineering, (b) the fragmented stage of the former should be overcome by a conscious interdisciplinary effort, and (c) such an effort is perhaps most effective if it is based on a concept of what we may call personal transdisciplinarity.

2.1 Human ecology as a trans-scientific endeavour

For the following discussion I make use of what I call the human ecological circle (see Figure 4.1). It suggests that for human beings there are three different levels of referring to or encountering the world around them, namely lifeworld, philosophy and science, and that there is or should be a connection between them. There is a 'primitive' closure between the three to the extent that every human being is, in a way, a philosopher (by thinking about him/herself and the world) and a scientist (by making use of what one may call 'people's science'). However, what is meant here are philosophy and science in their 'elevated' sense, as the result of centuries of rational discourse. We can then see the sequence of lifeworld, philosophy and science as a differentiation which, roughly speaking, runs parallel to the development of Western humanity through three stages, namely the archaic, the political and the economic stage, about which more will be said further on.

In the archaic stage human beings experience themselves as part of nature or they see analogies between nature and themselves. Religious principles transpire in every aspect of life. Special rituals may serve to appease or magically to influence holy forces or, more generally, to reproduce or restore the world order. As philosophy and science do not yet exist in the sense that we now know them, we therefore have for the archaic situation a circle which closes the lifeworld on itself in an unmitigated way. Thus a religious worldview is very directly self-referential in that it is based on experiences (collectively made over long periods of time) gained in the lifeworld which are interpreted in such a way as to bestow enduring guidance to the right kind of actions in the same lifeworld.

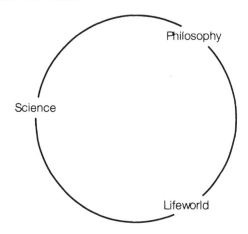

Figure 4.1   The 'human ecological circle': Three levels of world reference which should be connected to each other

The emergence of politically organized societies and urban settlements during the neolithic revolution forms a milieu in which an 'awakened' human mind starts to detach itself from purely religious thinking and to reflect upon the world, nature and human beings in rational terms. This marks the birth of philosophy. The prototype example for this development can be associated with the ancient Greeks, notably with Plato and Aristotle. To the extent that science in our sense does not yet exist or is still part of philosophy, knowledge is important for the sake of knowing, and the main function of *theoria* is one of contemplation (Gadamer 1988). It should be clear, however, that philosophy does not pervade the whole of society: its place is in the *polis*, whereas archaic principles of life very much persist in the areas outside the towns.

The onset of enlightenment is also the start of the development of modern science as an exercise in rational empiricism. As pointed out by Meyer-Abich (1988), it is eminently technologically oriented because of the kind of questions it asks: questions that must be answered by experiments. Science, therefore, is immediately associated with actions affecting the environment; its theories are meant as guides to practice rather than as instruments with which to understand the world and, consequently, they lead to manipulation. As a result, the technical products of such a science encroach totally on the everyday human lifeworld and make it lose much of its naturalness, a fact already recognized and deplored by Husserl (1935/77) in the thirties. On the other hand, science consciously and carefully distinguishes itself from the speculative metaphysics common in philosophy, that is, from the kind of questions which there can be no hope of ever answering on the basis of observation. This is serious because science claims for itself an ability to tell the truth about this world. But, as Picht (1973) points out, 'a kind of knowledge which manifests itself in the destruction of that which should be known cannot be true'.[4] How can this be if admittedly science produces certainty through reproducible facts? It obviously means that scientific certainty refers to half-truths only, and half-truths taken as the complete truth allow destruction.

This does not mean that science should be abandoned. However, to overcome the partiality just mentioned it must subject itself to a guiding supervision from philosophy on the one hand and take the realities of the lifeworld seriously rather than dismissing them as banal. Two points would seem to be of importance here:

1   The human ecological circle must be effectively closed again by supplementing any rational links between the three levels with emotional ties created by a new kind of (non-fundamentalistic!) religious attitude.[5]
2   With respect to the need for solutions we should not fall into the old instrumentalistic trap of the scientific expertise which provides a recipe as and when needed. We are past the times when the experts could come to an agreement anyway, and this is another clear indication of the fact that science does not suffice; as pointed out by Meyer-Abich (1988) scientists cannot argue about that which is certain, only about that which they do not know.

In this sense a general human ecology should not be seen as an undertaking designed to join the ranks of the experts, but rather one that develops a capability of establishing a conceptual framework within which structured conversations about problems can take place and a new kind of consciousness can develop. Meanwhile we may resort to some of the conventional measures indicated in the beginning, as long as they help to buy time for getting at the problem in a more fundamental fashion.

2.2 Human ecology against the background of a new cosmology

While worldviews are being discussed by philosophers in explicit terms, they are being lived by 'ordinary' people in everyday life. In other words, to the extent that a particular kind of worldview prominent in a certain culture influences the beliefs, values and norms that are held in this culture, its members will internalize it and act accordingly in an implicit way. Today such actions have become environmentally destructive, hence the importance of a new kind of cosmology with ecological qualities. It seems that such a new cosmology, which may be called evolutionary, is indeed emerging as a sort of amalgamation of two older worldviews, the morphological and the mechanistic. In the past there has been some degree of competition between the latter two. It is for this reason that there is talk about the 'two cultures' (C.P. Snow 1969) or a 'split of the worldview' (Riedl 1985). Morphological thinking has to some extent always remained important for the humanities, while mechanistic ideas have been the main credo in the natural sciences. However, as the instrumentality of the latter has been the decisive factor for the modern societal development, it is surely not misleading to speak about a dominance of the mechanistic cosmology in the immediate past.

The following is a coarsely simplified characterization of the three worldviews, the morphological, the mechanistic, and the evolutionary (more can be found in Steiner et al. 1989).[6] In all three cases there is an agreement on the point that the world can be seen as a sequence of strata, whereby smaller entitites make up larger entitites (see the discussion on hierarchies or 'levels of integration' in Young 1989: 20 ff). Speaking in modern terms, atoms are constituent parts of molecules, molecules (in organisms) of cells, cells of organs, organs of whole organisms, organisms of social communities. However, the three cosmologies differ in the way they see the causal connections between these strata. The situation is depicted in Figure 4.2. Some components (circles at the lower level) interact or stand in relation to each other (horizontal lines between circles). They form a new wholesome entity (circle at the higher level) only if these interactions or relations complement each other in a certain way, that is, if they all together make sense, if they are

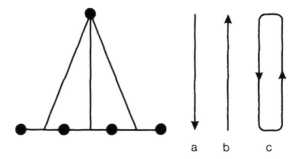

a   b   c

Figure 4.2 Schematic representation of the relationship between parts on a lower level of reality and a whole on an upper level of reality, with the direction of assumed causality as seen in (a) the morphological worldview, (b) the mechanistic worldview, and (c) the evolutionary worldview

supportive of a certain purpose, or if they comply to a certain form (hence the connections between the horizontal lines and the upper circle).[7]

In a morphological worldview forms have primacy over substance, which means that the stratification of the world just mentioned has the character of a regulative hierarchy. The whole determines its parts, hence there is a direction of causality from above to below (Figure 4.2a). The properties of the whole correspond to pre-existent patterns, such as Plato's ideas. In other words, there is assumed to be a teleological principle of organization preceding the parts, which then are realized in material terms. Notions of harmony and questions of the 'right place' of things are associated with this. For the ancient Greeks this means that, beyond a philosophy of nature, their thinking is eminently concerned with the right political behaviour of people which then would result in the right kind of state. To the extent that the whole precedes and governs the parts' notions about the world in a morphological worldview are holistic. If this relation between wholes and parts is applied to scientific disciplines associated with levels of reality it leads to what one might call an upward reductionism. As an example, consider the claim that what human individuals do is governed by societal structures and, hence, psychology reducible to sociology.

The term 'mechanistic cosmology' points to the fact that the principles of mechanics as established by Newton have been the prototype for this type of thinking. On this basis, the world is seen by the natural, later even by some of the social sciences as consisting of particles or bodies which, literally or otherwise, are in motion, a situation which can be described by a mathematical formalism. Of interest is, therefore, the question of how these components relate to each other and how they form a larger entity (e.g. the planets and the sun form a solar system). In this way the hierarchical structure of the world is felt to be a constitutive one, that is, the causal direction is from below to above (Figure 4.2b). The 'functioning' of an existing larger entity (such as an organism) can therefore be understood if it is broken down into its parts (atomism). This should reveal how the properties of the parts govern their relations, and how these relations make up the larger whole. The latter may show properties different from the parts, but they are derivable from them and, consequently, mere epiphenomena. A downward reductionism is assumed to be valid which, in terms of disciplines, means that sociology can be reduced to psychology, psychology to biology, biology to chemistry, and chemistry to physics.

A complementary aspect of the mechanistic cosmology is the 'machine metaphor', that is, the perceiving of entities as machines. The notion is that parts can be fitted together to form a new entity in such a way that this entity shows a particular kind of behaviour. Its design is derived from a pre-given purpose and thus the thinking in terms of machines can be regarded as a carry-over from a morphological cosmology. Indeed the pioneers of mechanistic cosmology imagined the world to be a huge machine set in motion by God.

The decisive point, however, is that a machine, unlike an organism, can be broken down into its components and put back together again, which is clearly a mechanistic notion. Also, it is associated with a belief in the powers of control and prediction. While thinking in this fashion is, up to a certain point, adequate in the field of engineering,[8] it becomes highly questionable and even detrimental if it is applied to non-artificial entities, in particular organisms, ecosystems, human beings, and human societies. An analysis at the level of parts may not tell everything there is to know about the whole. No doubt the relative unease with which many look upon the advent of genetic engineering, clearly still a child of the mechanistic age, is not accidental.

A decade ago Capra (1982) diagnosed a 'turning point' in our thinking about and our attitudes towards the world. Meanwhile the recognition that a new cosmology, which may be called 'evolutionary', is emerging has become commonplace. It can be seen as a reconciliation between the two 'one-way streets' typical of the competing mechanistic and morphological traditions and as combining them into a two-way connection between the level of parts and the level of the corresponding whole. This means that we have now a dual hierarchy: not only do the parts play a constitutive role for the superimposed wholes, but the wholes also have a regulative influence on the subordinate parts.[9] In other words, causation takes on a circular form (Figure 4.2c). The notion of circularity implies a switch from a more static to a more dynamic thinking. As pointed out by Sheldrake (1989) the morphological and the mechanistic worldviews have something in common, namely the belief in eternal, transcendental phenomena (that is, phenomena beyond our space–time), the ideal patterns on the one hand and the unchanging laws of nature on the other. In contrast, the one constant in the new evolutionary cosmology seems to be the principle of self-organization, which means that forms and possibly also the laws of nature are themselves products of evolution. In other words, everything has its history. We may be able to explain plausibly what happened in retrospect, but we will not be able to predict the future. This is an insight derived largely from the mathematical language of self-organization, that is, nonlinear system theory with its phenomena of bifurcation and chaos (see, for example, Nicolis and Prigogine 1977).[10] Also, new phenomena are now understood as emergent realities with their own causal powers.[11] In fact, the different levels of the stratified world are all seen as separate realities, none of them reducible to any other level, be it downwards or upwards.

2.3 Human ecology as an exercise in personal transdisciplinarity

Without in the sequel forgetting that an exclusively scientific approach to the human–environment problem is one-sided and partial, we do want none the less to turn now to such an approach and to the question of which form it could possibly take. As indicated earlier the perspective employed should have its centre within the human sciences. Here we are immediately confronted with the difficulty that they are fragmented into a multitude of disciplines, and within disciplines into a multitude of schools, each with its own object of study and its own vocabulary. Interconnections are obviously necessary between relevant disciplines, enabling the establishment of a genuinely interdisciplinary human ecology. However, in referring to inter-disciplinarity we should distinguish a weak and a strong variant, and it is the latter that I think is desperately needed.

The weak variant is the common case of an interdisciplinary research project: scientists who are specialists in different disciplines join forces and try to answer particular questions related to a common goal. A prerequisite for success in such a situation is, however, the existence of a 'strong' component, at least one person who takes on the role of a coordinating generalist. And as this same scientist comes him/herself from a specialist background, it may also mean that his or her discipline becomes the lead discipline. A fairly successful example of a project carried out on such a basis is the Swiss MAB[12] program in which aspects of the human–environment problem were studied in a number of mountain test sites and in which geography took on a leading function (Messerli 1989). One drawback was that an interdisciplinary theor-etical framework was not available at the time this program was carried out; what served as a basis for integration was a pragmatic–empirical approach which combined a concrete problem orientation with a corresponding data base.

A self-reinforcing process between a theoretical framework and improved coordination and integration between participating scientists can unfold with a strong variant of interdisciplinarity in which the strong component men-tioned above becomes the rule rather than the exception: the people con-cerned strive consciously for a certain degree of interdisciplinarity in person, or personal transdisciplinarity, as we may call it, by venturing from the familiar grounds of the home discipline on to the neighbouring grounds of other relevant disciplines. It amounts to what Boyden calls 'comprehensive scholarship': 'It is essential that there should be a central core of individuals whose main interest ... is the comprehensive and integrative process itself' (Boyden *et al.* 1981: xiv). The communication between disciplines can then rely heavily on intrapersonal integration. Human ecology of this type becomes grounded as much in scientific theory as in personal involvement.

3 A THEORETICAL FRAMEWORK FOR A GENERAL HUMAN ECOLOGY

On the basis of the previous reflections we can now start thinking about the establishment of a theoretical (scientific) framework for a general human ecology. This task can be seen as a twofold exercise in integration: on the one hand we need to find connections between relevant theories and knowledge of different disciplines, on the other hand we should be able to reconstruct relevant aspects and phenomena associated with the evolution of human societies. We will approach the former issue by means of the notion of recursive systems which recognizes the different levels of reality of this world as being related to each other by a circular causality. If we then turn our attention to the transformation of existing systems and the emergence of new systems, we deal with the latter issue. The two ensuing perspectives can be called 'extended ecological' and 'evolutionary' respectively. Their description, which we will attempt in the following, must remain rather sketchy. As the state of knowledge in the human sciences is one of utter fragmentation, the difficulty is obvious, but we hope nevertheless to be able to recover some fragments with fitting edges.

3.1 Linking recursive systems: an extended ecological perspective

The term 'recursive' is known in mathematics to indicate a function of the form $x_n = f(x_{n-1})$ which is calculated stepwise such that the function value xn calculated in the n-th step becomes the argument of the function for the calculation in the (n+1)-th step. Using the notion of a recursive system in a non-formal way, as we do here, means imagining a system in which events take place as interactions between parts and this in turn allows further events to happen. More precisely this involves a circularity between the interactions and the elements of the system in that the interactions make possible an enduring existence of the elements and the elements continue their very existence by further interactions. As an example, consider the notion of the autopoietic organization of organisms by Varela (1979: 13): 'An autopoietic system is organized (defined as a unity) as a network of processes of production (transformation and destruction) of components that produces the components that ... through their interactions and transformations continuously regenerate and realize the network of processes (relations) that produced them.' For this to be possible it is necessary that the interactions exhibit some degree of lawfulness or that they follow some rules. Only this can guarantee a recurrence of events such that a system is indeed organized. Each time an interaction takes place the corresponding rule is re-established or in some cases gradually transformed. Seen this way, therefore, we can also describe a recursive system as one in which there is a circular causality between its parts and its organization or its structure.[13] This, of course, is exactly along the line of thinking within the framework of an evolutionary cosmology in which there are interdependencies between different levels of reality such that none of them can be reduced to any other for the purpose of explanation.[14]

The 'human ecological triangle' shown in Figure 4.3 now serves to structure the discussion about a theoretical framework for a general human ecology. First, we note that it indicates three different levels of reality, namely the person (P), the society (S) of which P is a member, and the (biophysical) environment (E) in which P is living. Second, we can try to locate the position of relevant disciplines, at least of their traditional orientations, in the triangle: obviously psychology and physical anthropology can be associated with the corner P, sociology and cultural anthropology with the corner S, and geography and biology with the corner E. Third, we claim that basically we are dealing with three recursive systems, namely, person–society (P–S), person–self (P–P, we will consider different levels of reality within human beings below), and person–environment (P–E). The remaining connection, that is, the S–E side of the triangle, does not represent a recursive system, but rather a structural coupling between S and E.

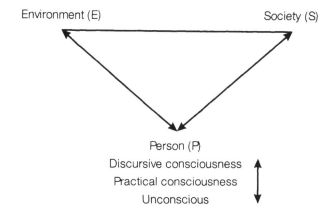

Figure 4.3   The 'human ecological triangle': it links up three recursive systems (shown by arrows) and one structural coupling (shown by a normal line)

If, as we claimed previously, the ecological crisis is really a human crisis, then a theoretical comprehension of the P–S recursive system must play a paramount role for a general human ecology. We find that such a comprehension is very adequately offered by the theory of structuration of society by Giddens (1984).[15] Several aspects of it will be discussed in a number of contributions in this volume, in particular in the papers by Lawrence and Nauser (Part III) as well as Werlen and Söderström (Part IV). Here it suffices to point out that the theory involves the notion of an agency–structure duality: human agents engage in social interactions and practices and by so doing reproduce (or transform) the social structures which govern these interactions and practices in a constraining but also enabling way.[16] The structures can be understood to be collections of rules.[17] A particular aspect of such rules is that they make sense. Sense is created by the fact that human societies are always language communities: the recurrent use of language as a means of communication by speakers in such a community establishes a network of meanings.[18] This, however, enables the development of a social reality which is detached from and possibly in contradiction to the reality of the surrounding (natural) environment on which ultimately it should be based. Indeed, words do not, as Wittgenstein thought originally, refer primarily to entities in the external world; instead, as he recognized later, they acquire their meaning only within the context in which they are spoken (Störig 1985). Here, therefore, we have a root of the present ecological crisis.

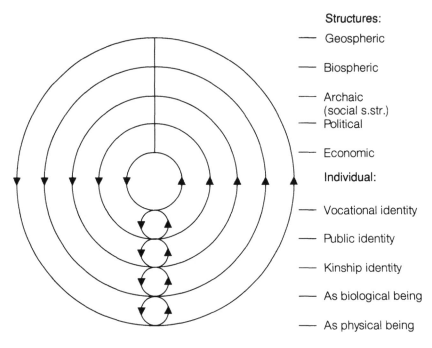

Structures:

—— Geospheric

—— Biospheric

—— Archaic
   (social s.str.)
—— Political

—— Economic

Individual:

—— Vocational identity

—— Public identity

—— Kinship identity

—— As biological being

—— As physical being

Figure 4.4 A model of the evolution understood as a sequence of recursive systems embedded within each other and applied to the present human situation. The arrows indicate causal relationships, the simple lines purely structural relationships

A P–P recursive system occurs because human beings as persons comprise components from different levels of reality of this world (see the evolutionary interpretation in Figure 4.4). The so-called mind–body problem is an expression of the fact that humans can be looked at in terms of two phenomena which cannot be reduced and explained one by the other. The human mind apparently is an emergent phenomenon that on the one hand is only possible on the basis of the material substrate of neural processes (and presumably, for its full development, on the existence of a social context), but on the other hand acquires its own causal powers. Also, the 'highest' form of consciousness, which involves self-awareness, apparently has grown out of 'lower' forms of consciousness closer to the organic processes of the body. It seems to be appropriate to follow Giddens (1984) and to distinguish the three levels of the unconscious, the practical consciousness and the discursive consciousness.[19] A human being operating in the last kind of consciousness is capable of verbally formulating and expressing his or her thoughts. In contrast, practical consciousness is a state of mind in which things are simply being done without agents being able to tell what exactly they are doing or why they are doing it. Typical examples of human actions belonging here are bodily skills such as riding a bicycle or a behaviour patterned by some social norms such as greeting somebody on the street. The notion of practical consciousness is at the core of Giddens' structuration theory, since it makes a routinized character of everyday life possible which in turn can give rise to a feeling of ontological security of agents. Werlen (this volume) discusses how a person, through a contradiction-free acceptance of social rules and an engagement in corresponding social practices with the co-presence of others, can acquire a cultural identity. The unconscious, finally, is the source of motives which rarely impinge in a very direct way on everyday courses of action. Rather they 'tend to have a direct purchase on action only in relatively unusual circumstances, situations which in some way break with the routine' (Giddens 1984: 6).[20] Nevertheless, we may assume that all different levels of the psyche are recursively connected to each other. As they can be associated with different types of knowledge, the question of possible interplay between them should be important in a human ecological context. We will come back to this point in Section 4.

We now turn to the P–E recursive system. E is a reality that, as long as it is a natural environment, can be described in terms of ecosystemic structures such as those relevant for alimentary and reproductional relations between organisms. The spatiality of such relations means that there are also spatial structures in the environment. For example, the way food resources are distributed may result in the establishment of territories, migration routes, etc. At the level of pure instincts, organisms show a genetically programmed and hence collective, that is, non-individualistic behaviour towards the environment as well as towards other members of the same species. In this sense, even if the animal in question is a social animal, there is no real separation between a 'society' and the environment. Ecosystemic structures relevant to an organism have become genetically internalized such that it 'knows', for example, what food to eat and where and when to find it. This is not to suggest, however, that there is an exclusive one-way street of causation; even at the purely instinctive level of existence organisms constantly influence their environment. As pointed out by Freese (1988) organisms always have cumulative and irreversible effects on the resources of the ecosystem concerned. With reference to Jantsch (1980: 70) he says: 'Living organisms themselves create the conditions for their own further evolution.'

At a certain point during biological evolution the possibility of ontogenetic learning starts to supplement the instinct basis and thus to make behaviour more individualistic. Such learning happens in direct contact with the environment (and includes such things as learning by imitation from one's fellows) and it requires the availability of what we called practical consciousness above. Animal–environment interaction at this level is the topic of Carello's contribution in this book. She describes it in the tradition of Gibson's ecological psychology as a case of 'direct perception'. This means that perception does not function as a kind of information transfer involving a mapping of environmental features into the animal's psyche, but that it is rather an active business involving the whole body such that there is a continuous cycle between perception and action. Hence an animal in direct contact with environmental structures constitutes another example of a recursive system. It can walk around, avoid obstacles, recognize sources of food, find places for resting (the so-called 'affordances'); in short, it can continue to live. In such a state of practical consciousness things have what Polanyi (1962) calls an 'existential meaning', that is, a meaning in itself, unlike the meaning of an intermediary in the form of a sign pointing to an object. There is a continuation for this kind of relationship to the environment for human beings: individuals may develop an attachment to certain aspects of their life-space independent from any socially generated symbolization (Weichhart 1990).

This, however, is different for components of the practical consciousness of human beings for which social rules[21] are constitutive. Such rules have detached themselves from genetically internalized structures with the consequence that a kind of separate reality is being created, which means that the transactions between a person and the environment acquire a social connotation. Evolutionarily speaking, this at first amounts simply to a social interpretation of natural features. For example, in an archaic society with a totemistic religion a particular animal is seen as a relative of human beings. Later the human-influenced environment becomes more and more socially altered and constructed such that it can be used as a 'locale' in order 'to provide the settings of (social) interaction' (Giddens 1984: 118). Lang (this volume) calls such an artificial environment an 'external memory' or a 'concretization of the mind': contents of the human psyche transform themselves into environmental (spatial) structures. Today this kind of materialization obviously has become excessive and disruptive for the natural components of the environment. It is important to note that the situation is again a recursive one: undoubtedly there is a feedback influence from these structures on the human mind and on human agency. We should distinguish, however, the social creation of environment emanating from the combined practical consciousness of the members of a community, leading to 'organically grown' cultural landscapes (such as those created by a long tradition of mountain farming and investigated in the Swiss MAB program mentioned earlier), from that kind of environmental transformation which is explicitly planned and is a construction of the discursive consciousness of some 'experts'.

The amalgamation of the environmental–spatial and the social obviously happens through a mediation of individual agency. As persons are engaged in a double duality with society and environment, they become automatically what we might call the 'locus of integration' (cf. Weichhart, this volume). It becomes clear that the presented theoretical approach structured on the basis of the human ecological triangle may be analytically useful, but that eventually it should be overcome in favour of a simultaneous consideration of all three recursive systems concerned. This need is very appropriately depicted by Guattari (1989), who talks about the 'three inseparable ecologies', namely the mental ecology, referring to relations of an individual to him/herself; the social ecology, referring to relations between different human beings; and the environmental ecology, referring to relations between humans and the environment.[22] They are inseparable because, as Guattari claims, the state of mind of humans is such that the way they behave towards and among themselves is invariably linked to the way they behave towards the environment. Obviously, the existence of causal links between person, society and environment does not indicate that the three are necessarily integrated, if by 'integration' we mean that they relate to each other in a mutually supporting way. In fact, as we live in an age of wholesale environmental destruction, we may, in the light of Guattari's diagnosis, see our task as one that asks for and investigates possible associated pathologies on the person–society side of the triangle. A corresponding discussion will appear below in Section 3.2.

3.2 Looking at the past as it unfolds to the present: an evolutionary perspective

A diachronic approach to human ecology is important in so far as the present can be understood only from its development from the past. In particular, as this present is characterized by a severe ecological crisis, the question arises as to the existence of a 'malevolution' or cultural degeneration[23] and its possible origins. Conversely, if we find that earlier human societies have lived more harmoniously and respectfully with regard to the natural environment, we can ask ourselves if there is something we have lost since and should recover for the purpose of a cultural regeneration.

Let us combine, therefore, the notion of recursive systems with an evolutionary perspective. We have noted earlier in Section 2.2 that evolution, first in its abiotic, then in its biological, and finally in its cultural form, can be understood as a process leading to a hierarchical sequence of different levels of reality such as subatomic particles, atoms, (bio-)molecules, cells, organs, organisms, societies. The consideration of the circular causative connections between these levels results in the additional interpretation of this sequence as one of emergent recursive systems (shown as a chain of small arrowed circles in the lower part of Figure 4.4). Evolution, however, seems to be more complicated than that. Following Jantsch (1980) it may be comprehended as a coevolution between a macro level sequence and a micro level sequence. While the hierarchy just mentioned corresponds to the latter, the macro level sequence refers to an order from the universe as a whole over galaxies and stars down to individual planets, such as Earth with continents, oceans, atmosphere, bioregions, and so on. We may assume that there are also recursive connections between micro- and macrophenomena (shown as a concentric pattern of large arrowed circles in Figure 4.4).[24] We remember that in each case a recursive system represents a duality between a level of subordinated components with some behaviour and a level of a superimposed whole with structures influencing this behaviour. It thus follows that the linear connections between the large circles in the upper part of Figure 4.4 must be interpreted as structural couplings. We note also that in the way Figure 4.4 is presented it describes the situation of human beings as (provisional?) 'end products' of evolution: they incorporate in themselves an accumulation of all different levels of evolutionary reality.

We now turn our attention to the cultural evolution of humankind after it has grown out of the physical–biological background,[25] that is, the two outer circles in Figure 4.4. Following Jaeger (1983, 1991) and Dürrenberger (1989) we can distinguish three emergent levels of societal development, namely the archaic, the political, and the economic, hence the three inner circles in Figure 4.4.[26] Each of these circles can be understood as connecting new kinds of individual agency to new kinds of structures. In each case the emergent phenomena are the results of 'revolutionary' changes: archaic societies follow the 'human revolution' (hominization, 4–2 million years ago), political societies follow the neolithic revolution (about 12,000–4,000 years ago), and the economic society follows the industrial revolution, beginning with the invention of the steam engine around 1780, but with roots further back in time in the Renaissance and the Reformation.

The structural characteristics of archaic societies are exclusively associated with direct relations among humans and between humans and the environment. People live in small local egalitarian groups in which they attain their identity according to kinship, sex and age. Their livelihood is based on nomadic hunting and gathering. All aspects of life are governed by an emotionally motivated religious world interpretation. The new phenomenon in political societies is, as the name suggests, a political organization with territorial and hierarchical aspects. Citizenship and/or class membership provide for the identity of individuals. Subsistence is now based on domesticated plants and animals, that is, agriculture and husbandry, and, since no longer all people are directly involved in production, must be guaranteed through a system of redistribution. People live in villages and cities. The world interpretation (of an élitarian part of the society) is of a rational speculative character, that is, philosophical. Finally, economic structures attain a status of overriding importance in the modern economic society. Production is industrially organized with entrepreneurs and salaried workers and employees, and is exchanged via markets. Also, it exceeds subsistence needs by far, which contributes to an excessive self-dynamic of the economic system. The identity of persons is now given by type of profession and the position within that profession. Other characteristics are: a high degree of division of labour, advanced technologies, high-energy throughput, mass consumption, global trade and traffic, and a general urbanization. The world interpretation is based on rational–empirical scientific thinking. We note that in each transition from an old to a new type of society the previous structures do not altogether disappear, but tend to become modified as a result of the newly emergent structures. Consequently, our modern Western society is a social system whose structures cover not only economic but also political and archaic aspects.

Consider Figure 4.4 now with respect to the question of 'malevolution': an inner large circle is evolutionarily speaking younger, an outer one older. The phenomena associated with a new circle, as they relate to an emergent level of reality, are by necessity to some degree emancipated from the reality represented by the surrounding circle. On the other hand, the newer reality is ultimately also dependent on the older reality as a basis for its continued existence. A special characteristic of human societies is, as already indicated, that they are language communities capable of creating their own sociocultural realities. 'The discrepancy between cultural images of nature and the actual organization of nature is a critical problem for mankind', says Rappaport (1979: 97). In accordance with Bateson (1972) he sees human societies as a case of cybernetic systems. Such systems generally 'attempt to maintain the truth value of propositions about themselves in the face of perturbations tending to falsify them. In systems dominated by humans, at least, the propositions so maintained (and the physical states represented by such propositions) may not correspond to, or may even contradict, homeostasis biologically or even socially defined' (Bateson 1972: 150). It is possible, therefore, that in Figure 4.4 the dynamics associated with an inner circle start to dominate those of an outer circle or disregard them altogether, and this, of course, sooner or later must lead to problems. In particular, the present ecological crisis can be seen as a situation in which the inner socio-cultural world has attained such a degree of detachment from the outer, more basic biophysical reality that destructive interactions between persons and the natural environment become possible or must even follow. In other words, the structural relationships between society and environment are incompatible, that is, they allow or even sanction such interactions. Generally we can diagnose such a state of affairs as one of overemancipation or, to the extent that the further existence of the society concerned is at stake, one of collective insanity.

In particular, we have such a situation with today's modern society in which the intransigent self-dynamic and the ensuing dominance of the economic component (the innermost circle in Figure 4.4) are overwhelming. From an evolutionary point of view we can say that it should be appropriately contained by the political system (the immediately surrounding circle in

Figure 4.4). Instead the latter behaves in a predominantly reactive rather than proactive way with respect to the economy. The political system in turn should be adequately influenced by cultural principles[27] with some archaic quality (the third circle from the centre in Figure 4.4), but this is obviously not the case. This necessity is demonstrated by the fact that the higher the degree of overemancipation the more we become aware of the limits set by nature. In the words of Schneider and Morton (1981: 286): 'We are certainly attempting to detach ourselves from Nature, but we are no more able to achieve that goal fully than were previous civilizations. The continuing paradox is that we are both detaching from and becoming more "embedded" in Nature.'

Among the degenerate results of this kind of development are the phenomena of anonymity, moral inversion, alienation, and the disunity of personal lives.[28] Here we will comment on the first two only. A discussion of the third phenomenon can be found in Jaeger (1991), who treats it as a case of a social practice whose goal has become external. The last aspect is addressed by Steiner et al. (this volume).

Anonymity appears in the course of the cultural evolution because, in the terminology of Giddens (1984), the integrative media holding a human society together turn from an originally exclusively social type into a predominantly system type. Social integration, the only means of integration in archaic societies, involves direct face-to-face contacts and is associated with situations of love (sexual or otherwise), sharing, cooperation, rituals, etc. In contrast, system integration occurs in an anonymous way via indirect relations, possibly over several relational links, between people who more often than not never see each other. Its medium in political societies is primarily power, in the economic society money. System integration enables the coordination of an increasing number of people, but only at the cost of growing anonymity with associated irresponsibility, that is, irresponsibility is almost unavoidable under the circumstances even with the best of intentions.[29] Appropriately Beck (1988) talks about *organisierte Unverantwortlichkeit* (organized irresponsiblity, the subtitle of his book). Mosler (this volume) concerns himself with possible countermeasures. He tries to develop ideas of how to institutionalize a kind of social trust which works in a situation of anonymity, that is, is not dependent on personal familiarity.

The 'process of moral inversion' (a wording used by Polanyi and Prosch 1975: 18) can be described as follows. In archaic societies the world interpretation is religiously inspired and is, by and large, ecologically sensible, that is, we have the case of a coupling (the S–E side of the triangle in Figure 4.3, the respective connection between the third and the fourth innermost circle in Figure 4.4) that provides for stability between society and environment. This is explicable by the fact that such societies have, by trial and error in their dealings with the environment over thousands of years, developed a cultural tradition that resides in the practical consciousness of their members. In addition, one may assume that such a tradition by necessity cannot be too

64

far from the instinctive basis of biological existence. Thus, the moral orders of archaic societies lead to respectful relationships with not only entities of the environment such as plants or animals but also whole parts of landscapes, such as rivers, woods or mountains.[30] At the other end, in our present economic society, the process of enlightenment has ended up in a degree of individualism never attained before, an individualism whose assets consist in the gain of political freedom and civil rights, but whose drawbacks are the eventual total loss of integrating moral orders of a collectively binding quality. Archaic principles become subdued and morality becomes a matter of personal relativism, that is, empirically observed preferences establish largely material-istic values *ex post facto*. For Meyer-Abich (1988) this is a *Kulturschwäche* (a weakness of culture), for Rappaport (1979) a case of 'usurpation': the economy, if it were to function sensibly, would have to be a subsystem at the service of the social system at large. Instead it occupies the 'sacred' level once reserved for religious principles and by so doing dominates the whole society in such a way that the other parts of it are at its service instead of the other way round.

4 TOWARDS A POST-ECONOMIC SOCIETY, BUT WHAT KIND?

There is much talk about the emergence of a post-economic (generally called post-industrial) society which, as materialistic values are the distinguishing mark of the economic society, is supposed to be grounded on a kind of post-materialistic values. As reported by Milbrath (1989) a comparative ques-tionnaire study carried out in the United States, England and West Germany in the early 1980s points to the appearance of a new 'environmental paradigm' encompassing a high valuation on nature, a generalized compassion toward other living beings, other humans, and other generations, a preference for low-risk technology, a recognition of the limits to growth, and an emphasis on participation, cooperation, and consultation in political affairs. However, this is not to say that the desperately needed 'ecological revolution' is now really under way or that it will all happen by itself. By and large there seems to be still a wide discrepancy between attitudes as those mentioned above and actual actions. It seems to be obvious, however, that a societal transformation is in progress, but where it will lead to remains uncertain. This should remind us that we are all agents in this process and, consequently, have the responsibility of reflecting about possible available options with respect to courses to take. Accordingly, the development towards a new environment-friendly society may be furthered or hampered.

If there is really a 'revolutionary change' to a post-economic society, does this mean that, seen in the graphical form of Figure 4.4, the result would be an evolutionarily new circle appearing within the one representing the economic society? A new circle would stand for a new recursive system, that is, a new level of structures within which human agents can interact and find an identity. What could this be? A first observation in the negative is that obviously environmental economics, as one of the conventional answers to the ecological crisis mentioned in Section 1, does not lead to such new structures. Its basic idea is that the destruction of the environment will be halted if the externalities created by the economic system are somehow internalized into the market mechanism. The simplicity of this approach is criticized by Pillet (this volume). He intends to show a way from 'environmental' to 'ecological economics' by, first, considering the environment to be a quasi-sector of the economy and, second, using an energy quality hierarchy method which enables ecologic–economic bridged systems analyses to be done. This may indeed constitute an important step toward a considerable improvement, but the problem remains that, interpreted from an evolutionary point of view, the cart is still put before the horse, that is, we remain at a 'low' level on which economic thinking is the driving force, which amounts to a further sanctioning of the moral inversion (see Section 3.2). What we need in contrast is a framework provided by a truly ecological culture, one which allows the reintegration of the technical, more purpose-oriented rationality of the economic system into the communicative, more value-oriented rationality of the lifeworld of responsible, that is, morally accountable human beings (Ulrich 1987). In a similar vein, Etzioni (1988) holds that the economic neoclassical paradigm, which sees human beings as self-centred, free-floating and utility-maximizing individuals, should be discarded in favour of what he calls an 'I and We' paradigm, which in contrast recognizes that humans cannot advance their 'I' sensibly unless they are well anchored within a sound community which they can perceive as theirs, as a 'We'.

Another conventional answer mentioned in Section 1 is scientific expertise. Here the situation with regard to the possibility of a new recursive system is less clear. On the one hand, scientific thinking of the usual kind leads to a problem similar to environmental economics: an attempt at 'internalizing' entities of the natural environment into society by accumulating more explicit knowledge about it is again a 'low level' connection outside a genuinely ecological cultural orientation. On the other hand, many see the coming post-economic society as an 'information society'. Surely the explosive development of computer and communications technology and the ensuing regional and global networking may lead to informational structures with a totally new quality. Bell (1973), in his reflections on a post-industrial society, entertains the notion that theoretical knowledge will gain a central position as the basis for future planning and design. For him this is a logical continuation of the cultural evolution which can be interpreted as a shift of dominant activities from the primary over the secondary to the tertiary (and the quaternary)[31] sector with a corresponding change of the basis of technology, namely from material resources over energy to information. Accordingly, he expects the emergence of what he calls an 'intellectual technology' which will make it possible to base decisions more on algorithms and less on intuition.

66

Others such as Toffler (1970) and Huber (1982) supplement this picture on the technological side with their notion of 'superindustrialization'. They see a society which, in addition to the already mentioned information technology, would be based on genetic engineering, biomass processing, alternative energy technologies, recycling, and so on, and would have the capability to build a synthesis between industry and ecology. However, as long as such a development is associated with the expectation of or even the demand for a new wave of growth, the latter point of compatibility is very questionable. Also, one can, of course, imagine that a society ruled by expert systems and artificial intelligence would indeed constitute a new kind of recursive system with its own dynamics. However, I am probably not alone with the uneasy feeling that such a future scenario would more likely develop into the horror story of an ultimate overemancipation putting us out of the frying pan and into the fire for good.

It seems to me that the question of whether or not we have a new recursive system on the horizon is less important than the feeling that, whatever happens, we need to some degree a backward orientation, a reattachment to something we have lost, a *re-ligio* in the true sense of the word, resulting in a new kind of ecologically compatible structural coupling between society and environment (the S–E line in Figure 4.2, and the lines connecting the inner to the outer circles in the upper half of Figure 4.4). It amounts to a rediscovery of the 'primordial bond' suggested by Schneider and Morton (1981: 83): 'The power that Nature wields over man, in terms of both the limitations imposed and the opportunities offered, has shaped a bond that has held man to Nature for as long as we can determine. This primordial bond remains today, despite the struggles of many to break it'.

Regarding the possibility of a future information society, the question clearly is not whether we should have any science at all, but rather what role it should play. An evolutionary reattachment can be achieved if, as suggested by Weizenbaum (1976), we abandon the 'imperialism of instrumental reason' and try to establish a science that is governed by higher ethical principles. This could be the basis of a different kind of information society, one that maintains a balance between emancipation from and embeddedness within older structures. A prerequisite would seem to be that 'information' is meant to be more than simply explicit knowledge of the scientific kind, that is, all kinds of knowledge to which humans possibly have access. This relates to the three levels of consciousness discussed in Section 3.1.[32] As they constitute an evolutionary sequence, their combination can become ecologically significant by providing the necessary degree of backward orientation. Consider the importance of this combination as discussed by Polanyi (1962, 1974). He makes a distinction between explicit and implicit knowledge. The former derives from discursive consciousness; scientific knowledge is a special, more or less formalized type. Conversely, implicit (or tacit) knowledge can be associated with practical consciousness and the unconscious. The special point of interest in Polanyi's discussion is his claim that human knowledge has the same structure as the world around us. This means that both can be illustrated by the situation shown in Figure 4.2. In discussing the new evolutionary worldview we have stated that it sees an entity on an upper level as a new kind of whole emerging from some interplay of parts on a lower level, whereby the phenomenon of 'emergence' means that the properties and behaviour of this whole cannot be reduced to those of the parts. On the side of human knowledge this is paralleled by an incompleteness of explicit knowledge. If the properties of and the relations between elements of a system (the circles connected by horizontal lines in Figure 4.2) are formulated in explicit terms, sense and meaning associated with the whole (the lines connecting to the upper circle in Figure 4.2) are lost. There is complementary tacit knowledge with which it can be recovered, but its activation requires a defocusing away from the system parts.[33] In a similar vein, Bateson (1972) points to the fact that consciousness is necessarily selective and partial and he goes as far as saying that 'mere purposive rationality unaided by such phenomena as art, religion, dream and the like, is necessarily pathogenic and destructive of life' (Bateson 1972: 146).[34]

With the partiality of explicit knowledge and the dependence of a sense of wholeness on complementary implicit knowledge, it should be obvious that with respect to the human–environment problem the latter is of great significance. Take the example of traditional mountain farming investigated by the Swiss MAB program mentioned in Section 2.3. Over the centuries the farmers have been able to develop a sustainable system of utilization and to reach a condition of ecological stability. The related knowledge is 'stored' in the members of the communitiy and passed on from generation to generation. And this is, of course, the problem: if the tradition discontinues, this knowledge is gone as well. As Messerli (1989: 12) says: 'it is . . . unimaginable that science can ever replace the local knowledge and experience which generations of mountain farmers have accumulated over many centuries of concrete work on nature.'[35] A corresponding argument is developed by Josefson in this volume in the context of hospital care: she points to the irreplaceability of 'knowledge by experience' of traditionally trained hospital nurses. Finally, to the extent that the collective unconscious comprises knowledge acquired in the course of our phylogenetic past it may also be a source of 'information' of great human ecological significance. The problem presumably is that most of us have lost an ability to tap this source. But perhaps emotions, which in their genuine form emanate from the unconscious, can be a guide. This for one leads Meyer-Abich (1988) to emphasize the importance of *erkenntnisleitende Gefühle*, (that is, feelings guiding cognition), whereas Bateson (1972) describes the situation as follows: 'the problem of grace is fundamentally a problem of integration and that what is to be integrated is the diverse parts of the mind – especially those multiple levels of which one extreme is called "consciousness" and the other the "unconscious".

For the attainment of grace, the reasons of the heart must be integrated with the reasons of the reason' (Bateson 1972: 129).

Going back to the question of how we, as responsible individuals, can possibly support a desirable development, we can now see that the problem with which we are faced has to do with the opposition between implicit and explicit instruments of regulation (cf. Lawrence, this volume). The conventional answers to the ecological crisis mentioned earlier belong to the latter category and, as we have learned, can be of limited help only. What we have to rely on is the implicit regulation developing within the self-organizing circle between human agency and socio-cultural structures. Does this mean that all is supposed to happen by itself? Not quite, provided that our present society can become open enough not only to allow passively but also to encourage actively alternative thinking and alternative lifestyles. In other words, if it is possible to set right external conditions, then Ladeur's (1987) paradox of the 'outside organization of self-organization' should come into force. The whole society should embark on a course of 'fundamental social learning' (cf. Milbrath 1989, and Section 3.2 in Steiner *et al.*, this volume) or, in the terminology of Bateson (1972), of 'learning III', a kind of learning that questions established routines. Out of it should emerge a new tradition, one that provides for ecological sustainability by means of a structural coupling between society and natural environment on a truly cultural level.[36] Such a tradition, however, should not become rigid again but should remain flexible and allow for continuous learning, such that material growth becomes replaced by human growth and the newly emerging society is a 'learning society' as proposed by Harman (1986: 3): 'The primary resource of future society is information, knowledge, learning, wisdom.'

A new tradition should be able to recover some components of the archaic quality of life without losing the positive aspects of the differentiation which humankind has acquired since primeval times.[37] Such components include: a renewed unity of personal lives, reinforced relations in the form of social integration, a reversal of alienation (for example, in work), a re-establishment of social trust, a renewed egalitarian relation between women and men, a strengthening of local and regional self-reliance and competence, increasing self-control of living conditions, and self-help. The point that the experience of developing one's faculties in several ways and of doing more things oneself may be an important factor in the formation of ecological ethics is stressed by Schubert (1989), while Jaeger (1991) notes that a successful sustainable development on a global scale presupposes the existence of regions that are capable of functioning as innovative milieux (cf. Introduction to Part IV). All of this may help to provide settings amenable to the re-establishment of respectful attitudes towards nature, but obviously it will involve considerable changes in lifestyles away from the still predominant consumerist attitudes. For all we know, however, this may mean not a loss but a substantial gain of quality of life.

To conclude let us consider one more point that would seem to be of importance. Any attempt at developing a new environment-friendly tradition must ultimately be founded upon an implicit kind of morality. Not only is this clearly absent at the present time, but also, as described by Macintyre (1985), the moral discourse is in great disorder. He demonstrates that any project of an explicit rational derivation of ethics is doomed to failure as long as a notion of human self-fulfilment or *telos* is lacking from our culture. Such a notion existed in the Aristotelian philosophy of antiquity and the middle ages, but the roots of a functional concept of human beings are much older; they lie in the religiously regulated lifeworld of human societies of the past. Their implicit morality derives from religious feelings, felt individually, but mitigated ritually in a collective fashion. The question arises as to whether this archaic principle could possibly have any significance for us modern humans. Steindl-Rast (1984) definitely thinks that we all have the capability to become overwhelmed now and again, to experience the world as a wonder, to develop a sense of belonging to it, to recognize that it is full of relatives. Feeling this way is a mindset which can be seen to be akin to the notion of 'identification' in Naess' (1989) ecophilosophy. To identify with the things around one means that one's relations to them become internal and thus a part of one's identity, and it means further that the distinction between facts and values vanishes. Such a reattachment to the foundations of our past may bring about progress in an ecological sense. As Gary Snyder puts it: 'In ecology we speak of "wild systems". When an ecosystem is fully functioning, all the members are present at the assembly. To speak of wilderness is to speak of wholeness. Human beings came out of that wholeness, and to consider the possibility of reactivating membership in the Assembly of All Beings is in no way regressive' (Snyder 1990: 12).

5 NOTES

1 From the movie *Why has Bodhi-Dharma left for the East?* by Yong-kyun Bae, South Korea 1989.
2 Quote translated from the German original by D.S.
3 The term 'worldview' means a 'grand theory' that is aimed at an understanding of the whole of the world. It 'affords a frameword within which we speak and think about how the things in this world are connected to each other, how they belong to each other, and what they mean for us human beings. Thus a theory in this sense has not only a scientific aspect. . ., but also a philosophical and religious dimension' (Jaeger 1988: 3, translated from the German original by D.S.). Consequently, a worldview always contains speculative elements. A worldview is what Whitehead (1978) calls a cosmology.
4 Translated from a quotation in German by Meyer-Abich (1988: 80) by D.S.
5 We also note that a genuine human ecology must be self-referential: it has itself a place in the circle (cf. Huppenbauer, this volume). One consequence is that a human ecologist cannot be merely a scientist, but his or her science must be combined with an altered attitude towards life and the world in general.

6   Compare also with the four types of worldview distinguished in the recent psychological literature and referred to by Weichhart (this volume). The following correspondences with our distinction would seem to hold: trait and interactional = mechanistic, organismic = morphological, and transactional = evolutionary.

7   Both the concept of 'mental processes' by Bateson (1979) and the speculative notion of 'morphic fields' by Sheldrake (1989) have a bearing on the upper part of Figure 4.2: 'the aggregate is a mind and, . . . if I am to understand that aggregate, I shall need sorts of explanation different from those which would suffice to explain the characteristics of its smaller parts' (Bateson, 1979: 91); 'by their very nature morphic fields interrelate and interconnect elements to integral wholes. They give meaning to the elements through their interconnections into such higher-level wholes' (Sheldrake, 1989: 201).

8   We say 'up to a certain point' because some of the modern large-scale technology has reached such a degree of inscrutable complexity that, as the book *Normal Accidents* by Perrow (1988) amply testifies, catastrophic events may become the rule rather than the exception.

9   See the discussion on this dual hierarchy in Harré *et al.* (1985).

10 Nonlinear system theory in the form of synergetics and in a social context is discussed by Mosler (this volume).

11 Applied to the world at large this notion implies that its hierarchical levels mentioned earlier must be regarded as a collection of separate irreducible realities. We note at this point that this is in line with the position established in a brand of philosophy of science known as 'transcendental realism' (see Bhaskar 1975, Harré *et al.* 1985).

12 The abbreviation MAB stands for 'Man and the Biosphere', an international program initiated by UNESCO in the 1970s.

13 Shotter (1984: 195 ff) refers to a situation in which action 'is structured both as a product and as a process, or better, is both structured and structuring' as 'duality of structure', and he sees it as a general evolutionary principle: 'an evolving world can be thought of as being full of agencies, as containing everywhere structurizing activities or formative causes'.

14 We note at this point that there is a relation between this notion of irreducibility and the one of complementarity. The term was originally coined by the physicist Niels Bohr who used it to refer to the duality of appearance of electromagnetic radiation: depending upon the method of observation it takes on either particle properties or wave properties (a holistic feature), but not both simultaneously. He later also applied it to point out that life forms cannot be reduced to chemical and physical processes, because whenever one tries to get to the bottom of 'things', the organism to be studied gets killed. Of particular significance for our discussion in connection with Figure 4.3, however, is the transfer of a corresponding notion to the realms of psychology and sociology by Devereux (1972): the two are complementary disciplines which cannot be reduced one to the other.

15 For similar theoretical ideas developed by Harré see Weichhart (this volume).

16 As can be seen, this theory overcomes the controversy between atomistic (mechanistic) approaches which claim that society is a mere epiphenomenon of the actions of individuals (individualism) and holistic (morphological) approaches which conversely see these actions as always determined by given structures (structuralism). Consequently, it fits into the framework of an evolutionary cosmology.

17 More precisely, Giddens regards structure 'as rules and resources recursively implicated in social reproduction; institutionalized features of social systems have structural properties in the sense that relationships are stabilized across time and space' (Giddens 1984: xxxi).

18 The aspect of sense and meaning is discussed by Giddens under the label of 'signification'. From here links can be found to social theories dealing specifically with it such as those by Schütz (1981) and Berger and Luckmann (1966). See also the discussion of 'intersubjective meaning contexts' by Werlen (this volume).

19 See also the special treatment of the same subject with respect to its relevance to the understanding of problems associated with the concept of 'environmental concern' or 'environmental awareness' by Nauser (this volume). We further note that this distinction can be paralleled not only with that of other authors, for example, with the 'deep structure of the mind', the 'behavioural routines', and the 'conscious awareness' of Harré et al. (1985), but also with the 'heart', the 'hands' and the 'head' of the Swiss school pioneer Johann Heinrich Pestalozzi (1746–1827).

20 Giddens (1984) restricts himself to a (critical) elaboration of Freud's notion of the personal unconscious. He does not consider the so-called collective unconscious as proposed by Jung. For our purposes we attach some importance to the latter. We also note that there are authors (as Gustavsson 1990: 5) who speak about a 'collective consciousness', a concept that reaches beyond the idea of a collective unconscious. It 'rests on the notion that all people are united through a transcendental field of consciousness, and that influence from this field is not dependent on contact between the people'. This speculative idea may be akin to Sheldrake's (1989) concept of 'morphic fields' as applied to human societies. Somewhat more tangible, if still unintelligible to a Western mind, is a phenomenon referred to by Bohm (1991: 83): 'Anthropologists who have studied hunter-gatherer groups say many of these people sit around in a circle without making decisions, and then they seem to know what to do'.

21 The 'social orders', if we wish to use a term employed by Harré et al. (1985).

22 Cf. Söderström (this volume).

23 For a lengthy treatment of processes of cultural degeneration as instances of 'common madness' see Jaeger (1991).

24 Some of the proposed connections of this kind may be the result of rather speculative ideas such as the Gaia hypothesis by Lovelock (1979) which suggests that Earth is some kind of superorganism.

25 For a biohistorical approach which investigates specifically the interrelationship between biophysical and cultural variables see Boyden (this volume, and 1987).

26 Other authors use different names. For example, Giddens (1984) talks about tribal, class-divided, and class (capitalistic) societies, and Boyden (this volume, and 1987) about the primeval, the early farming/early urban, and the high-energy phase. The double characterization of the middle phase indicates that, in accordance with a distinction made by Childe (1936/65), the neolithic revolution comprises two phases: an earlier agricultural and a later urban revolution.

27 Here and in the following, the word 'culture' is restrictively meant to indicate the state of mind of a society which expresses itself in a certain cultural orientation and also, in particular, in some form of moral orders.

28 All of those are prominent characteristics of today's economic society and contributory factors to a destruction of the natural environment in one way or another. Further important aspects of a malevolution whose roots go back to the neolithic revolution are organized warfare and patriarchy. In this context we note that Capra (1982) sees both the feminist movement and the peace movement as companions of the green ecological movement and thus as merely different expressions of the same underlying problem, namely that of a society having gone astray.

29 Also, of course, it becomes relatively easy to take advantage of the situation in a malevolent way. As an example, take the role of money in our society and the recurring scandals attached to it.

72

30 As one of many surprising examples that have been reported in the ethnological literature consider the mythology of the Desana, an Indian tribe living in the northwestern part of the Amazonian region, which amounts to a kind of biological variant of the energy conservation law, so to speak. As the creator of the world has created a limited number of living beings the total life energy in the universe is bounded. As a consequence any growth of the human population would mean a corresponding drop in the number of animals and plants and thus a threat to the livelihood of the tribe (Herbig 1985).

31 As a differentiation out of the tertiary, the quaternary sector stands for activities concerned specifically with the production, the processing and the distribution of information.

32 By referring to 'knowledge' in this broad sense we make an allowance for a blending of knowledge which is understood in the strict sense (the result of an act of cognition which is based on experience and insight and for which reasons can be given) and of belief, but then this is closer to the reality of human existence. A more extensive discussion of the human ecological significance of different types of knowledge can be found in Steiner (1991).

33 Similar notions regarding a structural likeness between world and human knowledge are related to the concept of a holographic world with its dualism of implicate and explicate order by Bohm (1980) and its transfer from the purely physical to the neurological and psychological realm by others (see Wilber 1982). Consider also the Buddhist idea that human beings normally perceive a world of apparently separate objects and they do this in a state of consciousness, whereas the real world is a 'nothing' in which everything merges into everything else, and it can be perceived as such only in a state of awareness (Baker-roshi 1991).

34 If this is true then the high status given to mathematics in our society in general and in science in particular should be questioned. In this respect see also Huppenbauer (this volume) and the Introduction to Part II.

35 Translated from the German original by D.S.

36 In a way we seek new answers to the old question asked in cultural ecology, namely, how do culture and natural environment relate to each other (for a more detailed account of this question see Steiner 1992).

37 Taylor (1972) talks about the need for a 'paraprimitive society'.

6 REFERENCES

Baker-roshi, R. (1991) 'Metamorphoses of simultaneous perception', lecture given during the 5th Cortona Week on 'Science and the Wholeness of Life', Cortona, Italy.

Bateson, G. (1972) *Steps to an Ecology of Mind. A Revolutionary Approach to Man's Understanding of Himself*, New York: Ballantine Books.

Bateson, G. (1979) *Mind and Nature. A Necessary Unit*, New York: E.P. Dutton.

Beck, U. (1988) *Gegengifte. Die organisierte Unverantwortlichkeit*, Frankfurt a/M: Suhrkamp.

Bell, D. (1973) *The Coming of Post-Industrial Society. A Venture in Social Forecasting*, New York: Basic Books.

Bennett, R.J. and Chorley, R.J. (1978) *Environmental Systems. Philosophy, Analysis and Control*, London: Methuen.

Berger, P.L. and Luckmann, T. (1966) *The Social Construction of Reality*, Garden City, NY: Doubleday.

Bhaskar, R. (1975) *A Realist Theory of Science*, Leeds: Leeds Books.

Bohm, D. (1980:) *Wholeness and the Implicate Order*, London: Routledge & Kegan Paul.

Bohm, D. (1991) 'Transforming the culture through dialogue', *Utne Reader* 44: 82–3.

Boyden, S. (1987) *Western Civilization in Biological Perspective. Patterns in Biohistory*, Oxford: Clarendon Press.

Boyden, S., Millar, S., Newcombe, K. and O'Neill, B. (1981) *The Ecology of a City and Its People: The Case of Hong Kong*, Canberra, London, and Miami: Australian National University Press.

Capra, F. (1982) *The Turning Point*, London: Wildwood House.

Childe, V.G. (1965) *Man Makes Himself* (first pub. 1936), London: Watts.

Devereux, G. (1972) *Ethnopsychanalyse complémentaristes*, Paris: Flammarion.

Dürrenberger, G. (1989) 'Menschliche Territorien. Geographische Aspekte der biologischen und kulturellen Evolution', *Zürcher Geographische Schriften* 33, Zürich: Verlag der Fachvereine.

Etzioni, A. (1988:) *The Moral Dimension. Toward a New Economics*, New York: Free Press, London: Collier Macmillan.

Freese, L. (1988) 'Evolution and sociogenesis. Part 1: Ecological origins', in E.J. Lawler and B. Markowsky (eds) *Advances in Group Processes* 5: 53–89, Greenwich, CT: JAI Press.

Gadamer, H.G. (1988) *Philosophisches Lesebuch*, 3 vols, Stuttgart and München: Deutscher Bücherbund.

Giddens, A. (1984) *The Constitution of Society. Outline of the Theory of Structuration*, Berkeley and Los Angeles: University of California Press.

Guattari, F. (1989) *Les trois écologies*, Paris: Editions Galilée.

Gustavsson, B. (1990) 'The effect of meditation on two top management teams', *Studies in Action and Enterprise* PP1990.4, Stockholm: Department of Business Administration, University of Stockholm.

Harman, W.W. (1986) 'The changing context of human and economic development', manuscript, Sausalito, CA: Institute of Noetic Sciences.

Harré, R., Clarke, D. and de Carlo, N. (1985) *Motives and Mechanisms. An Introduction to the Psychology of Action*, London and New York: Methuen.

Herbig, J.(1985) *Im Anfang war das Wort. Die Evolution des Menschlichen*, München and Wien: Carl Hanser.

Huber, J. (1982) *Die verlorene Unschuld der Ökologie. Neue Technologien und superindustrielle Entwicklung*, Frankfurt a/M: Fischer.

Husserl, E. (1977) *Die Krisis der europäischen Wissenschaften und die transzendentale Phänomenologie* (first pub. 1935), Hamburg: F. Meiner.

Jaeger, C. (1983) 'Wirtschaftswachstum – eine Fehlevolution?', *Alemantschen* 3: 125–35.

Jaeger, C. (1988) 'Theorie und integrative Ansätze in der Geographie', in D. Steiner, C. Jaeger and P. Walther (eds) 'Jenseits der mechanistischen Kosmologie – Neue Horizonte für die Geographie?, *Berichte und Skripten* 36, Zürich: Department of Geography, ETH.

Jaeger, C. (1991) 'The puzzle of human ecology. An essay on environmental problems and cultural evolution', habilitation thesis, Zürich: Department of Environmental Sciences, ETH (publication forthcoming).

Jantsch, E. (1980) *The Self-Organizing Universe: Scientific and Human Implications of the Emerging Paradigm of Evolution*, Oxford: Pergamon Press.

Ladeur, K.-H. (1987) 'Jenseits von Regulierung und Ökonomisierung der Umwelt: Bearbeitung von Ungewissheit durch (selbst-)organisierte Lernfähigkeit – eine Skizze', *Zeitschrift für Umweltpolitik und Umweltrecht* 10 (1): 1–22.

Lovelock, J.E. (1979) *Gaia. A New Look at Life on Earth*, Oxford: Oxford University Press.

MacIntyre, A. (1985) *After Virtue. A Study in Moral Theory*, London: Duckworth.

Messerli, P. (1989) *Mensch und Natur im alpinen Lebensraum. Risiken, Chancen, Perspektiven*, Bern and Stuttgart: Paul Haupt.

Meyer-Abich, K.M. (1988) *Wissenschaft für die Zukunft. Holistisches Denken in ökologischer und gesellschaftlicher Verantwortung*, München: C.H. Beck.

Milbrath, L.W. (1989) *Envisioning a Sustainable Society. Learning Our Way Out*, Albany, NY: State University of New York Press.

Naess, A. (1989) *Ecology, Community and Lifestyle. Outline of an Ecosophy*, trans. and rev. by D. Rothenberg, Cambridge: Cambridge University Press.

Nicolis, G. and Prigogine, I. (1977) *Self-Organization in Non-Equilibrium Systems*, New York: Wiley.

Perrow, C. (1988) *Normal Accidents. Living with High-Risk Technologies*, New York: Basic Books.

Picht, G. (1973) 'Der Begriff der Natur und seine Geschichte', lecture notes, winter semester 1973/74, Heidelberg: University of Heidelberg.

Polanyi, M.(1962) *Personal Knowledge*, Chicago, Ill.: University of Chicago Press.

Polyani, M. (1974) *Knowing and Being*, Chicago, Ill.: University of Chicago Press.

Polanyi, M. and Prosch, H. (1975) *Meaning*, Chicago and London: The University of Chicago Press.

Rappaport, R.A. (1979) *Ecology, Meaning, and Religion*, Berkeley, CA: North Atlantic Books.

Riedl, R. (1985) *Die Spaltung des Weltbildes. Biologische Grundlagen des Erklärens und Verstehens*, Berlin and Hamburg: Paul Parey.

Schneider, S.H. and Morton, L. (1981) *The Primordial Bond*, New York and London: Plenum Press.

Schubert, H.J. (1989) 'Zum Zusammenhang von Ethik und Macht am Beispiel Eigenarbeit', in B. Glaeser (ed.) *Humanökologie. Grundlagen präventiver Umweltpolitik*, Opladen: Westdeutscher Verlag.

Schütz, A. (1981) *Der sinnhafte Aufbau der sozialen Welt. Eine Einleitung in die verstehende Soziologie*, Frankfurt a/M: Suhrkamp.

Sheldrake, R. (1989) *The Presence of the Past*, London: Fontana / Collins.

Shotter, J. (1984) *Social Accountability and Selfhood*, Oxford: Blackwell.

Snow, C.P. (1969) *Two Cultures: And a Second Look*, Cambridge: Cambridge University Press.

Snyder, G. (1990) *The Practice of the Wild. Essays*, San Francisco, CA: North Point Press.

Steindl-Rast, D. (1984) *Gratefulness, the Heart of Prayer*, Ramsey, NJ: Paulist Press.

Steiner, D. (1991) 'The human ecological significance of different types of knowledge', in A. Kilchenmann and C. Schwarz (eds) *Perspectives of Human Ecology*, Berlin: Springer.

Steiner, D. (1992) 'Auf dem Weg zu einer allgemeinen Humanökologie: Der kulturökologische Beitrag', in B. Glaeser and P. Teherani-Krönner (eds) *Hummanökologie und Kulturökologie. Grundlagen–Ansatze–Praxis*, Opladen: Westdeutscher Verlag.

Steiner, D., Furger, F. and Jaeger, C. (1989) 'Mechanisms, forms and systems in human ecology', contribution to Third International Conference of the Society for Human Ecology, San Francisco, 7–9 Oct. 1988 (expected to appear in print in a conference volume).

Störig, H.J. (1985) *Weltgeschichte der Philosophie*, Zürich: Ex Libris.

Taylor, G.R. (1972) *Re Think*, London: Martin Secker & Warburg.

Thélin, G. (1987) 'Was bringt die Raumbeobachtung der Landschaft?', in H. Elsasser and H. Trachsler (eds) *Raumbeobachtung in der Schweiz*, *Wirtschaftsgeographie und Raumplanung* 1, Zürich: Department of Geography, University of Zürich.

Toffler, A. (1970) *Future Shock*, New York: Random House.

Ulrich, P. (1987) *Transformation der ökonomischen Vernunft. Fortschrittsperspektiven der modernen Industriegesellschaft*, Bern and Stuttgart: Paul Haupt.

Varela, F.J. (1979) *Principles of Biological Autonomy*, New York and Oxford: North Holland.

Weichhart, P. (1990) 'Raumbezogene Identität. Bausteine zu einer Theorie räumlich-sozialer Kognition und Identifikation', *Erdkundliches Wissen* 102, Stuttgart: Franz Steiner.

Weizenbaum, J. (1976) *Computer Power and Human Reason. From Judgment to Calculation*, San Francisco, CA: Freeman.

Whitehead, A.N. (1978) *Process and Reality. An Essay in Cosmology*, New York: Free Press.

Wilber, K. (ed.) (1982) *The Holographic Paradigm and Other Paradoxes*, Boston, MA: Shambhala Publications.

Young, G.L. (1989) 'A conceptual framework for an interdisciplinary human ecology, *Acta Oecologiae Hominis* 1, Lund.

5

HOW DOES THE PERSON FIT INTO THE HUMAN ECOLOGICAL TRIANGLE?

From dualism to duality: the transactional worldview

Peter Weichhart

1 A BASIC PROBLEM OF HUMAN ECOLOGY: BRIDGING SELF-CREATED CONCEPTUAL GAPS

The 'human ecological triangle' (cf. Figure 4.3 in Steiner, this volume, and also Jaeger and Steiner 1988: 137) may well serve as a symbol of the main objectives and of the fundamental – and yet unsolved – problems with which the scientific endeavour we call 'human ecology' is confronted. The relationships, interactions and interdependencies between society, person and environment constitute the ground and framework not only of our global (and local) ecological crises, but also of some of our global (and local) social crises. Traditional scientific approaches in the various disciplines concerned are evidently not appropriate to elaborate the complex models that are required for an adequate understanding of these relationships, and that must be regarded as an indispensable precondition for any solution of the problems. It is, however, much easier to demand an integrative approach than to carry it out. Asking the appropriate questions is a first and necessary step in the right direction, yet it is not a guarantee for getting the desired answers. One of the basic problems in this context arises from the need to find an adequate way to conceptualize the *relationships* between the three corners of the human ecological triangle.

The concepts traditionally employed for this task reveal a very specific style of thinking. When we discuss such ecological or social problems, we first try to identify the units or subsystems which, as single elements, factors, structures, forces, agents, etc., are perceived as relevant in constituting the problem. We use the scalpel of analytic decomposition in order to grasp the 'atoms' of the phenomena under consideration. This holds for any level of resolution or scale of investigation.

It seems to the author that owing to the traditions of scientific reasoning we

77

somehow become forced to come to terms with our problems by carefully identifying the very elements of reality. By doing this, we circumscribe the boundaries of these elements and isolate them conceptually against other elements (cf. Reichert 1992, and this volume). Even on the metatheoretical level to which the 'human ecological triangle' refers, we distinguish analytically between person, society and environment as the relevant 'entities' or elements to be dealt with when we define the subject matter of human ecology. Perhaps this way of thinking may be regarded as the consequence of general human concept formation: to identify an object and to deal with it as a cognitive concept seems possible only if we recognize the outer margins of this object, and define its boundaries against other objects and therefore are able to differentiate it from other elements of the world.

Yet the distinction between person and society and between both of these and environment logically implies that we also produce some sort of dichotomy or dualism between these elements. We thus consider the person as opposed to the environment and to society, the environment as a counterpart to human beings. Consequently, these polarities turn into implicit aspects in all of the following theoretical concepts and empirical projects. And yet a new, self-created problem arises, namely the task of bridging the gap that we have produced by our conceptual starting point. As Steiner (1988: 157) has mentioned, not the corners but the sides of the triangle symbolize the basic problem of human ecology.

Along with this text, several contributions to this book focus on human beings as persons and parts of larger social structures (cf. Carello and Mosler in Part II, Nauser and Lang in Part III). They look at the human ecological triangle primarily from the perspective of the individual. With regard to this point of view, the task of resolving the above mentioned dichotomies is closely related to the question of how the individual person fits into the larger context of environment and society. How can we find a conceptual and empirically useful approach which 'unifies' or integrates person and environment, person and society? Is it possible to consider the person as 'part' of the environment and society, or to interpret environment/society as a 'part' of the person?

2 STEPS TOWARDS INTEGRATIVE CONCEPTS OF 'PERSON AND ENVIRONMENT': LOOKING FOR ECOLOGICAL UNITS OF ANALYSIS

Claudia Carello's chapter centres on the first dichotomy; that between person and environment. Her arguments refer to one of the 'ecological' approaches in psychology, namely to the 'realist theory of perception', a tradition established by James Gibson. In order to avoid a redundant reiteration of Carello's description, only some of the more general theses and implications of this approach will be recapitulated here.

Her starting point is the reciprocity of perceiver – be it animal or human being – and environment. Perceiver and environment only represent the two aspects of one and the same irreducible system. Hence, there exists 'no such thing as *the* environment' that could be described regardless of a specific organism (cf. Weichhart 1979). Describing a part of the world as environment involves at the same time a description of the creature living in this environment. The complementarity between organism and environment is clearly articulated by the concept of affordance: 'the affordances of the environment are what it offers the animal, what it provides or furnishes, either for good or ill' (Gibson 1979: 127). Owing to its affordances an environment is directly related to an organism; the attributes of an environment constituting an affordance can be specified only in reference to a specific perceiver or user. Conversely, the perceiver's structural and functional characteristics must be understood as correlates of the supplying aspects and features of the environment. Affordances thus cut across the traditional objective–subjective dichotomy in scientific thought (cf. Heft 1981).

This approach obviously involves some very constructive proposals to overcome the first of our conceptual gaps. Without doubt the Gibsonian tradition offers some most interesting ideas which can be usefully implemented in human ecology. Nevertheless, when applied to this field, this approach seems to have some disadvantages and limitations. The author, therefore, suggests modifying some of its axioms and restricting their extent of validity. This seems particularly necessary for the axiom of realism.

By emphasizing the realism of Gibson's approach, Carello points out that environmental features are perceived directly without the mediation of mental or cognitive processes. However, if the animal under consideration is a human being, cognition must also be regarded as a factual feature of the organism–environment interaction whose evidence cannot be denied. Why should it not be possible to look at the relationship between the (realist) concept of perception and the (constructivist) concept of cognition under the aspect of complementarity – as the 'København school of quantum physics' did when dealing with the problem of the dual nature of light (as wave and particle)?

As Heft (1987: 179) has pointed out, Gibson's perspective suggests a reformulation of the standard view on the relationship between perception and cognition. Both should be seen as autonomous and complex processes serving different functions in the interplay between organism and environment. Yet it is no longer practical to consider perception as subordinate to cognition. The context of specific actions and events will determine whether perceptual or cognitive processes are useful. In the case of finding one's way, for instance, the pick-up of visual information over time in the perceiver's active movement through the environment does not depend on a cognitive map as a mental representation of the environment. On the other hand, tasks such as imagining new routes in a familiar environment or anticipating the possible route which a partner in social interaction might choose, presupposes the command of conceptual knowledge *qua* cognitive constructs.

The relation between perception in the sense of Gibson and cognition may also be paralleled with the relation between tacit knowledge and propositional knowledge (cf. Josefson, this volume) or with Giddens' (1984) concepts of practical and discursive consciousness.

A second aspect concerns the question of subjective, group- or culture-related meaning ('Sinn'). According to the axiom of realism, affordances are ecologically real or 'objective'. They are by no means mental representations of the environment. At the same time affordances are defined by the functional possibilities they offer to the perceiver. As Wohlwill and Heft (1987: 285) have stressed, this includes also subjective qualities which the perceiver imposes on sensory inputs – qualities that are related to values and meanings. The entire structure and contents of meaning are part of the specific social system to which an individual belongs. Mainly as a result of subjective internalization, meaning becomes implanted as an individual configuration of values in the course of the process of socialization. This process represents one of the central means by which individuals are bound to and integrated into a specific social and cultural context. It is obvious that the utilization of the affordances of the environment – and thus the affordances themselves – are fundamentally a function of the particular structure of values. Perceiving trees, for example, refers to very different affordances, depending on whether the perceiver is a conservationist, a forester, a speculator, or a vacationist: the characteristics of the affordances are a function of the cultural and socio-economic system to which the perceiver belongs. In the case of human beings, therefore, affordances are basically founded on the system of values and the context of meanings which are typical of the respective social system and which reveal a great variety across time and social groups. Consequently, in the case of human ecology, a deep concern with the value aspects of human existence is required in order to be able to deal adequately with the concept of affordance.

Subjective and social meanings, symbols and the whole system of culture are phenomena which cannot be conceived exclusively by a theoretical structure centred on visual perception. The transaction between these aspects of human environment and the individual is not a question of the way in which light interacts with the surface of physical objects; behavioural possibilities of human beings cannot be adequately and completely described as being exclusively determined by the surface layout of the environment. (The history of the discipline of geography has illustrated how a human geography depending on the 'physiognomic principle' was doomed to fail; likewise the classical German 'social geography' based primarily on a visual interpretation of the 'cultural landscape'.) This leads to the question of how Gibson's approach copes with the social dimensions of ecological problems. In spite of Carello's suggestions as to how to include social issues in the framework of

Gibson's approach, it seems somewhat doubtful that we would then be able to deal with the more complex social phenomena, especially with respect to the genesis of value systems.

In order to demonstrate and explain the specific and genuine tradition of Gibsonian ecological psychology, Carello emphasizes the differences in comparison to other ecologically oriented approaches of psychology. Evaluating the matter not from an insider's point of view but from an outsider's, it seems more appropriate to stress the evident general accordance of some of Gibson's central propositions with other examples of the 'contextual revolution' (Little 1987: 210) in psychology. In this respect Gibson's approach turns out to be a perception-orientated instance of the transactional worldview, which might turn out to be the most challenging development in modern psychology for human ecology and which reveals some interesting parallels and connections with other recent concepts such as structuration theory.

3 THE TRANSACTIONAL WORLDVIEW

In their discussion of philosophical backgrounds and underpinnings of psychology, Altman and Rogoff (1987) elaborated a fourfold taxonomy of worldviews which may be used to describe research and theory in this discipline. As representations of different cosmologies, these worldviews include contrasting assumptions relating to units of analysis, the role of the environment, temporal factors, concepts of causation and the role of the observer. Table 5.1 summarizes some characteristics of the four worldviews.

The authors do not claim that the various approaches of psychology fit perfectly into the categories described by this typology, which therefore is not to be regarded as a rigid classification of the whole field of research. They rather stress the fact that 'pure' examples are rare and that theories in psychology often contain ideas from more than one worldview. Nevertheless we may identify some approaches which are more or less typical of the four orientations. Classical instinct theories exemplify the trait orientation. The mechanistic perspective of the interaction approach, which may be considered as the dominant paradigm in contemporary psychology, is represented, for example, by modern personality theory, cognitive dissonance, social comparison or altruism theories (cf. Altman and Rogoff 1987: 16). Research on crowding, proxemics, as well as a large amount of the research on environmental perception and cognition, also reflect an interactional approach. The basic rationale of this worldview is the machine-analogy, and it is marked by a generally analytic orientation. Complex phenomena are analytically decomposed into separately existing elements, which are assumed to be related to one another by general laws.

In summary, interactional worldviews emphasize the separate existence of contexts and settings, person factors, psychological processes, and

81

temporal variables. Interactional research and theory ... describe dimensions of separate entities, examine their interactions, and attempt to understand antecedent–consequent causal relationships between variables. Environments are usually treated as independent predictor variables; psychological functioning is treated as a dependent or outcome variable.

(Altman and Rogoff 1987: 18–19)

Unlike interactional approaches, the organismic perspective does not concentrate on the elements of a phenomenon but selects the integrated system as a unit of study. The organism represents the basic metaphor. Although this paradigm considers 'the whole' as the proper unit of analysis, which possesses properties that are not directly derived from the properties of the elements comprising 'the whole', the elements are thought to be independently definable and functioning. Examples of this approach are: balance theory (Heider 1958); some aspects of Piaget's theory of development (1952); some of the research in environmental psychology (cf. Altman and Rogoff 1987: 22–4). The organismic view also corresponds to various conceptual and theoretical developments in other fields, such as systems theory, or the theory of living systems (cf. Miller 1978). Without doubt, approaches of this kind have promoted some important advances in the field of human ecology.

Nevertheless, the transactional worldview seems to be the most promising one for human ecology. Although the units of analysis also are holistic entities, the transactional whole is not composed of separate elements. It is rather a 'confluence of inseparable factors that depend on one another for their very definition and meaning' (Altman and Rogoff 1987: 24). The basic category of analysis of this paradigm is the event, which is conceived as a complexly organized unity constituted by a spatial and temporal confluence of human beings, settings, and activities.

There are no separate actors in an event; instead there are acting relationships, such that the actions of one person can only be described and understood in relation to the actions of other persons, and in relation to the situational and temporal circumstances in which the actors are involved. Furthermore the aspects of an event are mutually defining and lend meaning to one another, since the same actor in a different setting (or the same setting with different actors) would yield a different confluence of people and contexts ... the transactional worldview does not deal with the relationship between elements, in the sense that one independent element may cause changes in, affect, or influence another element. Instead, a transactional approach assumes that the aspects of a system, that is, person and context, coexist and jointly define one another and contribute to the meaning and nature of a holistic event.

The relations among aspects of the whole exist ... in their very definition, not in the influences of separate variables on one another. Relations among the aspects of a whole are not conceived of as involving mutual influences or antecedent–consequent causation. Instead, the different aspects of wholes coexist as intrinsic and inseparable qualities of the whole ... the transactional worldview incorporates temporal processes in the very definition of events. The transactional view shifts from analysis of the causes of change to the idea that change is inherent in the system ... Change is viewed more as an on-going, intrinsic aspect of an event than as the outcome of the influence of separate elements on each other ... Change ... may result in ... outcomes that are variable, emergent, and novel.

(Altman and Rogoff 1987: 24, 25)

We may infer from this lengthy quotation that transactionalism is compatible to general recent worldviews outside psychology, for instance, to that of modern physics, whose theories focus on the concept of 'field'. These theories suggest that in the case of subatomic high-velocity phenomena so-called particles (which in this perspective are revealed not to be 'real' particles) may be comprehended as momentary and changing bonds of energy and activity. Without going into detail we may trace certain similarities to synergetics and chaos theory. As the reader will certainly have noticed, it should be stressed that transactionalism also exhibits strong parallels to Giddens' structuration approach, which perhaps may be recognized as a sociological instance of this paradigm.

As we can see, Gibson's ecological realism fits perfectly into the transactional worldview (cf. Lombardo 1987). Apart from that, several other approaches identified by Altman and Rogoff (1987: 27–9) exemplify many aspects of the transactional perspective. Lewin's concept of life space (1964), the ethogenic or situated action approach (Ginsburg 1980, Harré 1977), ethnomethodology (Garfinkel 1967, Weingarten et al. 1979), Goffman's 'dramaturgical' social psychology (1959, 1971), perhaps Harré's (1983) concepts of personal psychology and other approaches resemble various transactional orientations. Undoubtedly, one of the most prolific examples is the ecological research done by Barker and his associates (1968, 1978). This school proposes the thesis that human conduct is inextricably linked with the physical and social environment in a continuous flow. Contrary to Carello's evaluation, this author argues that the central concept of this tradition – the behaviour setting – may be paralleled with Gibson's affordance concept in so far as both reflect a transactional view. In a more recent definition, a behaviour setting is conceived of as a 'bounded, self-regulated and ordered system composed of replaceable human and non-human components that interact in a synchronized fashion to carry out an ordered sequence of events called the setting program' (Wicker 1983: 12). Behaviour settings have a specific life-history; they proceed from formative or convergence phases to

Table 5.1 General comparison of trait, interactional, organismic and transactional worldviews

	Trait	Interactional	Organismic	Transactional
Definition of psychology	The study of the individual, mind, or mental and psychological processes.	The study of the prediction and control of behaviour and psychological processes.	The study of dynamic and holistic psychosystems in which person and environment components exhibit complex, reciprocal, and mutual relationships and influences.	The study of the changing relations among psychological and environmental aspects of holistic unities.
Unit of analysis	Person, psychological qualities of person.	Psychological qualities of person and social or physical environment treated as separate underlying entities, with interactions between parts.	Holistic entities composed of separate person and environment components, elements or parts, whose relations and interactions yield qualities of the whole that are 'more than the sum of its parts'.	Holistic entities composed of 'aspects', not separate parts or elements; aspects are mutually defining; temporal and spatial qualities are intrinsic features of wholes.
Causation	Emphasizes material causes, i.e. causes internal to phenomena.	Emphasizes efficient causes, i.e. antecedent–consequent relations, 'push'-ideas of causation.	Emphasizes final causes, i.e. teleology, 'pull' toward ideal state.	Emphasizes formal causes, i.e. description and understanding of patterns, shapes, and form of phenomena.

Observers	Observers are separate, objective and detached from phenomena; equivalent observations by different observers.	Observers are separate, objective and detached from phenomena; equivalent observations by different observers.	Observers are separate, objective and detached from phenomena; equivalent observations by different observers.	Relative: observers are aspects of phenomena; observers in different 'locations' (physical and psychological) yield different information about phenomena.
Selected goals	Focus on trait and seek universal laws of psychological functioning according to few principles associated with person qualities; study predictions and manifestations of trait in various psychological domains.	Focus on elements and relations between elements; seek laws of relations between variables and parts of system; understand system by prediction and control and by cumulating additive information about relations between elements.	Focus on principles that govern the whole; emphasize unity of knowledge, principles of holistic systems and hierarchy of subsystems; identify principles and laws of whole system.	Focus on event, i.e. confluence of people, space and time; describe and understand patterning and form of events; openness to seeking general principles, but primary interest in accounting for event; pragmatic application of principles and laws as appropriate to situation; openness to emergent explanatory principles; prediction acceptable but not necessary.

Source: Modified from Tables 1.1, 1.2 (pp. 9, 12–13) of Altman and Rogoff (1987)

operating ones and then to dissolution or divergent phases. They show a nested hierarchy which results from different types of couplings and linkages (cf. Wicker 1987). As 'configurations of actors, activities, and physical and social contexts that change in emergent and contextually linked ways' (Altman and Rogoff 1987: 30), behaviour settings are compatible with a view that accentuates the cognitive, culture- and value-related aspects of meaning.

4 FURTHER STRUGGLE FOR TRANSACTIONALISM: THE OUTSIDE AND THE INSIDE MIND

Alfred Lang, in his contribution to this volume (Part III), proposes a new heuristics for a more adequate understanding of the relationship between human beings and their environments. He criticizes the common models of humans in modern psychology, which he classifies as 'pre-Copernican' in the sense that they suffer from a restrictive subject–object dichotomy, and his criticism applies to other social sciences as well (cf. Lang 1988). By centring on the human individuals and their intrinsic psychic processes, pre-Copernican psychology has apprehended the environment exclusively from the perspective of the human subject. Therefore traditional psychology has never achieved the more complex and more appropriate model of an integrated subject–environment whole. In most cases the environment has been lost out of sight or at best has been considered something outside or opposed to the subject. The main aim of Lang's contribution is to encourage consideration of a more balanced view, a proposal for a more 'symmetrical' model of human beings and their environment by decentring the research object away from the individual and moving it towards person–environment units. The 'Copernican revolution' which he urgently demands is nothing but a plea for a trans-actional worldview.

Lang's arguments focus on a specific facet of the dualism between person and environment, namely on cultural artefacts and their relation to human beings. He starts out with the observation that the disciplines studying these phenomena usually do not consider artefacts to be 'part and parcel' of persons. One usually reflects on artefacts under the aspect of their objective existence, disregarding the fact that they are also created, possessed and appropriated by people, that they bear specific meanings and values, are loved or hated. In other words, Lang stresses the point that in the case of artefacts the boundaries between people and things are not as impermeable and stable as taken for granted in common scientific thought.

Perhaps the most important idea in Lang's argumentation is his conception of a kind of complementarity between the corporeal–personal whole of individuals and their connection and unity with the environment. Owing to this duality, both the separation of individuals from and their connection with the surrounding world are constituted and secured over the flow of time. Lang's central assumption refers to the concept of structure formation, which

qually applies to biological (cells), psychological (individuals) and socio-
)gical units (societies, cultures; cf. Lang 1988: 89–100). Structure formation
1ay be specified as a dynamic storage of the specific history of the respective
nit, which can be used or activated as a system of self-reference. These units
ave the ability to create a 'trace of their own history' (ibid., p. 99), which
llows them to relate past events and episodes to the present and even to
ossible events in the future. These storage systems may be described as
lynamic memories'. In the case of the cell, the genom works as a generative
nemory, by which phyletic history is stored; in the case of individuals it is
heir ontogenetic memory that allows the reproductive reference to past
vents and their speculative projection into an anticipated future. At the level
)f social units all cultural products, that is, material and symbolic things,
)uildings and the 'cultural landscape', may be interpreted as a kind of storage
)r memory which contain the history of the respective (sub-)culture and
vhich, owing to their relative durability, influence its future. In the terms of
Halbwachs (1967), cultural products reflect the 'collective memory' of a
ociety. With respect to the individual we may also assume the existence of
ome kind of 'external memory' such as things (as opposed to objects) or
)laces (as opposed to space), which Lang circumscribes as a 'concretization of
he mind'. By referring to the external memory system, which remains
·elatively stable over time, the specific units are able to preserve and assure
:hemselves of their identities despite temporal changes.

In summary, Lang's approach stresses the heuristic possibility of conceiving
naterial objects and cultural artefacts together with their inherent symbolic
neanings as an extension of persons: the dichotomy between 'inner' and
outer' world can thus be dissolved; persons and aspects of their surroundings
`orm a joint 'ecological unit' in their own right.

Needless to say, Lang's proposals for a 'concrete mind-heuristics' are in
ιccordance with the transactional worldview as it is present, for instance, in
:he works of Wapner (1981, 1987), Rapoport (1977, 1982), Altman and
Gauvain (1981), Stokols (1987), and their associates (cf. Altman and Rogoff
1987: 29–32). The necessity of changing the perspective, and not focusing on
:ither person or environment, but rather on person in environment, has been
·ealized, especially within the research context of identity development, place
dentity and residential environments. Some of the first to stress this
:onnectivity were geographers using a phenomenological approach and
)sychologists working on home environments: 'People are their place and the
)lace is people, and however readily these may be separated in conceptual
:erms, in experience they are not easily differentiated' (Relph 1976: 34). 'In
nany cases the boundary between self and the environment is far from
listinct. It may move outward, with part of the environment incorporated as
)art of the self. It may become increasingly permeable, so that the distinction
)etween self and environment becomes unclear or ... disappears ... the
:nvironment is experienced as an important part of one's self, as an integral

component of self-identity' (Ittelson 1978: 202–3; cf. Proshansky 1978: 155, Frey and Hausser 1987, and Weichhart 1990). It is self-identity, self-concept, and specific social identities of the individual that seem to be the core and focus of the unresolvable unity between person and environment.

5 THE HUMAN AGENT AS PERSON AND PART OF LARGER SOCIAL SYSTEMS

Hans-Joachim Mosler and Markus Nauser (both this volume) emphasize the other side of the human ecological triangle. They discuss various aspects of the question as to how the human agent as an individual person is part of the surrounding social structure and how this social system is reflected by and incorporated in the individual. Both authors are mainly concerned with the possibilities of establishing a politically practicable strategy that would induce people to be more responsible for the environment and to direct their behaviour in a way that is ecologically sound. The environmental crisis is identified primarily as a social problem: the 'tragedy of the commons' and the consequent exploitation and destruction of the natural environment result from social conflicts of interests. Traditional strategies of problem solving relate to governmental regulations and to the improvement of public attitudes by motivation campaigns. Both concepts are confronted with an insufficient individual acceptance; both rely on a mechanistic and causalistic concept of human beings and overrate the chances of manipulation and heteronomy through which the proposed regulations might acquire a touch of eco-fascism. The authors complain about the shortcomings of the traditional approaches and make some interesting proposals for a more appropriate conceptual framework: Mosler refers to synergetics and Nauser to the theory of structuration.

Essential topics of both papers are personal value-systems, personal responsibility, moral judgement, conscience, social confidence, and the question of how these personal and subjective systems of values and moral order are related to the superordinate social structures. Mosler discusses the development of public opinion concerning specific attitudes towards the environment and ecologically relevant behaviour by using the concepts and terminology of synergetics. He points out the interplay and feedback between collective attitudes and the formation of subjective attitudes of individuals as they are also revealed by processes of socialization. Via the pressure of conformity, the former influence the latter. The individual, striving for social acceptance and trying to avoid social isolation, is forced and is willing to accept and to internalize public opinion. However, by doing this, the individual reproduces and reinforces the social value system. The whole process may be interpreted as an example of the duality of structure, which, in Giddens' terms, describes the inseparable interplay between the individual agent and society that is to be seen as two aspects of the same reality. By nploying the concepts of stabilization, fluctuation, phase transition or aving principle, synergetics serves as the background for the description of ıe formation of collective and individual value structures in the course of me. Mosler's strategic proposals refer to a synergetic model based on the ıacrovariables of self-determined responsibility and obligation, public con-·ol, and trust.

Nauser, too, accentuates the social frame of reference for individual actions, ıus directly applying Giddens' theory of structuration. His arguments bear stronger relation to action and agency than Mosler's. He emphasizes the nportance of routines in everyday practice, of unacknowledged conditions nd unintended consequences of environmentally relevant conduct, and lustrates that the reproduction of structures in the context of specific actions ıay restrict and limit the possibilities of actions within a different context. Iis proposals for problem solving also refer to the key concept of moral ıbligations and subjective responsibility as the necessary condition for the esired 'ecologization' of agency.

Without a doubt, both papers fit into the framework of the transactional vorldview. They underline formal causes, actions and events, conceptualize ndividuals and social structures, not as separate parts but as mutually lefining aspects of one and the same reality and handle temporal and spatial ıualities as intrinsic features of the wholes under consideration.

However, from the viewpoint of the 'person corner' of the human cological triangle, the most important point seems to be the following: the luality of human agent and social structure, the inseparable connectedness ıetween the individual and the social system, person and society as mutually lefining aspects of the holistic reality of human existence – all referring to a pecific feature of the human being. Once more, it is ego-identity or the ıersonal self which seems to be the focus or the pivot upon which the duality ests (and is brought to work) and which is also the inner core of agency and ıf the episodic structure of human action.

6 THE PERSONAL SELF AS A FOCUS OF A DOUBLE DUALITY

ĭy what means do individual human beings acquire the ability and the power o act as more or less autonomous agents (even to the extreme possibility of uicide)? How do they manage the self-referential structure of the stream of onsciousness that assures their ontological security? By what means are ıuman beings certain of their identity, of being the same persons in the ourse of life-events from birth to death? What gives them the competence to nake moral decisions, to accept and obey a specific moral order? How is the levelopment of the self-referential personal structure that we call 'con-cience' possible? To answer questions such as these, individual psychology ıas developed a great number of theories of personality as well as a set of

concepts used for a more or less differentiated and profound circum-
scription of that inner core of human beings which seems to be the pivot
and medium for human attributes and abilities, such as responsibility,
conscience, or reflexivity; these concepts include 'ego-identity', 'self', 'self-
concept', 'I' and 'Me', Freud's trinity of 'id', 'ego' and 'super-ego', and even
more complex conceptions such as Allport's (1955) 'proprium', which is
conceived as including all aspects of personality that make for a sense of
inward unity.[1]

The author has the impression – but as a geographer he is far from being
competent in assessing the case – that despite the great efforts in psychology
no secure foundation and no common agreement or consensus have yet been
reached with regard to these basic concepts of personal being. He is not even
sure whether these notions of self refer to ontologically relevant facts, or
'actual entities', or make sense merely on a purely conceptual level of
constructive reflections. However, and be it only on the grounds of lay or
common-sense psychology, which in the course of everyday life and social-
interaction evidently presupposes the existence of entities such as the self or
the individual mind,[2] it seems inevitable to turn our interest to this inner core
of the human being as an individual person if we intend to make some
progress in the comprehension of the relations between person and environment and society.

In geography up to now such an endeavour or project was far beyond
general interest. Especially within German-speaking social geography, the
individual person has not been accepted as a meaningful topic of study.
Already in 1962 Bobek advocated the view that the individual as an object
matter of geography was an absurdity because only groups (so-called 'social–
geographical groups') were the relevant agents of the processes under consideration. In the meantime his notion has developed into a dogmatic axiom,
which also applies to the more recent studies of the field. Criticism of this
dogma and proposals to establish a more complex and differentiated concept
of humans based on concepts of psychology are effectively refused by
referring to the boundaries and competence of the discipline ('Well, but
would that really still be geography?').

In the social geography of the English-speaking world, we may distinguish
a group of socially committed approaches dealing with a geography of social
problems and characterized by a more or less emancipatory attitude and a
group of behaviourally oriented approaches. For the first group no need
exists to focus on the individual because of the general orientation towards
social systems, classes and power relations. For the second group one would
expect an explicit thematic emphasis on the individual because subjective
perception, assessment, and behaviour constitute the central object of
research. A more thorough examination reveals, however, that even the
most recent and most progressive approaches, which are dedicated to an
organismic–cognitive orientation far beyond a purely behaviourist con-

ception, in fact do not deal with the complex and holistic identity of single human beings, but only with some sort of subpersonal components of mental systems. In this case, the relevant elements of specific models are perception, cognitive structure, attitudes, emotions, aspiration level, preference, decision process, etc. These subpersonal components are regarded as if they represented some kind of modules or mental organs which, in accordance with the root metaphor of a machine-analogy, work together in a mechanistic and causalistic way. Yet by no means does the functionality of such input-processing modules stand for a meaningful and appropriate approximation to the psychic unity of a human individual (cf. Harré 1983: 9–20).

Only in the case of action–theoretical and phenomenological approaches does human geography accept the view of human beings as holistic and autonomous personalities. But even then some questions remain open, some problems unsolved. Thus, with the action-oriented approaches we may notice an underestimation of the role of the individuals as loci of agency and of their significance for the emergence of conduct. Werlen (1987: e.g. 22) stresses several times the notion of social analysis. Within this framework the basic categories and facts of the social world are not the persons, but the actions they generate. Therefore, the most important task of a modern social geography would be the comprehension and explanation of human actions as they emerge within a specific social context. The concept of a pure 'geography of the individual', he infers, would make it impossible to show how social structures are constituted.

In general, the author agrees with Werlen's propositions and conclusions, which are to some degree in accordance with the transactional worldview. Yet it seems that this concept does not go far enough and is not complete or sufficient enough because it neglects the problem of how and by what means the social context is related to or incorporated in the individual agent. So we may conclude that the personal context of human actions should also be of interest if we intend to explain the episodic flow and the outcomes of human conduct. Again Giddens' concept of the duality of structure is on the agenda.

Contemporary social sciences are marked by two contrasting lines of thought concerning human existence. A thoroughgoing individualistic conception, mainly represented by theorists of cognition, conceives human action as the genuine result of individual mental processes. The opposing concept of collectivism relates human action entirely to the surrounding social system: actions are the outcome of a web of intentional processes carried out by human subjects whose minds are structured and determined by an interpersonal social reality. 'For individualists the deepest problem is how intersubjectivity is possible and their great philosophical problem that of our knowledge of other minds; for collectivists, the deepest problem is how individuality is created and sustained in so thoroughly social a world. For the former, individual being is given and social being constructed; while for the latter, collective being is given and personal being is an achievement' (Harré 1983: 8). Against the background of the theory of structuration, the controversy between methodological individualism and methodological collectivism loses its relevance (Giddens 1984: 207–21; cf. also Lenk 1977, who points out the necessity for a complementary interrelation of the two views). Nothing like social structures could exist without individuals, who reproduce them in the course of their intentional actions. These actions are reflections and outcomes of the individual's consciousness in the sphere of practice and they in turn shape consciousness. Society is produced and reproduced in the episodic process of conduct, the performance of the individual's actions. Without social structure, which comprises constraint and enablement, the existence of the individual whose 'inner core' of self-consciousness emerges as a result and product of social interaction is impossible – as we will try to discuss in the following.

Let us refer to a recent theory of personality which may also be convincing for someone who supports the view of a definite methodological collectivism, namely to Harré's theory of personal being which reveals a considerable degree of comparability to structuration theory. In accordance with Giddens, he stresses the inseparability and mutual interdependence of the consciousness and mindedness of the human individual, which result in the basic human characteristics of 'having/possessing agency', and the surrounding social structure, which on the one hand shapes and determines the individual person to a very large extent but which in turn is the outcome of human conduct and therefore on the other hand is reproduced in the course of actions. Like Giddens he even uses the concept of duality to refer to this mutual conditionality.

Harré's main thesis might sound quite astonishing and provocative for the methodological individualist: 'mind is no sort of entity, but a system of beliefs structured by a cluster of grammatical models. . .the very structure of our minds (and perhaps the fact that we have minds at all) is drawn from. . .social representations. . .a person is not a natural object but a cultural artefact. A person is a being who has learned a theory, in terms of which his or her experience is ordered' (Harré 1983: 20). 'Our personal being is created by our coming to believe a theory of self based on our society's working conception of a person' (ibid., p. 26).[3] We must keep in mind that various other prominent writers, such as Mead (1934) or to some degree already Cooley (1902), also argued that both the self and consciousness were products of social processes and that 'being' or 'becoming' a person might result from taking the social role of persona. From this notion we can infer that in non-Western societies other concepts of individual beings may occur, resulting in comprehensions of self different from ours (Harré 1983: 29–30; cf. also Harré 1985).

Harré developed a complex theory concerning the processes and frameworks of how selfhood is learned by the individual human being. Similar to other authors in the social sciences (e.g. Luhmann 1984), he bases his theory on the axiom that in a 'social analysis of human life, conversation becomes the basic reality' (Harré et al. 1985: 71). The individuals' personal worlds of thought and feelings are largely 'reflections of the social world created by the conventions to which they adhere in interacting one with another ... The personal psychological world is created by appropriation of various conversational forms and strategies from that discourse. In so far as individual people construct a personal discourse, they become complex "mental" beings with individual "inner worlds"' (ibid., pp. 71–2). The 'language game of self-ascription' which also refers to moral order and the assignment of responsibility has its origin in a basic process of social interaction beginning in the earliest stage of ontogenetic development and which is called 'psychological symbiosis'. 'Psychological symbiosis is a permanent interactive relation between two persons, in the course of which one supplements the psychological attributes of the other as they are displayed in social performances, so that the other appears as a complete and competent social and psychological being' (Harré 1983: 105). The mother–infant dyad serves as the most relevant example. Investigations on the talk that mothers direct at their infants revealed that 'mothers address their infants as if the infants had well-developed psychological repertoires of intentions, wants, feelings and powers of reason from the moment of birth. The mother treated the child as if it was, in reality, what the nature of her speech implied she thought it was ... The mother does not interact with the child as it actually is but rather with a being of her own invention to whom she has ascribed quite sophisticated thoughts and feeling' (Harré et al. 1985: 75). In its further development the child is forced to appropriate the contents and meanings of this language game: 'One who is always presented as a person, by taking over the conventions through which this social act is achieved, becomes organized as a self' (Harré 1983: 106). Harré points out that by learning to conceive ourselves as personal beings we are driven to look for substantial centres of our experience: 'The "centred" model of experience ... is a permanent feature providing a perennial temptation to believe in an extra being, a noumenal self deeply embedded within our persons. Paradoxically, it is only by believing in such an inner active core of self that *our* psychological attributes and *our* moral order can be realized' (ibid., p. 108; emphasis Harré's).

Let us assume that Harré's theory is proven and that the personal self is in fact nothing else but the product of social practice and language usage or speech acts, is the result of the application of a common social theory of the self to oneself. In that case such 'emulation' of a personal self (but again, who or what is the subject of this process?) could not be comprehended simply as a reification but acquires the ontological status of an existential fact. Personal beings are real although they are the product of theoretical activity. What are, therefore, the central aspects of this emerging 'self' of the individual person?

Harré distinguishes three 'modes' of individuality. The first, consciousness, circumscribes the personal being as a formal unity. Its reflexive expression, self-consciousness, may be seen as an abstraction of the public person-concept. It is used as an 'explanatory concept with which we organize our experience, specifying its natural kind as a unity of spatio-temporally continuous point of view and point of action' (Harré 1983: 167). In a similar way he conceives agency, the second mode, which describes the personal being as practical unity, as the reflexive employment of a social theory with the active and willing self as its basic theoretical concept: 'In possession of a theory that I am an agent capable of acting against the tide of my inclination, capable of getting myself up and going etc., I have the means for readjusting the many means–end hierarchies which are involved in the preparation for action' (ibid., p. 193). Autobiography (the third mode) relates to the personal being as an empirical unity, which is based on the corporeality of human existence. Autobiography as a further basis of one's personal sense of identity refers to the continuity of one's location in space and time relative to one's point of view. The experiential continuity works as a system of reference in which a person's present actions are to be located, representing the past and anticipating the future, assuring a sense of a permanent self.

It is the interplay between these three modes of individuality from which the unity of the self emerges as it is experienced as an existential fact by the individual and as it is accepted, forced and presupposed by the social system. This unity of the self – however socially determined it may be – is the necessary and indispensable supposition that individual human beings are able to play their role as the autonomous agents who are hypothesized by structuration and action theory. The unity of the self is also the basis for the individuals' capacity of actually doing what the moral order requires, of fulfilling their commitments, of making moral decisions. Only by this unified self may individuals be said to have confidence, to be trustworthy, to possess a conscience, to be responsible. It may also be regarded as the preliminary condition of the specific human kinds of interpersonal interaction – in the good and in the evil sense.

Metaphorically speaking and in analogy to the idea of Dawkins (1976; cf. Lang 1988) that organisms are 'instruments' or 'survival-machines' of the genes, we may conclude that social systems 'construe' or 'manufacture' the individual self as a means for reproducing their own structure.[4]

As a result of this lengthy (yet hopefully helpful) digression into a recent theory of personality we now may be able to progress a step further and arrive at a conclusion concerning the problem we have been trying to deal with: how can we bridge the gaps, or resolve the dichotomies which are inherent in the human ecological triangle? Considering Harré's theory, we may hypothesize that from the very beginning the position of the person within the triangle has perhaps prevented us from seeing that a medium of integration already exists. It is the unity of self which, socially constructed as a main principle of human existence, functions as the inner core of the individual person and which is the focus or pivot of a double duality. The first duality, emerging in the events of human conduct, relates to the mutual conditionality of agent and social structure in the sense of Giddens. The second one emanates from the confluence of agent and environment which is described for instance by Lang's article.

By actively performing 'identity projects' (cf. Harré 1983: 256–87), the self-creation of a distinct person by an idiosyncratic transformation of the unity of self, individual persons may take control of their own development. Once possessing the unity of self an autopoietic process may begin to function. Now the human beings are able to work out actively and consciously the specificity and singularity of their personality, stressing the subtle differences to other persons.[5] In the course of this transformation the double duality of the inseparable connection with environment and social structure includes the constraints and the enablements affecting possible developments.

Our reflections on the personal self as conceived in Harré's theory are not in contradiction with methodological collectivism because it has been made especially clear that the origin of the self is deeply grounded in the social structure and the social practice of human conduct, or more precisely in the practice of conversation and speech acts. And at the same time the emergence of that self-referential deep-structure of the person we call the human self, however it may come into existence, forms the necessary precondition for the constructive mutual interplay between the agent and the social structure as hypothesized by structuration theory. Simultaneously, the self – which therefore may be conceptualized as a process, not as a structural entity – is established in its relation not only to the material but also to the symbolic structures of the environment, the things, the places and the cultural artefacts that are appropriated, loved or hated and in this way become part of the self. The referential connection to this material world as one of the most important means of conveying ontological security is also a basis for maintaining self-identity and the stability of autobiography. Together with the routines of day-to-day life (cf. Giddens 1984: 34–7) the physical environment and its connotations determine the *durée* of existential experience.

It seems quite useless, therefore, to try to come to terms with the phenomenon of the individual self by dealing with it as an entity in its own right because its mere existence and emergence are the consequence of an active and episodic flow of communication and interrelation between aspects of reality, which only together and in the course of relational processes constitute the existential facts of human beings.

Speculations such as those above may lead to another idea. Perhaps not the triangle, but the moving circle[6] (or the spiral?) with the unity of self in its centre should be regarded as the adequate visualization or symbolic representation of human ecology, its scope, and its basic problems.

7 NOTES

1 The aspect of personality included within the 'proprium' are bodily sense, self-identity, ego-enhancement, ego-extension, rational process, self-image, propriate striving, and the knower.

2 As we will see later, these common-sense theories about the human self may have a more than trivial significance for the solution of the problem.

3 Let us leave out of consideration that even this rather radical conception implies some hidden 'unit' or agent who 'applies' this social theory of the self to itself, which reveals some circularity in the argumentation. Perhaps Dagmar Reichert (cf. 1988) could help in thinking about this problem.

4 As we have seen, this metaphor must be restricted to our Western societies.

5 One might even speculate whether the history of humankind could not be interpreted reasonably well through the context of individual identity projects.

6 Envisaged as the trace of person, society, and environment in the course of action and events.

8 REFERENCES

Allport, G. (1955) *Becoming*, New Haven, CT: Yale University Press.

Altman, I. and Gauvain, M. (1981) 'A cross-cultural and dialectic analysis of homes', in L. Liben, A. Patterson and N. Newcombe (eds) *Spatial Representation and Behavior Across the Life-Span: Theory and Application*, New York: Academic Press.

Altman, I. and Rogoff, B. (1987) 'World views in psychology: Trait, interactional, organismic, and transactional perspectives', in D. Stokols and I. Altman (eds) *Handbook of Environmental Psychology*, 1, New York: Wiley.

Barker R.G. (1968) *Ecological Psychology: Concepts and Methods for Studying the Environment of Human Behavior*, Stanford, CA: Stanford University Press.

Barker, R.G., Barker, L.S., Fawl, C.L., Gump, P.V., Halstead, L.S., Johnson, A., Ragle, D.D.M., Schoggen, M.F. Schoggen, P., Wicker, A.W., Willems, and Wright, H.F. (1978) *Habitats, Environments, and Human Behavior*, San Francisco, CA: Jossey-Bass.

Bobek, H. (1962) 'Kann die Sozialgeographie in der Wirtschaftsgeographie aufgehen?', *Erdkunde* 16: 119–26.

Cooley, C.H. (1902) *Human Nature and the Social Order*, New York: Charles Scribner's Sons.

Dawkins, R. (1976) *The Selfish Gene*, Oxford: Oxford University Press.

Frey, H.-P. and Hausser, K. (1987) 'Entwicklungslinien sozialwissenschaftlicher Identitätsforschung', in H.-P. Frey and K. Hausser (eds) *Identität: Entwicklungen psychologischer und soziologischer Forschung*, Stuttgart: Enke.

Garfinkel, H. (1967) *Studies in Ethnomethodology*, Englewood Cliffs, NJ: Prentice-Hall.

Gibson, J.J. (1979) *The Ecological Approach to Visual Perception*, Boston, MA: Houghton Mifflin.

Giddens, A. (1984) *The Constitution of Society: Outline of the Theory of Structuration*, Cambridge: Polity Press.

Ginsburg, G.P. (1980) 'Situated action: An emerging paradigm', in L. Wheeler (ed.) *Review of Personality and Social Psychology*, 1: 295–325, Beverly Hills, CA: Sage.

Goffman, E. (1959) *The Presentation of Self in Everyday Life*, New York: Doubleday.

Goffman, E. (1971) *Relations in Public*, New York: Harper & Row.

Halbwachs, M. (1967) *Das kollektive Gedächtnis* (with a foreword by Prof. Dr. H. Maus), Stuttgart: Enke.

Harré, R. (1977) 'The ethogenic approach: theory and practice', in L. Berkowitz (ed.) *Advances in Experimental Social Psychology*, 10: 284–314.

Harré, R. (1983) *Personal Being: a Theory for Individual Psychology*, Oxford: Blackwell.

Harré, R. (1985) 'The language game of self-ascription: a note', in K.J. Gergen and K. Davis (eds) *The Social Construction of the Person*, New York: Springer.

Harré, R., Clarke, D. and de Carlo, N. (1985) *Motives and Mechanisms: an Introduction to the Psychology of Action*, London and New York: Methuen.

Heft, H. (1981) 'An examination of constructivist and Gibsonian approaches to environmental psychology', *Population and Environment: Behavioral and Social Issues* 4: 227–45.

Heft, H. (1987) 'Environmental cognition in children', in D. Stokols and I. Altman (eds) *Handbook of Environmental Psychology*, 1, New York: Wiley.

Heider, F. (1958) *The Psychology of Interpersonal Relations*, New York: Wiley.

Ittelson, W.H. (1978) 'Environmental perception and urban experience', *Environment and Behavior* 10: 193–213.

Jaeger, C. and Steiner, D. (1988) 'Humanökologie: Hinweise zu einem Problemfeld', *Geographica Helvetica* 43: 133–40.

Lang, A. (1988) 'Die kopernikanische Wende steht in der Psychologie noch aus! Hinweise auf eine ökologische Entwicklungspsychologie', *Schweizerische Zeitschrift für Psychologie* 47: 93–108.

Lenk, H. (1977) 'Struktur- und Verhaltensaspekte in Theorien sozialen Handelns', in H. Lenk (ed.) *Handlungstheorien interdisziplinär IV: Sozialwissenschaftliche Handlungstheorien und spezielle systemwissenschaftliche Ansätze*, München: Wilhelm Fink.

Lewin, K. (1964) *Field Theory in Social Science*, New York: Harper.

Little, B.R. (1987) 'Personality and the environment', in D. Stokols and I. Altman (eds) *Handbook of Environmental Psychology*, 1, New York: Wiley.

Lombardo, T.J. (1987) *The Reciprocity of Perceiver and Environment. The Evolution of James J. Gibson's Ecological Psychology*, Hillsdale, NJ and London: Erlbaum.

Luhmann, N. (1984) *Soziale Systeme: Grundriss einer allgemeinen Theorie*, Frankfurt a/M: Suhrkamp.

Mead, G.H. (1934) *Mind, Self, and Society, from the Standpoint of a Social Behaviorist*, Chicago and London: University of Chicago Press.

Miller, J.G. (1978) *Living Systems*, New York: McGraw-Hill.

Piaget, J. (1952) *The Origins of Intelligence in Children*, New York: International Universities Press.

Proshansky, H.M. (1978) 'The city and self-identity', *Environment and Behavior* 10: 147–69.

Rapoport, A. (1977) *Human Aspects of Urban Form*, Oxford: Pergamon.

Rapoport, A. (1982) *The Meaning of the Built Environment*, Beverly Hills, CA: Sage.

Reichert, D. (1988) 'Writing about circularity and self-reference', in R.G. Golledge, H. Couclelis and P. Gould (eds) *A Ground for Common Search*, Goleta, CA: Santa Barbara Geographical Press.

Reichert, D. (1992) 'On boundaries', *Society and Space* 10 (1), 87–98.

Relph, E. (1976) *Place and Placelessness*, London: Pion.

Steiner, D. (1988) 'Das Dreieck und der Kreis', in D. Steiner, C. Jaeger and P. Walther (eds) 'Jenseits der mechanistischen Kosmologie – Neue Horizonte für die Geographie?', *Berichte und Skripten* 36, Zürich: Department of Geography, ETH.

Stokols, D. (1987) 'Conceptual strategies of environmental psychology', in D. Stokols and I. Altman (eds) *Handbook of Environmental Psychology*, 1, New York: Wiley.

Wapner, S. (1981) 'Transactions of persons-in-environments: some critical transactions', *Journal of Environmental Psychology* 1: 223–39.

Wapner, S. (1987) 'A holistic, developmental, systems-oriented environmental psychology: some beginnings', in D. Stokols and I. Altman (eds) *Handbook of Environmental Psychology*, 2, New York: Wiley.

Weichhart, P. (1979) 'Remarks on the term "environment"', *GeoJournal* 3(6): 523–31.

Weichhart, P. (1990) 'Raumbezogene Identität. Bausteine zu einer Theorie räumlich-sozialer Kognition und Identifikation', *Erdkundliches Wissen* 102, Stuttgart: Franz Steiner.

Weingarten, E., Sack, F. and Schenkein, J. (eds) (1979) *Ethnomethodologie. Beiträge zu einer Soziologie des Alltagshandelns*, Frankfurt a/M: Suhrkamp.

Werlen, B. (1987) 'Gesellschaft, Handlung und Raum. Grundlagen handlungstheoretischer Sozialgeographie', *Erdkundliches Wissen* 89, Wiesbaden and Stuttgart: Franz Steiner.

Wicker, A.W. (1983) *An Introduction to Ecological Psychology*, New York: Cambridge University Press.

Wicker, A.W. (1987) 'Behavior settings reconsidered: Temporal stages, resources, internal dynamics, context', in D. Stokols and I. Altman (eds) *Handbook of Environmental Psychology*, 1, New York: Wiley.

Wohlwill, J.F. and Heft, H. (1987) 'The physical environment and the development of the child', in D. Stokols and I. Altman (eds) *Handbook of Environmental Psychology*, 1, New York: Wiley.

6

PHILOSOPHICAL REMARKS ON THE PROJECT OF HUMAN ECOLOGY

Markus Huppenbauer

My philosophical remarks on this volume are concerned in the first part with the problem of integration and in the second part with a critique of the practical importance of human ecology.

1 THE THEORETICAL FRAMEWORK OF THE HUMAN-ECOLOGICAL TRIANGLE

Let me start with the human–ecological triangle (cf. Steiner, this volume), the relationships between person, society, and environment. This triangle presupposes that the fields which it tries to integrate have been dealt with separately by scientific analysis. Therefore, in talking about the human-ecological triangle we also mean the search for a new, that is, integrated form of science. But what does 'integration' mean in this context? Suppose we know what the notions 'person', 'society', 'environment' mean: what is it that connects such different things as personal emotions, social symbols and processes of the biosphere? It is obvious that these things interact in the real world, but do we have a theoretical framework that enables us to conceive of this interaction as an expression of a unity which connects all three aspects?[1] If we are not able to conceive of this unity, we do not know what we mean by the human–ecological triangle.

1.1 The limits of previous attempts at integration

In this volume there are attempts not only to demand but to execute such an integration: 'eMergy analysis' (Pillet, this volume) and 'technometabolism' (Boyden, this volume) are the names of such concepts. They try to achieve the integration by describing the interactions in the human–ecological triangle as amounts of energy. I consider these models as useful because we know at least vaguely what we are talking about, even if the different transfers of energy are of a high complexity. Such statements can at least be measured or tested. However, one may also accuse these models of reductionism: how can we compare and integrate, for instance, statements on amounts of energy and

99

statements on social symbols? Do we have a satisfactory description of the ecological crisis, knowing to what extent energies are moved by persons and society? There is yet another question which I think will be hard to answer: how is it possible that things (state of affairs) which do not consist of energy and matter, such as social structures, rules, symbols and theories, are able to transform and use matter and energy? In our everyday life we are no longer surprised by the fact that we can travel by train from Zürich to Basel. Yet do we have any theoretical frameworks which enable us to see that our inorganic and organic environment is moved and structured by human activity and human symbols (cf., on this problem, Weichhart 1989)?

1.2 New forms of science

I am concerned here with a very specific aspect of integration. At the centre of my attention is the problem of a self-referential human ecology. In my opinion, human ecology knows itself only if it is capable of depicting itself as an integrated part of the human–ecological triangle.[2] If it is not capable of doing so it reproduces those analytic distinctions which it attempts to overcome. Yet how does human ecology integrate itself into the human–ecological triangle?

There seems to be an agreement amongst several contributors to this volume on the following point: the so-called 'high-energy phase' of human history causes the ecological crisis. It has also been assumed that the modern form of science is an essential condition of this industrial phase. Therefore, the fight against the ecological crisis has often been linked with the search for new forms of science.[3]

What does 'new form' mean? I would like to mention some key terms which by now are widely used: it has been tried to conceive of knowledge, and those who have knowledge, as part of that which is known, in order to avoid the subject/object distinction; and instead of reducing wholes to their parts, the attention has been turned to the interaction of the parts and their integration into a more comprehensive whole. Are such strategies sufficient to establish a new form of science? Is it enough for this purpose to study the interactions within the human–ecological triangle without reducing one part to another one? I guess that by doing this our objects and methods are becoming more complex than they already are. However, the form of our human–ecological theories has not fundamentally altered. How do I reach this conclusion? To answer this question we have to look back at the history of philosophy: since Plato, human knowledge (or science) is characterized by the stating of universal structures and laws. For this purpose modern science has developed mathematical and quantitative models. Since Descartes, philosophical thinking has been faced with the problem of how we ought to justify and interpret that basic form of science. It is well known that Kant referred back to the difference between subject and object in order to explain why mathematics-oriented science is able to describe nature properly. A look at Prigogine and Stengers (revised edition, 1986) shows that these questions are not obsolete. Their main problem is the question of how science can depict irreversible processes, bearing in mind that the mathematical structure of classic science cannot portray this kind of irreversibility. Is there an unbridgeable gap between the mathematically structured science and the world of change with its variety? Does human ecology not reproduce this gap if it takes the shape of a universal and mathematical theory?[4] This question is important for the following reason: the thought that nature is characterized by various individual and temporal processes seems to become more and more important, but the application of the universal and mathematical forms mentioned above implies that science loses its touch with nature. That is to say, science fails to adapt itself to the natural environment; and this is certainly one reason why the ecological crisis is increasing. We are only able to depict human ecology as an integrated part of nature if we find adequate theoretical forms to describe the individual and temporal processes of nature. The fact that the relation between mathematical models and reality has become unclear, in climatology and meteorology for instance, is just a start (nevertheless, most scientists might still consider a mathematically structured science as an ideal despite the difficulties they face). To give an example: Prigogine and Stengers (1986: 300) do not question the mathematical formulas they use for describing individual processes – which becomes clear when we look at the graphs: time is still perceived as a linear parameter. How do we connect the idea of an internal time of individual processes introduced by Prigogine and Stengers with this universal parameter? Was it not the idea of a linear time-parameter that made it possible for mathematics to be applied to natural processes?[5] There are more questions to that. The point is that it seems to me far from being clear how these questions should be discussed by philosophers and scientists so that they become relevant for science.

If one could develop – which I think is unlikely – a science that is not mathematically structured, how would it be distinguished from art or other ways of depicting the world? By its renunciation of pictures? In other words: what is there to be said against the view that literary forms of art, such as novels and poems, are perhaps even more complex ways to depict the world? It is unlikely that thinkers after Wittgenstein deny that such forms have truth-value, but do we know how to determine their truth (especially the truth of poetic fiction)?

2 THE PRIMACY OF PRACTICE

In this volume several different theories are put forward under the title of human ecology. The legitimacy and unity of the human ecological project seems mainly to be due to what I like to call the 'primacy of practice'.

2.1 The modern concept of human beings and the function of religion

Many of the presented arguments deal with the same problem: it is demanded and presupposed that something should be done to get a grip on the ecological problem of our society. Further, several contributors consider the discrepancy between what we know and what we really do as the main problem. Who is ready to accept constraints on his or her mobility, for instance? How can we handle this discrepancy?

First of all, the mentioned primacy of practice needs to be questioned self-critically (cf. Summerer 1989). It has to be noticed that the primacy of practice also belongs to the industrial society. Scientific knowledge is primarily judged by its practical usefulness. Almost all energies are invested in technical activity by which we control nature. This dynamic rests mainly on the concept of human beings which has been advocated during the last two hundred years: humans understood themselves as practical beings.[6] The emphasis which is put on human practice in this volume is shared by most members of the industrial society.

I think that we have to criticize that concept of human beings in order to make progress. I am not arguing against technology in general, I am arguing only against a certain concept of humans which has a great impact on our everyday life. However, I cannot think of a public ecological discussion which has not dealt with questions such as: what is the point of our discussions? and what can we do? By asking these questions we reproduce the concept of human beings which is partly responsible for the ecological crisis.

The primacy of practice is more and more questioned by our knowledge about the unforeseeable consequences of our technological and economic practice, and maybe also of the actions we take against these consequences. In addition, there is no doubt that environmentally harmless behaviour implies refraining from acting, especially using less energy and raw materials. These are, of course, pragmatic objections to our primacy of practice. The more fundamental question of whether there are things related to life which are not at our disposal should also be asked. Inspired by the sociologist Niklas Luhmann some German philosophers of religion and some theologians consider it the function of religion to deal with those things we cannot control. Such things (some are talking about contingencies) are, for instance, birth, climatic and biological factors of our environment, and also certain social and psychological factors. Persons cannot escape these things and this applies especially to death. At this point human activity reaches its limits. According to certain theorists of religion, what is beyond these limits has been called God. Nowadays we cannot use the term 'God' without making a great hermeneutic effort. The basic limits of human activity remain, even though the technology of industrial societies might have changed them to some extent. It might be the awareness of these limits which saves us from the illusion that everything is feasible and at our disposal. The insight into the limits of humans is – since Kant – one central aspect of the philosophical enlightenment. I do not know what follows from the enlightenment for the demands of environmentally harmless actions. Above all, we cannot refrain from acting. However, it might be very important to give more weight to contemplation.[7] The same applies to the things we subsume under the title of 'art'. Art changes nature but does not destroy it. Could these things not be examples to us of how we should deal with our environment?

2.2 The theological context of the ecological crisis

These thoughts about the discrepancy between what we know and what we actually do have parallels within Christian theology.[8] The New Testament is already familiar with that discrepancy, although its historical context is different. One might say that one of the most fundamental insights in the New Testament can be formulated in the following way: human beings know what is good for them[9] but do not succeed in obtaining it since they want to reach it on their own. Passages from Paul and John make clear that they regarded it as naïve that human beings want to overcome the discrepancy between knowing and acting on their own.[10] In their view, believing in God means being aware of the fact that the most essential things in life are given to us rather than being made by us.[11] The New Testament calls the life-giving source 'God'. I think it is necessary to analyse the ecological crisis in this theological context: on the one hand the ecological crisis could thus be seen as a consequence of the human attempt to realize ourselves through techno-logical practice, ignoring all limits. On the other hand, the ecological crisis could be regarded as a consequence of the human inability to leave things as they are. Instead of discovering their own value and beauty we see them as a threat to human life and identity. We therefore tend to destroy them through technological practice.

This perception implies the acceptance of the limits of human activity. Human ecology is not what it could be as long as we do not take these limits into account. Once it was religion which reminded us of our limits.

3   NOTES

1 A depiction of such an integration might be possible within the framework of a general theory of systems.
2 Something like an 'ecology of mind' (Bateson 1972) is required.
3 The Zürich Human Ecology Group has developed its work for a number of years under the title of 'Theorie und integrative Ansätze'; cf. the report of the workshop held in Zürich, 21 December, 1987 (Steiner et al. 1988).
4 See, to that effect, the following contributions in this volume: Boyden talks about 'fundamental principles' and 'identifying principles and recurring patterns'; at least in a rudimentary way, Pillet seems to refer to amounts of energy which can be formulated in mathematical terms; Carello refers to mathematical relations.

5 Cf. Picht (1979). I agree with Picht that the problem of mathematics in science is part of a more comprehensive problem. It might be called, in connection with our philosophical tradition, the problem of universals. In the context of human ecology, we also have to determine the nature of conceptual forms of universality (for instance, concepts and 'fundamental principles'). Are they merely conventions or structures of nature itself? Are we really able to depict individual processes within the framework of universal forms of science?

6 I do not mean practice in the Aristotelian sense, rather in the modern everyday sense: as technology and human beings as *Homo faber*.

7 To mention an example, human love is likely to appear as a present and not as something which has been made.

8 By Christian theology I do not mean a metaphysical theory of God. Such a theory is not available any more. I mean rather a non-esoteric and understandable thinking about the things which are not at our disposal. The theology I mean can be called Christian theology because of its relation to the historical 'categories' of the New Testament. I owe these thoughts to the protestant theologian R. Bultmann.

9 The term 'good' must not be understood in a moral sense, that is, as compliance with moral demands. On the contrary, in the view of the New Testament the intention to comply with moral demands in order to come to grips with one's life is sin as such (cf. Gal. 3:10). The New Testament defines the good life – reformulated in an anthropological and modern way – as the openness of humans to the variety of life in time.

10 Cf. Rom. 7:16.

11 Rom. 3:28 formulates this, in a slightly unusual terminology, as the 'justification before and by God' (this should not be taken in a moral sense as I said above). Or, to take another example: we celebrate – or at least we used to – Thanksgiving, although we know how important our achievement in, for instance, agriculture is. We celebrate Thanksgiving because we are grateful to the things for which we are not responsible and which we cannot control: biological growth itself.

4 REFERENCES

Bateson, G. (1972) *Steps to an Ecology of Mind. A Revolutionary Approach to Man's Understanding of Himself*, New York: Ballantine Books.

Picht, G. (1979) 'Ist Humanökologie möglich?', in C. Eisenbart (ed.) *Humanökologie und Frieden*, Stuttgart: Klett-Cotta.

Prigogine, I. and Stengers, I. (1986) *Dialog mit der Natur – Neue Wege wissenschaftlichen Denkens* (5th edn), München and Zürich: Piper.

Steiner, D., Jaeger, C. and Walther, P. (eds) (1988) 'Jenseits der mechanistischen Kosmologie – Neue Horizonte für die Geographie?', *Berichte und Skripten* 36, Zürich: Department of Geography, ETH.

Summerer, S. (1989) 'Voraussetzungen einer Umweltethik', in B. Glaeser (ed.) *Humanökologie – Grundlagen präventiver Umweltpolitik*, Opladen: Westdeutscher Verlag.

Weichhart, P. (1989) 'Die Rezeption des humanökologischen Paradigmas', in B. Glaeser (ed.) *Humanökologie – Grundlagen präventiver Umweltpolitik*, Opladen: Westdeutscher Verlag.

Part II

THE IMPLICIT AND
THE EXPLICIT

7

INTRODUCTION TO PART II

In the Introduction to Part I we have noted that we cannot expect to find a 'grand theory' of human ecology. But despite all the seeming divergence exhibited by the concerns and viewpoints of the disciplines that we regard as being constitutive for a general human ecology (the collection of papers in this volume may be taken as an illustration), it is still possible to find a common ground. The papers contained in Part II may serve as an example of how this may be achieved. What they have in common can be seen in the following: all papers make reference to or make use of different (implicit and explicit) modes of consciousness and knowledge that are constitutive for the existence of human beings and their dealings with the (biophysical and social) environment. If one is willing to accept the argument that the person can be seen as a medium of integration,[1] it is obvious that we should try to do precisely this: explore the role of the different domains of human consciousness for the way human beings relate to the social surroundings and the biophysical environment. This is the topic of Section 1.[2] From here we will proceed to a discussion of the value of the most explicit form of scientific expression, namely mathematics, in a human ecology context (Section 2) and to an aspect of human life that traditionally had an implicit character, namely morality.

1 'HEAD, HANDS AND HEART'

Let us recall at this point the tripartition of human consciousness into three levels: discursive consciousness, practical consciousness, and unconscious (Steiner, this volume, based on Giddens 1984). In parallel we may identify three different kinds of knowledge[3] that can be acquired or are contained in each one of these states of consciousness. In an evolutionary perspective we can, broadly speaking, relate (genuine) emotions emanating from the unconscious to a religious existence, actions performed in practical consciousness to a lifeworld unfolding in a traditional setting, and reasoning and communication in discursive consciousness to a style of culture culminating in philosophy and science. In any real case, of course, these categories of consciousness may blend with each other.[4] For any individual, the different states of consciousness are not separate entities which are 'on' or 'off', but all three of them are in a constant mutual relationship to each other, a fact which is recognized in modern forms of psychotherapy such as 'focusing'. Here the therapeutic process is based on the (re)establishment of a holistic[5] nexus between the three modes of experience: mind, body, and emotion (Wild-Missong 1983). As we know, modern society is characterized by the fact that the first of these modes is held in higher esteem than the others and, consequently, is assigned a dominant status. As we also know, but do not readily admit, the other two modes may interfere by effecting seemingly 'irrational' components of behaviour. In fact, as pointed out by Etzioni (1988), the fundamental frame for human behaviour is first set by normative–affective factors and only within their confines is there some amount of room left for rational considerations. Given this role of the non-conscious parts of our psyche, human life in its totality can only be lived, but not described. Or can it?

Reconsider the notion of parts and wholes as a concept of hierarchical structure,[6] which may be applied to the description of the world around us as well as to that of human knowledge. With reference to Polanyi (1962) we noticed that explicit knowledge refers to the level of parts only, and that implicit knowledge is required in addition to perceive or create a whole. More precisely, only by focusing away from the explicit and keeping it as just a subsidiary awareness can the implicit take shape in the form of a whole. This is totally in line with modern psychological views on the interplay between the conscious and the unconscious, which can be described by the metaphor of foreground and background: if one puts one's full awareness on some object it moves from the back- to the foreground. And if this object recedes from the foreground again, it changes the qualities of the background (Wild-Missong 1983). In linguistic terms one could say that the level of parts provides the syntax, the level of the whole the semantics. Indeed, of course, a sentence spoken or written in a natural language is something that consists of words related to each other by some rules of grammar, but at the same time provides some meaning. Meaning is provided by the following facts: first, some of the terms used point to objects in the world around us; second, at a more abstract level, terms stand in some relation to other terms;[7] and third, a discursive communication unfolds as a practice against the background of some socio-cultural rules.[8] This sequence indicates a move from the more explicit to the more implicit. Conversely, it is probably correct to say that the more formal, that is, the more explicit, a linguistic expression becomes, the more its semantic content becomes a direct function of the symbols used, which means that its scope is limited to the level of parts only. If we use mathematical formulations, e.g. system theoretic approaches, to model human ecological situations, this aspect becomes crucial. It is for this reason that we want to delve into this matter at some greater depth in the next section of this introduction.

For a human individual to become socialized into a web of meanings as described above means that he or she has acquired some kind of internal knowledge of how things are done. Indeed much of human agency has a routinized character that does not require much reflection. In this sense, the two papers by Claudia Carello and Ingela Josefson deal with two different aspects of practical consciousness, that is, precisely a frame of mind within which human beings simply act in some way or other without being able to say in words what they do or why they do it.[9] Human agency of this sort may be directed towards the biophysical environment or unfold within a social setting in relation to other human beings. In the first case individuals accumulate knowledge on the basis of experience gained in repeated direct dealings with the environment, in the second case as a result of learning in the social context of an established cultural tradition.

Carello's contribution deals with the first case of practical consciousness. As it relates to speechless situations devoid of any social connotation, it first of all applies to animals. Their interaction with the environment is described as a continuous perception–action cycle providing a very immediate relationship to environmental features. It is a situation which, in Polanyi's (1962) terms, can be called existential: things in the environment do not have a symbolic meaning, but they mean something in themselves. The question arises of whether or not such a relationship to the biophysical environment also applies to humans.[10]

Josefson's paper refers to the second instance, in that it deals with the assets of practical knowledge of hospital nurses, acquired gradually on the scene under the tutorship of an older experienced nurse, and contrasts it to the theoretical knowledge transmitted in a discursive manner. The importance of this kind of 'knowledge by familiarity', as Josefson calls it, in a human ecological context is twofold. First, to be in good hands within a traditional setting, in which not only knowledge but also values are being transmitted, may raise the feeling of security of the actors concerned and may, indirectly, also affect their relationship to the biophysical environment. Second, if the kind of knowledge conveyed through apprenticeship concerns actions on the environment, such as traditional farming practices, then we deal with a combination of the two aspects of practical consciousness mentioned.

In the usual manner of scientific writing we use our discursive capabilities in a rather formal fashion to establish some distance to the object of our investigation. If our concern is with the non-conscious levels of the human psyche then we have the somewhat paradoxical situation that we use explicit means to describe what is implicit. In other words, we are in the difficult position of someone who talks about the foreground when he or she means the background. The paper by Dagmar Reichert is an attempt to overcome some of this difficulty in that she uses a style which is not in accord with the usual standards of scientific writing. It may be seen as one possibility to comply with the previously mentioned requirement that a general human ecology should incorporate trans-scientific components. Reichert makes use of the linguistic freedom otherwise accorded to literary writing to treat the human ecologically significant question of the boundedness or unboundedness of a person. This kind of boundary-crossing affects not only the mind but also the unconscious. Dreams, whether night- or day-dreams, cannot be described easily in terms of language. Conversely, however, language may, as in parts of this paper, be given a poetic quality, and poetry has the power to speak to our hearts. But this is just one aspect. More fundamental may be the fact that style and form blend with content.[11] This provides for a second boundary crossing or rather boundary elimination. The way in which the text is presented is a metaphor for the underlying message that an individual, who has a multitude of relations to his or her (biophysical or socio-cultural) environment, becomes a person with an identity not by positioning him/herself as a point within the network of relations, but rather by being the network itself. It suggests that an end should be put to the infamous split between subject and object.

It is presumably not exactly a coincidence that this paper has been written by a woman. One of the old metaphorical ways of perceiving the male principle is as something concentrated in the middle with definite shape and form and with sharp boundaries around it. The female principle is seen as a residual. As pointed out by feminist critique, however, this residual is all around the centre and stretches to infinity (Klinger 1991).[12] Consequently, Reichert's paper may also be read as a critical manifestation, as a deconstruction of the centre and its boundaries. Also this perspective clearly supports the notion expressed previously, namely that this text is an attempt at combining the explicit with the implicit.

2 MATHEMATICS AND HUMAN ECOLOGY

In their papers Gonzague Pillet and Hans-Joachim Mosler suggest the use of mathematical models to describe human ecological situations. There are assets and there are shortcomings, if not to say dangers, associated with such an application of mathematics. Let us start with the latter aspect first. It follows from the foregoing that common forms of mathematical formulations, being the most explicit among discursive communications, can refer to a level of parts only, and that in a very specific way.[13] In so far as this is equivalent to saying that mechanistic thinking still prevails, the posting of a caveat for using mathematical descriptions of human ecological problems is in order. It is in line with these mechanistic properties that, with respect to hierarchical structures, for a mathematical system to be consistent it must be closed to one level. Otherwise the logic may be disturbed by paradoxes occurring as a consequence of level-mixing in situations of self-reference.[14] On the other hand, as Gödel has shown, the logical consistency of a mathematical system cannot be proved within the system alone (see Delbrück 1986). More popularly this may be interpreted to mean that to establish whether something is true on one hierarchical level may require information from another, higher level.

What does this mean with regard to the papers by Mosler and Pillet? The former suggests using a synergetic approach to model social systems.[15] Synergetics has its origin in laser physics. It can be understood as comprising the stochastic variant of nonlinear system theory, which has become a fundamental instrument for the formal description of self-organizing systems. The notion of self-organization is, as pointed out by Mosler, important in that it may show the way to a 'third solution' of the environmental problem, the first and the second solutions comprising the conventional ideas with respect to an explicit regulation of society in general and of individuals in particular. It has been claimed that synergetic models are truly holistic because they provide a genuine connection between a micro and a macro level of a hierarchy. Indeed it is possible to demonstrate with such models that the behaviour of elements becomes constrained by aspects of the collectivity to which these elements belong. As a human society can exist only if its members are subject to some behavioural constraints, synergetic modelling may also be applicable, with some degree of plausibility at least, to social situations. At the same time one should recognize that the macro level in synergetic models simply consists of a statistical aggregation of micro level information. This becomes a critical point: such a kind of macro sociology is not the same as a sociology that looks at society in terms of organizing values, norms, and rules: 'Social institutions are not explicable as aggregates of "microsituations"' (Giddens 1984: 141); 'sociology is not concerned, as such, with large-scale mass or group behaviour, ... but (paradigmatically) with the persistent relations between individuals (and groups), and with the relations between these relations' (Bhaskar 1978). Formal system theory, therefore, whether synergetic or not, cannot be the sole panacea for the solution of (human) ecological problems. To the extent that a system simply represents a collection of relations between a (often large) number of individual elements, system theory amounts to a perpetuation of mechanistic ideas (Steiner et al. 1989, Primas 1992).

Of quite a different nature is the quantitative approach to ecological economics by Gonzague Pillet. He starts from the basic postulate of environmental economics: environmental damages are unpriced externalitites that must be internalized into the economic system, hence given a price. He shows that the two approaches developed so far, market internalization (concerning individual environmental items in the context of short-term situational analyses), and environmental internalization (the environment with its overall resource-providing and waste-assimilating capacities is being regarded as a component of the economic system under a longer-term perspective) are not sufficient. As a further step Pillet proposes the modelling of economic activities by means of energy flows. This has the advantage that direct links between human activities and biophysical aspects of the environment can be established in terms of a physical language. Surely this is still a reductionistic approach. However, the level to which the problem is reduced is part of the biophysical world, and we can imagine that this kind of model is capable of signalling 'limits to growth' kind of information,[16] in particular in a regional context.[17] In turn, this is not true for a modelling of the same problem in terms of money. There are no comparable limits to the growth of money, the less so as money becomes an increasingly abstract entity. Furthermore, money is a purely socio-cultural concept and can have no real bearing on components of the natural environment.

A simplification of the socio-cultural world to a level of physical–mathematical models obviously has its dangers. No harm is done if the limitations of such models are recognized at all times. However, the belief that only science which makes use of mathematics is good science is still widespread. Or at least this was the dominant belief some twenty years ago. At that time Harvey (1969: 182) could say: 'it is universally agreed that mathematics is the language of science.' It was the time of the so-called quantitative revolution in (human) geography.[18] Other social science disciplines had gone through a phase of mathematization earlier.[19] It became fashionable to borrow concepts from physics to model social processes, a mix of approaches which has become known as 'social physics'. For Harvey (1969: 187) this was a desirable development as 'the calculi useful to physics are probably better developed than any other calculi'. True, he remarked that considerable care in the use of these concepts would be required, but also that 'the rewards may be great'. Meanwhile there has been some disenchantment with quantitative approaches as the hope that they would become a very effective tool for problem solving has been thoroughly shattered. Many pioneers of the quantitative revolution have left the field and are doing something quite different today. Harvey is an excellent example: he has shed completely the positivistic stance he had in 1969 and is now a renowned expert on Marxist theory.[20] Others have turned to humanistic and phenomenological approaches (see, for example, Johnston 1986).[21] Geertz (1983: 3) puts it thus: 'Ten years ago, the proposal that cultural phenomena should be treated as significative systems posing expositive questions was a much more alarming one for social scientists . . . than it is now. In part, it is a result of the growing recognition that the established approach to treating such phenomena, laws-and-causes social physics, was not producing the triumphs of prediction, control, and testability that had for so long been promised in its name.'

None the less, social physics is by no means dead today, as is exemplified by the two papers by Mosler and Pillet. Of course, there is an essential difference in their respective use of mathematics. The application of synergetics to social problems as described by Mosler constitutes an analogue model, that is, a model that is based not on substantial theory about the social situation in question, but on borrowed concepts from physics and structural similarities between physical and social processes. The popularity which synergetic models have attained in the wake of the development of nonlinear system theory has led to the rather curious situation that physicists do not hesitate to write books about sociology these days (Weidlich and Haag 1983). In contrast, Pillet's kind of energy modelling is real social (and environmental) physics in the sense that the use of mathematics is directly related to a physical phenomenon occurring in a human ecological context. To be fair let us point out that a majority of people who today continue to use mathematical descriptions of (human) ecological situations (including nonlinear system models) probably do this with a fairly different kind of awareness compared to the one which was preponderant twenty years ago. They do not value mathematical formulations *per se*, but recognize that they have real assets only if embedded in a broader context.

The critical remarks made here should not suggest that there is no place for mathematical approaches in human ecology. To get rid of them would amount to throwing out the baby with the bathwater. However, as pointed out by Gould (this volume) in the context of using mathematics to model the diffusion of AIDS, one has to be careful not to let this use 'deteriorate into silly computer games of nonlinear systems'. If this is avoided then results obtained from quantitative analyses may be very useful indeed. We note the following positive aspects:

1 To some extent mathematical manipulation, such as in multivariate statistics, can take the place of the controlled laboratory experiment. It thus provides a tool to deal with problems of multicausality in nonexperimental situations.[22]

2 Quantitative analyses may help to detect empirical phenomena which otherwise might go unnoticed. The phenomenon of counterurbanization may be a good example (see Berry 1976, Ernste and Jaeger 1986/87).

3 In large systems the many elements that they contain may, for statistical reasons, indeed show a tendency for regularities at the collective level. Hence mathematical models (such as those used by Mosler or Gould in this volume) have a descriptive and to some extent perhaps a predictive value.

4 Mathematical modelling has a 'heuristic function' (Wilson 1990: 402): it may provide valuable assistance in theorizing exercises. As an example, consider the simple case of the game theoretic concept of the prisoner's dilemma for the evaluation of social choice mechanisms (Dryzek 1987). Also, system theoretic analyses may lead to counterintuitive results and hence to new insights (Forrester 1971). It is presumably in this area where Mosler's application of synergetics is strongest.

5 A special and, it would seem, particularly important case of theorizing occurs in connection with 'limits-to-growth' kinds of analyses. In the preceding text we have pointed to energy flow modelling (as in Pillet's

contribution) as an example. Another one is provided by the species-specific models for viable populations (minimum size, minimum density) developed in conservation biology (Soulé 1987).

6 To model quantitatively such things as, for example, urban development may be valuable in the sense that by so doing a standard can be established against which deviations can be measured. It is to be expected, however, that when dealing with a particular city, the deviations are more important than the standard (see Steiner 1981).

7 More generally, Harvey's claim that 'to mathematise large areas of social-science research is . . . generally healthy, simply because it demands a prior clarification of concepts and propositions about empirical phenomena' (Harvey 1969: 187) may still be valid, if not taken as a dogma that bans all other approaches from the scene.

To conclude our reflections about the role of mathematics we recall that mathematically formulated truths, if they are truths at all, can be half-truths only. And we recognize that the dream about a unification of science in general, and of human sciences in particular, brought about by a mathematization of the subject matter of the various disciplines, is just that, a dream. It is good to heed the advice given by Naveh and Lieberman (1984) in their treatise of landscape ecology: a mathematical description does not suffice, it must always be supplemented by statements formulated in natural language. On the other hand, Polanyi (1962) of all people, who otherwise stresses the importance of implicit knowledge, still postulates a superiority of theoretical knowledge in terms of objectivity: a theory is something other than myself, hence objective knowledge, he says. There is no contradiction here, because, at once, he sees such a theory in the wider context of an implicit background. Not only does he stress the important role of intuition in the scientific process, but he also maintains that the discovery of objective truth in science consists in the apprehension of a rationality which commands our respect and arouses our contemplative admiration. Polanyi defends this conception of objectivity against reproaches of Platonism: 'twentieth-century physics, and Einstein's discovery of relativity in particular, . . . demonstrate . . . the power of science to make contact with reality in nature by recognizing what is rational in nature' (Polanyi 1962: 6).[23]

3 THE MORAL DIMENSION

The quote from Polanyi provides a ready connection to the problem of environmental ethics. What role may the different states of consciousness play here? Does a cognitive act that provides us with insights about the wisdom of nature stir up a sufficient degree of reverence so that we feel inspired to establish a new respectful kind of relationship to this nature? Presumably, as a prerequisite, we would need a different kind of science to start with, one that leans more towards contemplation than manipulation, or, stated differently, one whose significance in terms of theoretical knowledge lies more in educational aspects than operational potentiality.[24]

An interesting question relates to the fact elaborated upon by Capra (1982), namely that there is a striking parallel between the insights of modern physics and the teachings of Eastern mysticism. Now the former obviously are the result of explicit thinking of a high degree of intellectuality within discursive consiousness, the latter an attempt at translating religious or mystical experiences made via a communication with implicit contents of the unconscious. Does it mean that there can be an equivalence between the explicit and the implicit? Perhaps yes in this one special case. It seems indeed that the nature of the outermost reality, in the sense of the recursive systems embedded within each other,[25] is such that it can be described completely in mathematical terms, that is, by the theory of quantum physics. 'Complete' means that it embraces the level of parts as well as the whole.[26] 'Quantum mechanics is the first logically consistent and mathematically formulated holistic theory' (Primas 1992: 11).[27] Of course, it does not follow that evolutionarily later realities must also be exhaustively describable in mathematical terms.

Even if the results are comparable, there is surely a difference between directly having a mystical experience and looking at equations of quantum physics. Not everybody is given to either of them anyway. At a more mundane level, there are a number of other aspects to consider in connection with the development of environmental ethics. In discursive consciousness, mathematical modelling of the kind used by Pillet may be of importance here as it has the capability to demonstrate limits set to us by the given biophysical reality.[28] And it may be that being reminded of limits brings us back to considerations of a religious nature. As Huppenbauer (1990: 122) puts it: 'In the final analysis humans cannot themselves produce the energy which they tap from nature. It is precisely in the realm of technology, where humankind seems to be most powerful, that the unavailable is encountered. Nature responds to the urge of technology, it grants energy to human beings.'[29]

Of course, science will not be a fruitful ground for the development of feelings towards nature as long as it adheres to the myth of being value-free. A step forward here is a situation in which the significance of cognitive insights provided by science are brought into a general discussion in the public at large, because a communicative political process must then decide about any value aspects. In principle, of course, a free discourse has been the ideal of modern and democratic society for a long time, except that the same type of society has succumbed to an unhealthy reign of the experts in the past. For many, therefore, Popper's (1945) conception of an open society, in which anything can be questioned at any time (except the premise that the society should be an open one) is a favourite model. However, this model fails with regard to the ecological crisis, because there is no guarantee at all that in such a society a decision to sustain the environment rather than to spoil it would ever be taken. In comparison, the kind of society envisaged by Habermas (1989), in which a communicative rationality would be used to arrive beforehand at basic principles (such as those concerning our responsibility towards the environment), has a definite advantage. Nevertheless, the role of discursive consciousness in shaping a new kind of ethics remains questionable. Can it lead to more than simply a legalistic kind of regulation? Here Mosler's idea of symbolic communication (wearing a badge) is interesting, as it may help the people concerned to develop trusting relations and emotional ties within a particular segment of society with a dedication to a common cause.[30]

It would seem that the hopes for a new morality should have better grounds in the potential regulative power of implicit norms, rules and experiences internalized in the psychic realm of practical consciousness. In the context of the ecological crisis, which one would be more important: a social attachment, that is, a sense of being embedded in a social network that provides for less alienative conditions of living, work, and social intercourse, or a place attachment, that is, a sense of belonging as a result of a continuous opportunity to experience directly natural phenomena of the environment? Most likely we require a combination of both. A possible example of this would be gardening, an activity in which implicit knowledge of the kind Josefson is talking about is acquired from more experienced people, but which at the same time provides for direct and personal contacts with the environment.[31] What we said about the role of childhood in this context may lead to the question of the extent to which we not only might have the ability to carry over childlike qualities into our adult life but also in fact do so.

To the extent that what we have just mentioned amounts to establishing a new tradition, we are confronted with an obvious problem: we do not have the time. The development of new implicit forms of regulation may require generations. As we cannot wait to take corrective and impeding actions with respect to the ongoing destruction of the environment, we are back to the necessity of forms of explicit regulation. However, we can say at this point that such regulation should proceed with a high degree of caution, and that the necessary prudence can come about only if cognitive insights are supported by corresponding feelings. It is Jonas (1984) who stresses this point: that such a combination is an indispensable prerequisite for the emergence of a new kind of ethics. A process of identification with the environment in the sense of Naess (1989; see also Steiner, this volume) may be what we require in this connection, except that most of us probably will not have any idea at all of what to make out of this requirement. Some clues may be found by reading Reichert's paper. At any rate, this seems to be clear: we need to steer a path between the Scylla of mere distanced and cool cognition and the Charybdis of a totally immersing and heated fundamentalism. Stated more simply: we require a balanced bridge between facts and values.

4 NOTES

1 Cf. Introduction to Part I, Section 4.

2 Cf. Note 19 in Steiner (this volume).

3 The term 'knowledge' is used in a broad sense here and may include such forms as might be more commonly called 'belief'.

4 The use of language is an obvious example: It is a prerequisite for the unfolding of a world of rational discourse at a level of discursive consciousness. However, in the evolutionary sense, language is in the first place an instrument of communication employed largely in an unreflected state of practical consciousness. This is true not only with respect to the learned bodily movements (of the lips, the tongue, and the larynx) involved, but also with respect to the observation of the rules of grammar, the selection of words, and the adherence to meanings. For a treatment of language as a tradition see Keller (1990).

5 The term 'holistic' as it is used here and in the following relates to concepts not of the old morphological worldview, but of the modern evolutionary worldview, that is, it is associated with the notion of circular causalities between parts and whole (for a short discussion of the different worldviews see Steiner, this volume).

6 Cf. Figure 4.2 in Steiner (this volume).

7 These first two levels of meaning are described by Sayer (1984) as words standing in reference relations and in sense relations, respectively.

8 Rappaport (1979: 156–7) describes this hierarchy of meaning as follows: 'Lower-order meanings, in particular the meanings of words *per se*, are those of definition and distinction, and as such are virtually synonymous with information in the technical sense. Low-order instrumental values can be specified adequately in words and numbers. Higher-order meaning is of a different sort. . . . information is radically reduced . . . But as information is decreased, meaningfulness is increased; for similarities, substantive or structural, between that which we seek to understand and that which we already "know", are made explicit. . . . Higher-order meaning is, then, not information in the digital sense but, rather, metaphoric. Although the discursive content of higher-order meanings may be less than that of lower-order meanings, higher-order meanings may well be affectively more powerful, which is to say more meaningful.' In a similar vein Jaeger (1991) stresses the importance of the fundamental difference between algorithmic instructions and socio-cultural rules.

9 However, partial contents of practical consciousness may by lifted on to a discursive level such as when, for example, an athlete watches his or her performance on a video recorder and discusses it with the coach.

10 We will come back to this question in the Introduction to Part IV, Section 2.

11 This kind of presentation shows some degree of affinity to the writings of such 'post-modern' authors as Gunnar Olsson. In the sixties Olsson was one of the pioneers of the quantitative revolution in geography (see, for example, Olsson 1965). Ten years later we find him being interested in the ambiguities created by a circular reality (Olsson 1975). For Geertz (1983: 23) this is not necessarily unusual, as he diagnoses the present state of the social sciences as one of 'blurred genres', i.e., one in which 'so much more of the imagery, method, theory, and style is to be drawn from the humanities than previously'. He also, however, expresses some cautionary concern: 'All this fiddling around with the proprieties of composition, inquiry, and explanation . . . is about as likely to lead to obscurity and illusion as it is to precision and truth. If the result is not to be elaborate chatter or the higher nonsense, a critical consciousness will have to be developed.'

12 A realization of this principle in geographical space would seem to be the spatial pattern of the Ancient Greek landscape: the *polis* as the centre of culture in the

middle, nature and the wild around it. Evolutionarily speaking, we think it is no coincidence that the emergence of city states at the beginning of Antiquity occurs in tandem with the establishment of patriarchal social structures.

13 We note that the situation changes, if the concept of infinity is allowed to enter (theoretical) mathematical considerations.

14 The paradoxes discovered in connection with the development of set theory are well known, such as the barber who shaves all the men of a village who do not shave themselves, an example provided by Russell (see, for example, Meschkowski 1984). Obviously, if the barber shaves himself he is somebody who does not shave himself, and if he is somebody who does not shave himself he shaves himself, etc. To avoid such phenomena Russell and Whitehead (1910) developed their typology, which requires a strict separation of hierarchical levels. It is clear, however, that what seems to be an obstacle to logical analysis does not present any problems to speakers of a natural language using such sentences. Natural languages are universal in character and can deal with level-mixing and self-reference.

15 For a discussion of other aspects of Mosler's paper cf. Weichhart (this volume).

16 Cf. Introduction to Part I, Section 2.

17 Cf. Introduction to Part IV.

18 Actually, at that time, this revolution was already over in North America, while it was in full swing in the United Kingdom and still in its infancy in continental Europe.

19 Typically enough, the first book on 'statistical geography' was written by sociologists (Duncan et al. 1961).

20 What Harvey has to say today about the quantitative orientation is of the following genre: 'Those who have stuck with modelling since those heady days have largely been able to do so, I suspect, by restricting the nature of the questions they ask' (Harvey 1989: 212).

21 Olsson was cited earlier as another typical example of a 'convert'.

22 Obviously, however, the more complex a situation is, the more decisive becomes the question of adequacy of mathematical description.

23 Two remarks come to mind at this point. First, in saying that a scientific theory transcends the experience of our senses and embraces the vision of a reality behind, Polanyi is in line with the conceptions of a new philosophy of science known as transcendental realism (Bhaskar 1978; cf. also the Introduction to Part IV). Second, the talk about rationality in nature is of interest in the context of the ecological crisis. Dryzek (1987), for example, develops a concept of ecological rationality and then investigates the extent to which social choice mechanisms are in agreement with it.

24 The character of modern science has become very much an object of feminist critique. It is argued that scientific truth becomes reduced to the technically implementable, that this reduction is very closely associated with fantasies of virility, and that consequently only an 'active' kind of science is real science (Scheich 1991).

25 Cf. Figure 4.4 in Steiner (this volume).

26 Actually it may be more appropriate to say that there is no level of parts at all: 'material reality is a whole, more precisely a whole that does not consist of parts' (Primas 1992: 10–11; translated from the German original by D.S.).

27 Translated from the German original by D.S. As the theory not only is holistic in a cosmological sense but also specifies that there can be no separation between the observer and the reality observed, it should not come as a surprise that it includes paradoxes. The paradoxical phenomena (for example, an electron or a photon exhibits particle-like properties in one situation and wave-like properties in another) that can be derived from the theory, however, are real in a very fundamental sense: they cannot be manipulated away by resorting to a different

logical structure.
28 Cf. Introduction to Part I, Section 2.
29 Translated from the German original by D.S.
30 To be fair to Habermas we should point out his notion of a basic kind of morality as a precondition for communicative action, a morality that has its roots in something deeper and more universal than any particular tradition, namely in a fundamental aspect of human existence: the urge together with the capability for social interaction, which makes human beings competent communicative agents in any society (Habermas 1990).
31 For a discussion of psychological benefits of gardening see Kaplan (1973).

5 REFERENCES

Berry, B.J.L. (1976) 'The counterurbanization process. Urban America since 1970', in B.J.L. Berry (ed.) 'Urbanization and Counterurbanization', *Urban Affairs Annual Reviews* 11: 17–31.

Bhaskar, R. (1978) 'On the possibility of social scientific knowledge and the limits of naturalism', *Journal for the Theory of Social Behaviour* 8: 1–28.

Capra, F. (1982) *The Tao of Physics*, London: Fontana/Collins.

Delbrück, M. (1986) *Wahrheit und Wirklichkeit. Über die Evolution des Erkennens*, Hamburg: Rasch & Röhring.

Dryzek, J.S. (1987) *Rational Ecology. Environment and Political Economy*, Oxford: Blackwell.

Duncan, O.T., Cuzzort, R.P. and Duncan, B. (1961) *Statistical Geography. Problems in Analyzing Areal Data*, Glencoe, Ill.: Free Press.

Ernste, H. and Jaeger, C. (1986/87) 'Neuere Tendenzen schweizerischer Migrationsströme. Teil 1: Eine Literaturübersicht zum Phänomen der Entstädterung; Teil 2: Entstädterung in der Schweiz', *Geographica Helvetica* 41 (3): 111–16, and 42 (1): 27–34.

Etzioni, A. (1988) *The Moral Dimension. Toward a New Economics*, New York and London: Free Press and Collier Macmillan.

Forrester, J.W. (1971) 'Counterintuitive nature of social systems', *Technology Review* (73): 53.

Geertz, C. (1983) *Local Knowledge. Further Essays in Interpretative Anthropology*, New York: Basic Books.

Giddens, A. (1984) *The Constitution of Society. Outline of the Theory of Structuration*, Berkeley and Los Angeles: University of California Press.

Habermas, J. (1989) *The Theory of Communicative Action, 1: Reason and the Rationalization of Society; 2: The Critique of Functional Reason*, Cambridge: Polity Press.

Habermas, J. (1990) *Moral Consciousness and Communicative Action*, Cambridge: Polity Press.

Harvey, D. (1969) *Explanation in Geography*, London: Edward Arnold.

Harvey, D. (1989) 'Notes on the project to "remodel contemporary geography"', in B. MacMillan (ed.) *Remodelling Geography*, London: Macmillan.

Huppenbauer, M. (1990) 'Poiesis als Problem einer Humanökologie', in H.-J. Braun (ed.) *Martin Heidegger und der christliche Glaube*, Zürich: Theologischer Verlag.

Jaeger, C. (1991) 'The puzzle of human ecology. An essay on environmental problems and cultural evolution', habilitation thesis, Zürich: Department of Environmental Sciences, ETH (publication forthcoming).

Johnston, R.J. (1986) *Philosophy and Human Geography. An Introduction to Contemporary Approaches*, London: Edward Arnold.

119

Jonas, H. (1984) *Das Prinzip Verantwortung. Versuch einer Ethik für die technologische Zivilisation*, Frankfurt a/M: Suhrkamp.

Kaplan, R. (1973) 'Some psychological benefits of gardening', *Environment and Behavior* 5 (2): 145–62.

Keller, R. (1990) *Sprachwandel. Von der unsichtbaren Hand in der Sprache*, UTB 1567, Tübingen: Francke.

Klinger, C. (1991) 'Was ist und zu welchem Ende betreibt man feministische Philosophie?', Interdisciplinary Lecture Series on Feminist Perspectives in Science, University and ETH, Zürich (will be published in book form by Verlag der Fachvereine, Zürich).

Meschkowski, H. (1984) *Was wir wirklich wissen. Die exakten Wissenschaften und ihr Beitrag zur Erkenntnis*, München and Zürich: Piper.

Naess, A. (1989) *Ecology, Community and Lifestyle. Outline of an Ecosophy*, trans. and rev. D. Rothenberg, Cambridge: Cambridge University Press.

Naveh, Z. and Lieberman, A.S. (1984) *Landscape Ecology. Theory and Application*, New York: Springer.

Olsson, G. (1965) *Distance and Human Interaction: a Review and Bibliography*, Bibliog. series 2, Philadelphia, PA: Regional Science Research Institute.

Olsson, G. (1975) *Bird in Egg – Egg in Bird*, London: Pion.

Polanyi, M. (1962) *Personal Knowledge*, Chicago: University of Chicago Press.

Popper, K.R. (1945) *The Open Society and its Enemies*, 2 vols, London: Routledge & Kegan Paul.

Primas, H. (1992) 'Umdenken in der Naturwissenschaft', *Gaia* 1 (1): 5–15.

Rappaport, R.A. (1979) *Ecology, Meaning, and Religion*, Berkeley, CA: North Atlantic Books.

Russell, B. and Whitehead, A.N. (1910) *Principia Mathematica*, Cambridge: Cambridge University Press.

Sayer, A. (1984) *Methods in Social Science. A Realist Approach*, London: Hutchinson.

Scheich, E. (1991) 'Die zwei Geschlechter in der Naturwissenschaft: Ideologie, Objektivität, Verhältnis', in D. Engfer, P. Fry, R. Gratzfeld, A. Scheller and S. Stalder Ghidossi (eds) *Im Widerstreit mit der Objektivität. Frauen in den Naturwissenschaften*, Zürich and Dortmund: eFeF-Verlag.

Soulé, M.E. (ed.) (1987) *Viable Populations for Conservation*, Cambridge: Cambridge University Press.

Steiner, D. (1981) 'Zur Mathematisierung der Geographie', in P. Hoyningen-Huene (ed.) *Die Mathematisierung der Wissenschaften*, Zürcher Hochschulforum 4, Zürich: Artemis Verlag.

Steiner, D., Furger, F. and Jaeger, C. (1989) 'Mechanisms, forms and systems in human ecology', contribution to Third International Conference of the Society for Human Ecology, San Francisco, 7–9 Oct. 1988 (expected to appear in print in a conference volume).

Weidlich, W. and Haag, G. (1983) *Concepts and Models of a Quantitative Sociology. The Dynamics of Interacting Populations*, Berlin: Springer.

Wild-Missong, A. (1983) *Neuer Weg zum Unbewussten. Focusing als Methode klientenzentrierter Psychoanalyse*, Salzburg: Otto Müller.

Wilson, T.P. (1990) 'Sociology and the mathematical method', in A. Giddens and J.P. Turner (eds) *Social Theory Today*, Cambridge: Polity Press.

8

REALISM AND ECOLOGICAL UNITS OF ANALYSIS

Claudia Carello

1 INTRODUCTION

Ecological psychology in the tradition of James Gibson (1966, 1979; see also Reed and Jones 1982) is a realist theory of perception. The approach is directed at understanding how animals (including humans) get around in their environments. How do creatures in different habitats, with widely varying locomotory styles (walking, swimming, flying, slithering) and differential reliance on seeing, hearing, smelling, and touching as well as thermal and electrical sensitivity, perceive essentially the same things: obstacles, openings, shelter, mates, edible things, and so on (see Figure 8.1)? This theoretical framework was developed to address everyday achievements of perceiving and acting; it has not been applied explicitly to problems that constitute the current ecological crisis. In this paper, a few core assumptions of the ecological approach will be identified. The last section of the paper will highlight those concepts that may prove useful for those who might elaborate this approach with respect to the effect of individuals and societies on the environment.

At the outset it should be noted that in the pursuit of a thoroughgoing account of how animals maintain contact with their environments, realism is the starting point; it is not a position to be proved. A realist theory of perception asserts that terrestrial objects and events, not mental representations of them, are perceived. The particulars of Gibson's ecological approach reflect those concepts, methods, and strategies that are consistent with realism; concepts, methods, and strategies that undermine realism must be eschewed (Turvey and Carello 1981). The punchline is twofold: (1) realism rests on a lawful relationship between perception and action, and (2) it is only by honouring the reciprocity of animal and environment, that is, by using ecological units of analysis, that ecological laws of perceiving and acting can be identified.

1.1 An overview of ecological realism

For Gibson, the source of the adjective 'ecological' for psychology reflects the assertion that an animal and its environment comprise an irreducible system. Perception cannot be understood as something that happens 'in' animals or

Figure 8.1 Creatures of different sizes, action capabilities, and perceptual sensitivity perceive many of the same animal–environment relations. Although each of the depicted animals has chosen a different path, all are locomoting through openings

brains; it is a fact of animal–environment systems. As we shall see, this means that descriptions of animals must reflect their environments and descriptions of environments must reflect the animals that get around in them. 'Organism–environment synergy does not suggest merely that an organism implies the existence of some environment (or vice versa) but, more strongly, that each component of an organism-niche system logically conditions the very nature of the other component' (Turvey and Carello 1981: 315). In this view, therefore, there is no such thing as the environment as a generic description that will suffice for every creature. This is, of course, consistent with some treatments of ecological niche (e.g. Elton 1927; see Whittaker *et al.* 1973, for a review). But, for our purposes, that concept must be elaborated to include the epistemic fit between organisms and environments (Alley 1985, Gibson 1966, Michaels and Carello 1981); perceptual systems are suited to the structured energy distributions that inform about animal–environment relations.

The way in which environment descriptors become animal-referential reflects a second core assumption of Gibson's ecological approach, namely, perception cannot be understood independently of action. The mutually con-straining relationship between perception and action requires that accounts of perception be with reference to the activities that it guides and that accounts of action be with reference to how it might be guided. This precludes an account in which a motor program is written on the basis of a representation which is the 'product' of perception. The perceiving–acting cycle (see Kugler and Turvey 1987) is such that perception constrains action by detecting the information by which activities are guided and action constrains perception by altering the animal–environment system and thereby the available information.

The coupling of perception with action necessarily takes perception out of the realm of statically imposed, momentary stimuli and into the realm of events. An event is defined as 'changes in the layout of surfaces, changes in the colour and texture of surfaces, and changes in the existence of surfaces' (Gibson 1979: 94). Depending on the nature of the style of change, certain aspects of a particular combination of substance and surface will be transformed while other aspects will remain invariant. Transformations such as translation, rotation, and reflection change the location or orientation of an object but do not change its structural integrity. A transformation such as growth alters the size and shape of facial features but can leave underlying identity invariant (Shaw and Pittenger 1978). Burning, melting, exploding, and so on change substance and surface dramatically. Combinations of substance and surface can be said to support certain kinds of transformations and not others. Because events are defined with respect to persistence and change, notions of space and time as separate and absolute are not useful to an ecological analysis. A third assumption, therefore, is that events, not time and space, are primary realities:

> The flow of ecological events is distinct from the abstract passage of time assumed in physics. The stream of events is heterogeneous and differentiated into parts, whereas the passage of time is supposed to be homogeneous and linear. Isaac Newton asserted that 'absolute, true, and mathematical time, of itself and from its own nature, flows equably without relation to anything external'. But this is a convenient myth. It assumes that events occur 'in' time and that time is empty unless 'filled'. We should begin thinking of events as the primary realities and of time as an abstraction from them. . . . Events are perceived but time is not.
> (Gibson 1979: 100)

The same kind of argument applies to space. It is not a vessel to be filled by objects. 'The persisting surfaces of the environment are what provide the framework of reality. . . . Surfaces and their layout are perceived, but space is not' (Gibson 1979: 100).

The final assumption that we will consider is that perceiving and acting are made possible by laws at the ecological scale. In approximate terms, these laws take the form 'Animal–environment relation X generates optical (acoustic, haptic, etc.) structure Y; optical (acoustic, haptic, etc.) structure Y specifies animal–environment relation X'. Such laws are demanded by realism because they do not entail epistemic mediation. Rather, they reflect the mutual compatibility between animal and environment, between perception and action, that makes direct epistemic contact between animals and environments possible.

The argument will be made that acknowledging the reciprocity between animal and environment and between perception and action yields ecological units of analysis which, in turn, reveal the laws that might otherwise be obscured. Orthodox perceptual theories, in not using ecological units of analysis, cannot find ecological laws – or, perhaps more accurately, do not consider laws for perceiving and acting to be findable – and, thereby, undermine realism. Gibson developed his ecological approach as an alternative to the perceptual orthodoxy that he considered unworkable:

> I should like to think that there is sophisticated support for the naïve belief in the world of objects and events, and for the simple-minded conviction that our senses give knowledge of it. But this support is hard to find when the senses are considered as channels of sensations; it becomes easy when they are considered as perceptual systems.
>
> (Gibson 1967: 167)

Channels of sensations are passive conduits by means of which input arrives at the brain where, finally, activity occurs. Their limitations (i.e. the kinds of inputs they are presumed to carry) separate organisms from the world to be perceived. Perceptual systems are themselves active; they 'orient, explore, investigate' the surroundings. The notion of perceptual systems underscores the goal of understanding perception in the natural circumstances of getting around.

Naturalness pervades the ecological themes to be developed here. Other scholars have also been concerned with the lack of concern for naturalness in the way psychology is studied typically. In order to avoid confusion, I will spend a little time to distinguish Gibson's ecological approach from that of others who have used 'ecological' to modify their brand of psychology, most notably Kurt Lewin (1944), Egon Brunswik (1956), and Roger Barker (Barker 1965, 1968; Barker and Wright 1949).

1.2 Other ecological approaches to psychology

The theme that distinguishes other ecological approaches from Gibson's is their assumption that the physical world is somehow 'transformed into the psychological world' (Barker and Wright 1949), an assumption that is at odds with realism. Lewin, for example, used 'psychological ecology' to refer to the relation between the psychological and non-psychological (material/cultural) worlds. He considered the problem of establishing connections between non-psychological facts and psychological situations or habitats to be the ecological problem. Brunswik used the term 'ecological' to denote a body of problems concerned with the description of the psychological world and its relation to behaviour. Ecological variables for Brunswik are things such as perceived size and distance as distinct from measured size and distance. He considered these to be probabilistically (not geometrically or lawfully) related. Stimuli indicated environmental facts but did not specify them. Barker's ecological psychology was an attempt to parallel ecology in applying field-study methods to the study of behaviour in naturally occurring non-experimental situations or psychological habitats. The hallmark of his approach was the behaviour setting which was considered to stand midway between the non-psychological milieu and the psychological habitat: 'The behaviour setting is not a part of the non-psychological milieu as such, nor a part of the milieu which has been transformed into psychological habitat for a particular individual; it is a part of the milieu as it is generally perceived' (Barker and Wright 1949: 137).

All of these accounts reflect a basic animal–environment dualism: the world as described by physics is the real objective world, different from the meaningful world that is a subjective contribution of perceivers. In positing something standing between an animal and its apprehension of its surroundings, the preceding approaches violate realism and, therefore, are untenable to an ecological psychologist in the tradition of Gibson. We will now turn our attention to how the assumptions identified in the preceding section become elaborated to yield ecological units of analysis that are sought in that tradition.

2 AFFORDANCES: ANIMAL- AND ACTIVITY-REFERENTIAL DESCRIPTIONS OF THE ENVIRONMENT TO BE PERCEIVED

Asserting the mutual relationship between animals and environments highlights Gibson's concern that orthodox perceptual theory has ceded the right to describe the environment to physics (see Carello and Turvey 1985). Representational/computational theories of perception are forced to propose computations over representations because meaningful experience of the environment cannot be squeezed out of descriptors such as intensity and wavelength. But, Gibson argued, intensity and wavelength are only appropriate descriptors if the problems of concern involve, for example, the design of telescopes or photoelectric cells. A scientist interested in how animals get around in their environments is duty-bound to develop the appropriate descriptive vocabulary for that class of problems. This vocabulary is no less rigorous and refers to properties that are no less real; it simply must be at the ecological scale of behaving animals.

The demand for a descriptive vocabulary that is animal-referential and activity-referential from the start is met to a significant degree by Gibson's concept of affordance: 'The affordance of anything is a specific combination of properties of its substance and its surface taken with reference to an animal' (Gibson 1977: 67). In particular, it refers to the behavioural possibilities of a surface layout for a particular animal. For example, a relatively horizontal, rigid, extended surface approximately knee-height off the ground is 'sit-on-able' to a creature with the dynamic capability of maintaining balance while lowering its centre of mass until the buttocks are supported by the surface. To the same creature with the intention of reaching something high, the same surface layout might afford stepping on. To the same creature just passing through, the same surface layout is simply an obstacle to locomotion. To a different creature with different dynamic action capabilities (or 'effectivities', Turvey and Shaw 1979) the same surface layout might afford sheltering. Regardless of the particular circumstances, it is the affordance that is perceived. One does not perceive a list of properties – horizontal, rigid, extended, knee-high – and infer that a surface can be sat upon. The invariant combination of substance and surface relative to a particular animal that renders something sit-on-able for that animal is the affordance. Given the immense variety of surfaces (e.g. Breuer and Barcelona chairs; rocks and tree stumps; tables and car hoods) that are sit-on-able for somebody, information about that affordance must be very complex. Let us examine some critical aspects of affordances which are implicit in the foregoing examples and which will have to be incorporated into a description of that information.

First, affordances are a way of describing the environment that is scaled to the perceiver. This can easily be seen with respect to the animal's size. A chair that is sit-on-able for an adult may not be so for a child. But other, lower, surfaces are sit-on-able for a child and the same invariant combination of animal–environment properties will specify sit-on-able in the two cases (Figure 8.2). This has been shown experimentally by Mark (1987) who found that people allowed only to look at a raised surface were able to see when it was at the maximum height to permit sitting upon in the ordinary fashion. This turned out to be a constant proportion of the perceiver's height. Moreover, changing a person's height (essentially by providing 10 cm thick soles for their shoes) changed the perceived boundary of sit-on-able surfaces appropriately. But scaling to perceivers is not simply a size problem. Under different dynamic conditions, the same geometric layout will be perceived differently by the same perceiver. For example, the perceived boundary of one's forward reach depends on the manner of reach, balance requirements, the presence of obstacles, and so on (Carello *et al.* 1989a). Information about affordances must include information about the dynamic consequences of activities for a given animal–environment system.

Second, it should be noted that affordances are not restricted to static surface layouts. On the contrary, the fact that objects and surfaces participate in events is rather crucial to the way in which they structure energy distributions. Certain properties of objects and surfaces are only revealed

Figure 8.2 What can be sat upon comfortably differs for people of different sizes but a sit-on-able surface can be described using an invariant ratio of leg length to surface height, suggesting that the information about this affordance is the same

through exploration (hefting, sniffing, scrutinizing); certain properties are only revealed in motions of the objects or movement by the observer. Information about both persisting and changing properties must be available in order for animals to maintain epistemic contact with their surroundings.

Finally, affordances are a way of describing the environment that entails meaning. The behavioural possibilities of the surface layout – what the animal can do in the environment – is really what the environment means to the perceiver. Meaning is not something that is added to a representation of the environment; it is not an attribution by the perceiver to the environment. Perceiving what is reachable, for example, does not mean that one perceives distance and then deduces whether or not reaching is possible given the circumstances; the affordance is perceived directly. Gibson argued that information about affordances is available in the structured energy distributions (reflected light and sound, chemical diffusion, tissue deformation) generated by surface layouts and events and by the animal's own actions.

In short, the system comprising an animal and its environment is full of relations that have meanings for the conduct of acts and, in Gibson's view, there is information specific to those meanings.

2.1 The human-made environment

Gibson's emphasis on the natural fit between organisms and their environments does not mean that altered environments are beyond the purview of the theory. The notion of affordances applies to the human-made environment as well. The example of sit-on-able surfaces shown in Figure 8.2 illustrates this. Indeed, much of affordance research takes place with respect to architectured surroundings (e.g. traversability of surfaces, E. Gibson *et al.* 1987; passability of apertures, Warren and Whang 1987). Tools, furniture, buildings, and so on can be seen as exploiting and refining affordances. If a flat rock of a certain size and heft relative to the hand affords pounding with, then one can search for rocks that are better pounders or one can alter available rocks until they do the job. A similar scenario can be offered for designing furniture and buildings. An intention to perform an act prompts exploring the surroundings for something that will do the job. Noticing that some things do the job better leads to a differentiation of affordances and, ultimately, to construction of objects or places that are 'tailor made'.

A delightful, multiply nested example of this treatment of tool use is provided by the nutcracking skills of nonhuman primates. Wild chimpanzees crack nuts with not only heavy sticks or rocks but also on tree stumps or flat rocks that might be considered natural 'anvils', matching the required hardness of the stick and anvil to the hardness of the particular nuts (Boesch and Boesch 1983). Moreover, although the naturalists at first assumed that the sticks were being used as hammers, the chimpanzees were using them as nutcrackers, anchoring the stick at one end and cracking the nut underneath with a lever action. As Boesch and Boesch discovered when they tried out the hammer method for themselves, the impact scatters the nut. The chimpanzees had settled upon a much more efficient method.

It is easy to see that nutcrackers could be refined by humans with the capability to machine tools. But the more technologically advanced constructions of humans are not perceived in a different manner from naturally occurring objects and places. This is not to say that the affordances of artefacts and natural objects will be equally readily apparent. A given affordance may be easier to perceive if the design somehow highlights the relevant properties. Other affordances of that object may be more difficult to perceive, however, as suggested by the occurrence of functional fixedness (Duncker 1945). Finally, considerations such as aesthetics may have the effect of obscuring behavioural possibilities (e.g. creations that elicit the familiar 'what is it?' response, for example, Figure 8.3).

3 ECOLOGICAL LAWS OF PERCEIVING–ACTING

A major step in Gibson's ecological development occurred with his resolution of a problem that had long puzzled early perception theorists: how

Figure 8.3 Design for a stool whose 'sit-on-able-ness' is not immediately apparent. The top surface, which is approximately at knee height, is composed of reasonably densely packed, rounded 'spikes' (adpated from Smets 1986)

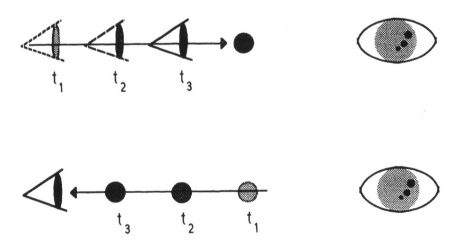

Figure 8.4 An age-old puzzle of perception: An eye approaching an object (*top left*) and an object approaching an eye (*bottom left*) are not distinguished in the retinal pattern (*right*). The successive retinal locations that are excited are indicated by increasing size of the projection (*right*)

movement of the self is distinguished from movement of something in the world. Under the standard analysis, the problem arises because in both cases the same retinal locations are stimulated successively (Figure 8.4). According to the prevailing view, the input is insufficient to distinguish the two circumstances and, therefore, epistemic mediation is necessary. As the story went, the ambiguous input must be compared to other inputs such as whether or not commands have been sent to the muscles to move. In the absence of muscle commands, the perceiver could assume that the retinal excitation was caused by movement of something 'out there'. Gibson noted, however, that no such comparison is necessary. At the ecological scale – the scale at which locomoting animals evolved, developed, and get around – movement of the self always generates global change in optical structure. Movement of an object toward a stationary observer generates a local disturbance in an unchanging global optical structure. The two circumstances are distinguished on the basis of optical information, not cognitive intervention (Figure 8.5).

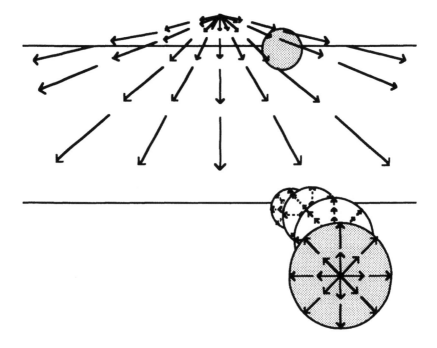

Figure 8.5 As Gibson pointed out, ego-movement and motion of an object are distinguished by the optical structure generated by the two events. *Top*: as an animal moves forward, light is structured in a manner that has been characterized as a global optical expansion. Light reflected from an object in the field of view is consistent with the global field structure. *Bottom*: an object approaching a stationary animal generates a local disturbance in the optic array. Parts of the background become occluded by the object and then come back into view.

The foregoing can be used to illustrate what is meant by ecological laws of perceiving and acting. Such a law takes the form shown in Figure 8.6 (Turvey and Carello 1986).

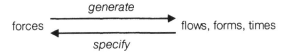

Figure 8.6   The form of an ecological law of perceiving and acting

The forces are generated by an animal engaging in activity in a cluttered surround (i.e. an ordinary environment, whether natural or architectured). The flows, forms, and times are properties of the optical structure (or acoustic structure, haptic structure, etc.) that always accompany those activities. Let us focus on the ego-movement/object motion example because there is a good deal of experimental evidence to draw on. When one moves forward, a pattern of global outflow is generated at the point of observation. It does not matter whether the point of observation is occupied by a walking human, a roller skating human, a flying insect, a galloping gnu, or whatever; forward locomotion generates global expansion. Nor does it matter what kind of ocular apparatus (e.g. chambered eye, compound eye) occupies the point of observation; the information generated will be the same. Therefore, the pattern of global optical expansion is said to specify the event that produced it and a properly attuned organism will, on detecting this pattern, perceive him/herself to be moving forward.

A growing number of experimental observations lend support to this assertion. When this animal–environment circumstance is simulated (e.g. by placing an observer in a room that can be moved as a whole, or in front of a screen on which flow patterns are projected), observers do, indeed, perceive themselves to be moving. Toddlers fall over (Lee and Aronson 1974, Stoffregan et al. 1987), houseflies modulate their wing beats (Wagner 1986a), and adults make postural adjustments to steady themselves (Lee and Lishman 1975, Stoffregan 1986). These adjustments have been shown to be synchronized to subtle modulations of the flow (Anderson and Dyre 1989). It has also been demonstrated that the form of the global flow (e.g. undulations, deformations) is specific to other aspects of locomotion such as walking versus moving smoothly over the ground (Stappers et al. 1989). The direction in which one is heading is perceived within better than 1° of accuracy on the basis of global radial outflow, even when the focus is not within view (Warren et al. 1988). It has been shown mathematically that information about the imminence of an upcoming collision is available (Lee 1976) and experimental investigations verify that perceivers are sensitive to this information, houseflies (Wagner 1986b) as well as humans (Lee et al. 1982, Lee et al. 1983; Todd 1981). It has also been shown that information is available about how hard an upcoming collision will be (Lee and Reddish 1981) and experimental investigations confirm that perceivers are sensitive to this information (Kim *et al.* 1989).

While many of the preceding experiments simply require perceivers to make categorical judgements (e.g. Which simulated object will contact first? Is an impending collision going to be hard or soft?), those that explicitly tie perception to action are especially instructive. For example, the speed of opening of a baseball batter's stance is linked to time-to-contact (Hubbard and Seng 1954), as is the mean acceleration during the forehand smash of expert table tennis players (Bootsma and van Wieringen 1990). These kinds of results reveal a tight link between information and the modulation of activity (Turvey 1977).

4 PERSON, ENVIRONMENT AND SOCIETY IN GIBSON'S ECOLOGICAL APPROACH

When discussing the interrelatedness of person, environment and society,[1] the obvious entry point for a perceptual psychologist schooled in the Gibsonian tradition is to consider person and environment as an instance of animal–environment mutuality. The terminology of animal–environment systems rather than person–environment systems is used by ecological psychologists to emphasize that perceiving meaningful properties such as edible or climbable or impassable does not require sophisticated intellectual machinery. If a housefly can do it, then it seems less compelling to attribute the ability to propositional thought (cf. Fodor 1975).

In asserting that describing environments is part of the task of perception theory, Gibson emphasized that that theory cannot be indifferent to what there is to be perceived. As was pointed out earlier, what is perceived is not simply an environment but an animal–environment relation as captured in the notion of affordance. Ironically, Gibson's concern for the long-neglected environmental component of perception – again, not treating perception as something that happens solely in animals but as an achievement of animal–environment systems – has led some to misinterpret his position as claiming that perception is all 'out there', external to animals (e.g. Shotter 1983).

> For although, as Gibson sees it, an ecological psychology is sensitive to the dynamic relations of mutuality and reciprocity between living entities and their environments, the beings in Gibson's world are depicted merely as observers, not as actors, i.e. not as beings able to provide for themselves, by their own action, conditions appropriate to support their action's continuation. They move about, but they do not act; thus rather than 'makers', they are presented as 'finders' of what already exists'.
>
> (Shotter 1983: 20)

The evidence for such an assertion is not at all clear. Gibson provided a functional taxonomy of movement systems that distinguished, among others, the orienting–investigating system in which 'adjustments of the head, eyes, mouth, hands, and other organs [are] for obtaining external stimulus information' and the performatory system in which 'movements occur that alter the environment in ways beneficial to the organism, such as displacing things, storing food, constructing shelter, fighting, and using tools' (Gibson 1966: 57). So, to reiterate, the environment is one component of a larger, irreducible system and this system is the proper unit of analysis at which to understand perception.

A consideration of animals as 'makers' of their environments has obvious relevance for understanding how society would fit into a Gibsonian perspective. But the notion of 'society' connotes a good deal that is uniquely human and, as such, has not received emphasis in the development of the ecological approach.[2] I would hesitate to include it as an entity separate from animals and environments; it seems to reside in both. In these final sections I will sketch how social issues fit into a Gibsonian ecological framework.

4.1 Society as environment

Most obviously, many of the products of society (e.g. tools, buildings, roads) constitute the human-made environment and we have already argued that that submits to an affordance analysis. Actualizing an affordance may alter the animal–environment system and thereby alter its affordances for me, for some other person, or for some other species. For example, to commuters a particular stretch of land may afford building a road to ease travel between two cities. That same stretch of land may also afford finding shelter and resources to a variety of nonhuman, perhaps endangered species. Building the road may eliminate crucial affordances for those other species but this is, in principle, no different from the disruption of an insect's econiche when a bird collects material to build a nest.

Because the effects of human activity are often quite far-reaching, of course, ordinary instances of actualizing affordances may have extraordinary consequences for the long-term health of the planet. Environmentalists would hope to forestall damage by getting people to understand the consequences of their actions. Ecological psychologists would consider this to be a problem in the education of attention. Affordances are often multiply nested and must be individuated by a perceiver with an intention. A chair is sit-on-able on the occasion of my needing to sit and stand-on-able on the occasion of my needing to reach something high. The information about both affordances is always available but a perceptual system must be attuned to that information in order to detect it. But let us suppose that more than one surface layout affords my intended activity, say, climbing to the second floor of a building. On what basis does one choose? Experiments have shown that, even when provided with optical information alone, perceivers choose as their preferred stairway that which would be energetically least costly given their particular biomechanical configuration (Warren 1984). In other words, part of perceiving the affordance is perceiving the future consequences of actions.

One could speculate that if humans' attention were educated to the affordances for other people or for other species or for society at large, then certain behaviours might be altered accordingly. Highways would not be built through wetlands, the design of public housing would be less sterile, chemicals that destroy the ozone layer would not be used. But, of course, the situation is not so straightforward because, for example, actions to preserve the ozone layer over the long term may compete with actions to preserve the viability of a farm in the short term. Or the connection between certain actions and their consequences may be more apparent than the connection between other actions and their consequences. Or the immediacy of certain consequences may simply be more compelling. But, again, these are standard problems in the issue of individuating affordances.

Less standard, perhaps, are affordances that are 'socially sustained'. Perceiving that a mailbox affords mailing a letter, for example, rests on an understanding of the postal system and this understanding does not arise from structured light but from

> the cooperative linguistic community that symbolically invests and sustains objects and their affordances. ... When we come to 'social' objects we leave the realm of purely biological–behavioural and enter the world of cultural rules and roles. In the human community there just is a vast proportion of perceptual experience that is sustained by cultural rule-given prerequisites. It is naïve, not to say one-eyed, for Gibson to pass lightly from the natural environment to the socially-sustained environment yet go on describing the objects of the latter as no different to the objects of the former, thereby failing to see the enlarging fact that the relationship between persons and objects is now complexly mediated by a normative structure that is not connected with the visible afford-ances of objects *per se*.
>
> (Noble 1981: 72–3)

A more accurate characterization of Gibson's position, I think, is not that the objects of the natural and socially sustained environments are the same but that the theory of information pick-up is the same:

> The theory of information pick-up ... closes the supposed gap between perception and knowledge. The extracting and abstracting of invariants are what happens in both perceiving and knowing. To perceive the environment and to conceive it are different in degree but not in kind. ... Knowing is an *extension* of perceiving. The child becomes aware of the world by looking around and looking at, by listening, feeling, smelling, and tasting, but then she begins to be *made* aware of the world

as well. . . . Extended or aided modes of apprehension are all cases of information pick-up from a stimulus flux'.

(Gibson 1979: 258)

What differs is the nature of the link between information and the event that is informed about. When structured light, sound, and so on are generated lawfully by an event, that structure is information in the specificational sense – the sense of information that is at the heart of ecological realism. When structured light, sound, and so on are linked conventionally to the event, that structure is information in the indicational sense – the sense of information that is required for socially sustained affordances. In both cases, the link between the structure and the event is systematic; the former is information about the latter. Information in the indicational sense, which is central to linguistic meaning, is predicated on information in the specificational sense (Turvey and Carello 1985). As Gibson observed, 'the learner has to hear the speech in order to pick up the message' (Gibson 1979: 258). To be sure, coming to terms with linguistic meaning in a way that satisfies the demands of realism is not trivial. But it is not impossible and is being pursued in exciting directions on a number of fronts (e.g. Barwise and Perry 1983).

Notions of socially sustained affordances and the cooperative linguistic community highlight the importance of the actions of individuals in co-operation with each other. It is to social cooperation that we now turn.

4.2 Society as agent: social interaction

Social cooperation requires perceiving the affordances of animated objects. The environment comprises various combinations of substances and surfaces. Some of those surfaces are attached to people and at that level it is easy to see how perceiving the affordances of animated objects need be no different from perceiving the affordances of inanimate objects. People can be obstacles to locomotion or they can be projectiles to be evaded (as is a charging middle linebacker in American football). But they also have what might be con-sidered more complex, social affordances. They can be cooperative or threatening or nurturing and, presumably, can structure energy distributions in such a way as to specify the relevant affordances (Baron and Boudreau 1987, McArthur and Baron 1983). Many social affordances, for example, are related to growth or ageing. Information about growth is made available by lawfully produced changes in craniofacial morphology (Shaw et al. 1974, Shaw et al. 1982). Craniofacial morphology, in turn, has been found to be relevant to perceived dependency and approachability (Berry and McArthur 1986), attractiveness (Carello et al. 1989b), and to the eliciting of protective (Alley 1983) and abusive behaviour (McCabe 1984). Social affordances need no more be considered inferential attributions than are non-social affordances such as walk-on-able or graspable.

Beyond characterizing the information about social affordances, how are we to understand the social interactions that ensue? Can the activities coordinated between two or more people submit to the same kind of analysis as the coordinated activity of a single person? The question arises because, unlike the coordinating limbs in locomotion, the components in between-person coordination do not share a single nervous system. One might imagine, therefore, that the coordinating basis is different. The ecological characterization of perception and action as being mutually constraining, however, suggests that even within-person coordination is information-based (Kugler and Turvey 1987). If this is true, then between-person coordination should show the same characteristics as analogous within-person coordinations (Schmidt *et al.* 1990). Although this conjecture has only been assessed with respect to fairly simple social interactions, it appears that they can be usefully characterized as coupled oscillators (e.g. Newtson *et al.* 1987) and benefit from theoretical developments in that domain. Indeed, optically coupled biological oscillators – two people who are instructed to swing one leg below the knee so that it maintains a particular phase relationship relative to the other person's leg – exhibit nonequilibrium phase transitions of a sort that are found generally in nature (Schmidt *et al.* 1990). Certain cooperative states are more stable than others, cooperation breaks down as the system is driven too far from equilibrium, fluctuations in the system increase as the transition is approached. These same characteristics have been found in within-person bimanual coordination (e.g. Haken *et al.* 1985, Kelso *et al.* 1986). In coordination, the components – be they the limbs of the body or the actions of two individuals – behave as if they were a single 'virtual' system (Kugler and Turvey 1987). Principles underlying cooperative phenomena do not distinguish what or who is cooperating.

5 CONCLUSIONS

The strategy pursued by ecological psychologists in the tradition of James Gibson is, most obviously, an effort not to treat problems as more local than they are. Perception is not something that happens inside animals. 'It is a keeping-in-touch with the world, an experiencing of things rather than a having of experiences. It involves awareness-of instead of just awareness' (Gibson 1979: 239). To understand a problem in perception, therefore, it must be addressed at the level of an active animal–environment system. What does the animal of interest need to perceive in order to get around its environment successfully? Presumably, its success is made possible by ecological laws of perceiving and acting. Calling these laws ecological is a reminder that they hold within circumscribed domains, not really all domains. Part of the ecological strategy, therefore, also involves identifying those bounded domains.

One of the hallmarks of the ecological approach that derives directly from the use of ecological units of analysis has been its capacity for handling perception and action as natural phenomena and not special, extraphysical, mental phenomena. A growing number of scholars working in the domain of complex systems share this view: 'All physical nature operates with only a few principles, but they require many forms and are expressed in a variety of emergent processes' (Iberall and Soodak 1987: 501). Results from experimental analyses of between-person interactions suggest that social phenomena may be characterized similarly. Iberall and Soodak (Iberall 1984, 1987; Iberall et al. 1980; Soodak and Iberall 1987) have been developing what they call 'the thermodynamic approach to the study of human society'. They argue that understanding the organized activities of a society will require, among other things, 'an account of the outside constraints, fluxes, and forces imposed on the culture, both from the geophysical environment and from. . .neighboring, interacting cultures. . .as well as the genetic species development occurring on the evolutionary scale' (Soodak and Iberall 1987: 467). The ecological unit of analysis for a society may be daunting but promises a lawful basis for understanding social action.

6 NOTES

1 This chapter was first presented as a paper at the international conference 'Person – Society – Environment', held at Appenberg, Switzerland, May 1989; the conference resulted in the present volume.
2 Interestingly, Gibson's early career was noted as much for his views on social psychology and social issues as for his work in perception (Reed 1988). The writing of a book on the socialization of behaviour was interrupted by World War II and he never really went back to it. His biographer speculated that this neglect was a combination of the increased importance of his work on visual perception and his alienation from mainstream social science as a result of the cold war climate in the United States: 'His response to the [intellectual freedom] cases seems to have been to abandon his own social science research in favour of helping colleagues in difficulty and to redouble his efforts in the study of perception' (Reed 1988: 113). Although many notes for the socialization book exist in the Gibson archives at Cornell University, these are from a period before his ecological approach to perception had matured. A paper he published in 1950 is a summary of these views (Gibson 1950).

7 REFERENCES

Alley, T.A. (1983) 'Growth-produced changes in body shape and size as determinants of perceived age and adult caregiving', *Child Development* 54: 241–8.
Alley, T.A. (1985) 'Organism–environment mutuality, epistemics, and the concept of an ecological niche', *Synthese* 65: 411–44.
Anderson, G.J. and Dyre, B.P. (1989) 'Spatial orientation from optic flow in the central visual field', *Perception and Psychophysics* 45: 453–8.
Barker, R.G. (1965) 'Explorations in ecological psychology', *American Psychologist* 20: 1–14.
Barker, R.G. (1968) *Ecological Psychology*, Stanford, CA: Stanford University Press.

Barker, R.G. and Wright, H.F. (1949) 'Psychological ecology', *Child Development* 20: 131–40.

Baron, R.M. and Boudreau, L.A. (1987) 'An ecological perspective on integrating personality and social psychology', *Journal of Personality and Social Psychology* 53: 1–7.

Barwise, J. and Perry, J. (1983) *Situations and Attitudes* Cambridge, MA: MIT Press.

Berry, D.S. and McArthur, L.Z. (1986) 'Perceiving character in faces: The impact of age-related craniofacial changes on social perception', *Psychological Bulletin* 100: 3–18.

Boesch, C. and Boesch, H. (1983) 'Optimisation of nut-cracking with natural hammers by wild chimpanzees', *Behaviour* 83: 265–86.

Bootsma, R. and van Wieringen, P.C.W. (1990) 'The timing of rapid interceptive actions', *Journal of Experimental Psychology: Human Perception and Performance* 16: 21–9.

Brunswik, E. (1956) *Perception and the Representative Design of Psychological Experiments*, Berkeley, CA: University of California Press.

Carello, C. and Turvey, M.T. (1985) 'Vagueness and fictions as cornerstones of perceiving and acting: A reply to Walter', *Cognition and Brain Theory* 7: 247–61.

Carello, C., Grosofsky, A., Reichel, F., Solomon, H.Y. and Turvey, M.T. (1989a) 'Perceiving what is reachable', *Ecological Psychology* 1: 27–54.

Carello, C., Grosofsky, A., Shaw, R.E., Pittenger, J.B. and Mark, L.S. (1989b) 'Attractiveness of facial profiles is a function of distance from archetype', *Ecological Psychology* 1: 227–51.

Duncker, K. (1945) 'On problem solving' (trans. L.S. Lees), *Psychological Monographs* 58: 5.

Elton, C. (1927) *Animal Ecology*, London: Sidgwick & Jackson.

Fodor, J.A. (1975) *The Language of Thought*, Cambridge, MA: Harvard University Press.

Gibson, E.J., Riccio, G., Schmuckler, M.A., Stoffregan, T.A., Rosenberg, D. and Taormina, J. (1987) 'Detection of the traversability of surfaces by crawling and walking infants', *Journal of Experimental Psychology: Human Perception and Performance* 13: 533–44.

Gibson, J.J. (1950) *The Perception of the Visual World*, Boston: Houghton Mifflin.

Gibson, J.J. (1966) *The Senses Considered as Perceptual Systems*, Boston, MA: Houghton Mifflin.

Gibson, J.J. (1967) 'New reasons for realism', *Synthese* 17: 162–72.

Gibson, J.J. (1977) 'The theory of affordances', in R.E. Shaw and J. Bransford (eds) *Perceiving, Acting and Knowing*, Hillsdale, NJ: Erlbaum.

Gibson, J.J. (1979) *The Ecological Approach to Visual Perception*, Boston, MA: Houghton Mifflin.

Haken, H., Kelso, J.A.S. and Bunz, H. (1985) 'A theoretical model of phase transitions in human hand movements', *Biological Cybernetics* 51: 347–56.

Hubbard, A.W. and Seng, C.N. (1954) 'Visual movements of batters', *Research Quarterly* 25: 42–57.

Iberall, A.S. (1984) 'Contributions to a physical science for the study of civilization', *Journal of Social and Biological Structure*, 7: 259–83.

Iberall, A.S. (1987) 'A physics for studies of civilization', in F.E. Yates (ed.) *Self-Organizing Systems: The Emergence of Order*, New York: Plenum Press.

Iberall, A.S. and Soodak, H. (1987) 'A physics for complex systems', in F.E. Yates (ed.) *Self-Organizing Systems: The Emergence of Order*, New York: Plenum Press.

Iberall, A.S., Soodak, H. and Arensberg, C. (1980) 'Homeokinetic physics of societies – A new discipline: Autonomous groups, cultures, polities', in H. Reul, D. Ghista and G. Rau (eds) *Perspectives in Biomechanics* 1(A), New York: Harwood.

Kelso, J.A.S., Scholz, J.P. and Schöner, G. (1986) 'Nonequilibrium phase transitions in coordinated biological motion: Critical fluctuations', *Physics Letters* 118: 279–84.

Kim, N-G., Carello, C. and Turvey, M.T. (1989) 'Optical information for the prospective control of collisions', paper presented at the International Society for Ecological Psychology, Hanover, NH, October.

Kugler, P.N. and Turvey, M.T. (1987) *Information, Natural Law, and the Self-Assembly of Rhythmic Movement*, Hillsdale, NJ: Erlbaum.

Lee, D.N. (1976) 'A theory of visual control of braking based on information about time-to-collision', *Perception* 5: 437–59.

Lee, D.N. and Aronson, E. (1974) 'Visual proprioceptive control of standing in human infants' *Perception and Psychophysics*, 15, 529–32.

Lee, D.N. and Lishman, R. (1975) 'Visual proprioceptive control of stance', *Journal of Human Movement Studies* 1: 87–95.

Lee, D.N. and Reddish, P.E. (1981) 'Plummeting gannets: A paradigm of ecological optics', *Nature* 293 (5830): 293–4.

Lee, D.N., Lishman, J.R. and Thomson, J.A. (1982) 'Visual regulation of gait in long-jumping', *Journal of Experimental Psychology: Human Perception and Performance* 8: 448–59.

Lee, D.N., Young, D.S., Reddish, P.E., Lough, S. and Clayton, T.M.H. (1983) 'Visual timing in hitting an accelerating ball', *Quarterly Journal of Experimental Psychology* 35A: 333–46.

Lewin, K. (1944) 'Constructs in psychology and psychological ecology', *University of Iowa Studies in Child Welfare* 20: 3–29.

McArthur, L.Z. and Baron, R.M. (1983) 'Toward an ecological theory of social perception', *Psychological Review* 90: 215–38.

McCabe, V. (1984) 'Abstract perceptual information for age level: A risk factor for maltreatment?', *Child Development* 55: 267–76.

Mark, L.S. (1987) 'Eyeheight-scaled information about affordances: A study of sitting and stairclimbing', *Journal of Experimental Psychology: Human Perception and Performance* 13: 361–70.

Michaels, C.F. and Carello, C. (1981) *Direct Perception*, Englewood Cliffs, NJ: Prentice Hall.

Newtson, D., Hairfield, J., Bloomingdale, J. and Cutino, S. (1987) 'The structure of action and interaction', *Social Cognition* 5: 191–237.

Noble, W.G. (1981) 'Gibsonian theory and the pragmatist perspective', *Journal for the Theory of Social Behavior* 11: 65–85.

Reed, E.S. (1988) *James J. Gibson and the Psychology of Perception*, New Haven, CT: Yale University Press.

Reed, E.S. and Jones, R. (eds) (1982) *Reasons for Realism: The Collected Papers of James J. Gibson*, Hillsdale, NJ: Erlbaum.

Schmidt, R.C., Carello, C. and Turvey, M.T. (1990) 'Phase transitions and critical fluctuations in the visual coordination of rhythmic movements between people', *Journal of Experimental Psychology: Human Perception and Performance* 16: 227–47.

Shaw, R.E. and Pittenger, J.B. (1978) 'Perceiving change', in H. Pick and E. Saltzman (eds) *Modes of Perceiving and Processing Information*, Hillsdale, NJ: Erlbaum.

Shaw, R.E., McIntyre, M. and Mace, W.M. (1974) 'The role of symmetry in event perception', in R.B. MacLeod and H.L. Pick (eds) *Perception: Essays in Honor of James J. Gibson*, Ithaca, NY: Cornell University Press.

Shaw, R.E., Mark, L.S., Jenkins, H. and Mingolla, E. (1982) 'A dynamic geometry for predicting growth of gross craniofacial morphology', in A. Dixon and B. Sarnat (eds) *Factors and Mechanisms Influencing Bone Growth*, New York: Liss.

Shotter, J. (1983) '"Duality of structure" and "intentionality" in an ecological psychology', *Journal for the Theory of Social Behavior* 13: 19–43.

Smets, G. (1986) *Vormleer: De paradox van de vorm*, Amsterdam: Uitgeverij Bert Bakker.

Soodak, H. and Iberall, A.S. (1987) 'Thermodynamics and complex systems', in F.E. Yates (ed.) *Self-Organizing Systems: The Emergence of Order*, New York: Plenum Press.

Stappers, P.J., Smets, G.J.F. and Overbeeke, C.J. (1989) 'Gaps in the optic array: Toward richer operationalizations of optic array and optic flow', paper presented at the Fifth International Conference on Event Perception and Action, Oxford, OH.

Stoffregan, T.A. (1986) 'The role of optical velocity in the control of stance', *Perception and Psychophysics* 39: 355–60.

Stoffregan, T.A., Schmuckler, M.A. and Gibson, E.J. (1987) 'Use of central and peripheral flow in stance and locomotion in young walkers', *Perception* 16: 113–19.

Todd, J.T. (1981) 'Visual information about moving objects', *Journal of Experimental Psychology: Human Perception and Performance* 7: 795–810.

Turvey, M.T. (1977) 'Preliminaries of a theory of action with reference to vision', in R.E. Shaw and J.B. Bransford (eds) *Perceiving, Acting and Knowing*, Hillsdale, NJ: Erlbaum.

Turvey, M.T. and Carello, C. (1981) 'Cognition: The view from ecological realism', *Cognition* 10: 313–21.

Turvey, M.T. and Carello, C. (1985) 'The equation of information and meaning from the perspectives of situation semantics and Gibson's ecological realism', *Linguistics and Philosophy* 8: 81–90.

Turvey, M.T. and Carello, C. (1986) 'The ecological approach to perceiving–acting: A pictorial essay', *Acta Psychologica* 63: 133–55.

Turvey, M.T. and Shaw, R.E. (1979) 'The primacy of perceiving: An ecological reformulation of perception for understanding memory', in L.-G. Nilsson (ed.) *Perspectives on Memory Research*, Hillsdale, NJ: Erlbaum.

Wagner, H. (1986a) 'Flight performance and visual control of flight in the free-flying housefly (*Musca domestica*), II. Pursuit of targets', *Philosophical Transactions of the Royal Society of London* B312: 553–80.

Wagner, H. (1986b) 'Flight performance and visual control of flight in the free-flying housefly (*Musca domestica*), III. Interactions between angular movement induced by wide- and small-field stimuli', *Philosophical Transactions of the Royal Society of London* B312: 581–95.

Warren, W.H. (1984) 'Perceiving affordances: Visual guidance of stair-climbing', *Journal of Experimental Psychology: Human Perception and Performance* 10: 683–703.

Warren, W.H. and Whang, S. (1987) 'Visual guidance of walking through apertures: Body-scaled information for affordances', *Journal of Experimental Psychology: Human Perception and Performance* 13: 371–83.

Warren W.H., Morris, M.W. and Kalish, M. (1988) 'Perception of translational heading from optical flow', *Journal of Experimental Psychology: Human Perception and Performance* 14: 646–60.

Whittaker, R.H., Levin, S.A. and Root, R.B. (1973) 'Niche, habitat, and ecotope', *American Naturalist* 107: 321–38.

9

ON SCIENCE AND KNOWLEDGE

Ingela Josefson

In recent years, developments in the research area of artificial intelligence have drawn attention to fundamental questions related to the nature of knowledge. Let me illustrate this with an example.

> The geriatric robot is wonderful. It isn't hanging about in the hopes of inheriting your money – nor of course will it slip you a little something to speed the inevitable end. It isn't hanging about because it can't find work elsewhere. It's there because it's yours. It doesn't just bathe you and feed you and wheel you out in the sun when you crave fresh air and a change of scene, though of course it does all those things. The very best thing about the geriatric robot is that it listens. 'Tell me again' it says, 'about how wonderful/dreadful your children are to you. Tell me again that fascinating tale of the coup of '63. Tell me again . . .' and it means it. It never gets tired of hearing those stories, just as you never get tired of telling them. It knows your favorites, and those are its favorites too. Never mind that this all ought to be done by human caretakers; humans grow bored, get greedy, want variety. It's part of our charm.
>
> (Feigenbaum and McCorduck 1984: 100ff)

This was written by Edward Feigenbaum, professor of computer science at Stanford University and Pamela McCorduck, professor of literature. The primary application of artificial intelligence research is in expert systems. The geriatric robot is an example of this. That is why Feigenbaum points out that the bottle-neck in this research area is in the area of knowledge-related issues. If these systems are to be developed, ways must be found to 'tap' the knowledge of human experts. The systems are based on experts' detailed descriptions of their professional knowledge; this knowledge is then reformulated in the uniform rules of logic and transferred to the machine. This has proved to be a much more difficult process than hitherto imagined. It is difficult to 'tap' the entire pool of professional knowledge that these experts possess.

Let us examine for a moment the question: what must a nurse know to care for a geriatric?

It goes without saying that he or she should have a wealth of factual knowledge – general medical knowledge. The nurse must know, for example, the symptoms of senile dementia, symptoms which he or she usually learns to recognize in the course of training. For the last ten years or so, nursing training has been classified as tertiary education, and comprises a three-year theoretical course. The kind of knowledge that a nurse acquires in his or her university-level education can be passed on in the form of statements. 'You should do this and that if you have a patient with this or that symptom.' I call this part of a body of professional knowledge 'propositional knowledge'. Equipped with this knowledge and a little practical experience, the nurse faces the daily challenges of the care sector. At best, he or she can deal with the everyday problems that are encountered, provided that the patient's symptoms agree with the description in the handbook. But the nurse is at a loss when confronted with a case which deviates from the description in the handbook. With his or her lack of experience the nurse is likely to adhere strictly to the rules that he or she has learned, no matter what form of reality is.

Let me illustrate this with the following example taken from an interview with a British nurse.

> A middle-aged man came to the emergency room with a suspected coronary. According to the handbook the patient must be kept abso-lutely still. But this man is upset and worried, he stubbornly insists that he wants to go to the lavatory. A young nurse refuses to let him, because it is against the rules she has learned. At this point an experienced nurse intervenes; she puts the man carefully in a wheelchair and helps him to the lavatory. As she said, 'I felt that there was less risk for the man if we let him do what he wanted and be calm, than have him lie absolutely still and become more and more upset.'
>
> (Josefson 1988: 17)

Daring to break the rules presupposes sound judgement and the ability to assess what the consequences may be. Moreover, in nursing, these judgements must often be made very quickly.

This is not the sort of knowledge that can be learned from a book, because we are dealing with the ability to make a judgement in an unusual, even unique, case. Instead, it is acquired from the variety of practical experience, from familiarity with the work. I therefore call this 'knowledge by familiarity'.

Let me give another example which illustrates the aspects of a nurse's professional knowledge which come under the heading of knowledge by familiarity. (Source: An interview conducted by the author in the course of a case study on nursing.)

A nurse in her forties described how, as a newly qualified nurse, she was working on a cancer ward in Stockholm. She was given the chance to work alongside an older, experienced nurse for the first few months of her time there. One morning, during their daily round of the wards, the older nurse stopped outside the room of a seriously ill patient and said, 'Go and telephone this patient's relatives, he has not long to live'. The young nurse obeyed, the relatives came, and the man died later in the day. The young nurse asked her colleague how she could see what was going to happen. She herself could detect no change in the patient; he looked the same as he had done for several weeks. She asked her colleague to describe in detail how she could see that death was approaching. The answer she got was: 'I can't describe how I can see it, and you can't learn it from a book either. But once you have seen it a few times, you know exactly what is going to happen. You learn to "see".' Twenty years later, this nurse says that she now has this knowledge.

Propositional knowledge and knowledge by familiarity are different aspects of professional knowledge. They are not separated by any clear boundaries; rather they are interdependent. It is practical experience which gives life to theoretical knowledge. Propositional knowledge may be learned from books; the knowledge of experience is acquired through practice. The tradition for expression is passed on; for example, from older to younger colleagues. It is developed with the help of reflection. This means that time for reflection is essential if this sort of knowledge is to develop. Time is scarce in the working life of today. Today, practising nurses are eligible for grants if they attend a theoretical course of training at university level. However, it would appear more difficult to make time at the workplace for people to reflect on their day-to-day experiences. In the care sector, there is a risk that if the difficult situations which are encountered are not processed, they may make people hard and cynical. Experience which is not processed is not converted into knowledge.

In our culture there is a clear tendency to equate propositional knowledge with other kinds of knowledge. Knowledge by experience is often called intuition which, at least in the Swedish language, evokes associations with something unspecified, something inherited, something which women in particular have been blessed with.

Let me give an example of this by quoting from a document on the nursing profession and information technology. At a congress in the USA held in 1976, Robin Parsons, the principal of a nursing college, made the following statement to the nursing delegates:

The nurse must, once and for all, learn to express her professional knowledge in exact terms. She cannot satisfy the demands of computer technology by using a vocabulary that has become part of the tradition of nursing. Imprecise, abstract terms such as *tender loving care*, *better patient care* and *caring for the whole person* are meaningless to computer technicians. They expect the nurse to be able to describe these concepts in a logical, scientific way, using a vocabulary that is free from the clichés and the esoteric rhetoric that is part of her traditional vocabulary. Her description must show how the result of her care can be measured quantitatively and evaluated.

If we try to describe 'tender loving care' in exact scientific terms we may be sure that the results will be nonsense. Love is expressed in action, but becomes trivial if defined in a scientific language to which it is not suited. The deep respect of the Western culture for scientific knowledge rests on a firm base. Socrates and Plato created a stable basis for our values. It may be refreshing to remind ourselves that this is not an eternal truth. In ancient Greece, craftsmanship was highly prized. Wisdom was seen as something practical and creative: the first philosophers were people with practical skills. They lived in the practical work they carried out and were not recluses who withdrew from the world and pondered abstract problems.

The high value placed on practical work was diminished more and more in the fifth century, the century of Socrates and Plato. This is closely linked to the religious conviction which emerged in this period. The soul is immortal, it existed first – the spiritual goes before the physical. The soul is imprisoned in the body: the body is, moreover, mortal. Effort must be made to turn the soul away from material things, the tangible and worldly things. In Plato's philosophy it is largely mathematics, the most abstract of all sciences, which may serve as a means of purifying the soul. It may even be seen as the key to the riddle of the universe. In the days of the revolution of the natural sciences, mathematics once again takes a central place. It is the precise, abstract science which is Galileo's starting point when he says in the 1600s that the book of nature is written in the language of mathematics.

Some philosophers of that time and of later centuries were captivated by the view of mathematics and logic as an instrument for describing ideas, and attempted to reduce as much of all human experience as possible to a purely logical calculation. The mathematicians of the seventeenth century attempted to create a universal language in which people could communicate more simply and speak with greater precision and clarity. Leibniz' plan was to create a logical language with the certainty of mathematics, a language which would encompass all areas of human problems, particularly metaphysics, theology and ethics. He said: 'Let us calculate, for this is the way that all religious and philosophical disputes may be mastered.'

This idea is based on the fundamental assumption that it is possible to reduce all knowledge to a mathematical language. These attempts have failed, but the dream of the exact language is still alive in, for example, the world of computer research.

Our culture's one-sided preference for propositional knowledge exists at the expense of the knowledge of familiarity. I see examples of this in developments of, for example, nursing training and training for the child care sector in recent years. Ten years ago these practical training courses were run under the auspices of our universities. The University Ordinance states that training must be on a scientific basis. What is largely practical know-how must be transformed into the exact language of science. The results of these efforts may be expressed in the following way. In a recent publication a

Finnish–Swedish professor in care science, who is well known in Scandinavia, writes: 'It is important to be able to give intellectual and spiritual content to the word "love" so that it may be understood and applied in the same way by different people'. It is the case that 'questions related to the meaning of life may be classified as the eternal questions of science, which means that they will always be the subject of research in different disciplines' (Eriksson 1987:43).

In a culture where abstract, precise science is valued more highly than practical common sense, there is a serious risk that people will be tempted to exceed the limits of what is meaningful in order to formulate in an exact language. There is reason to reflect on the observation made by Ludwig Wittgenstein (1982) in his *Tractatus logico-philosophicus* to the effect that we know that if all possible scientific questions have been answered, we have not yet touched on our life problems.

1 REFERENCES

Eriksson, K. (1987) *Pausen*, Helsinki: Almqvist & Wiksell.

Feigenbaum, E. and McCorduck, P. (1984) *The Fifth Generation, Artificial Intelligence and Japan's Computer Challenge to the World*, New York: Signet Book.

Josefson, I. (1988) *Från lärling till mastare*, Lund: Studentlitteratur.

Parsons, R. (1976) 'The future of the nursing profession. Computer technology and clinical nursing practice', *SA Nursing Journal* 43, 10.

Wittgenstein, L. (1982) *Tractatus logico-philosophicus*, Frankfurt a/M: Suhrkamp.

2 FURTHER READING

Göranzon, B. and Josefson, I. (1988) *Knowledge, Skill and New Technology*, London: Springer.

10

EXTERNAL EFFECTS AS A BRIDGE BETWEEN ENVIRONMENTAL AND ECOLOGICAL ECONOMICS

Gonzague Pillet

1 INTRODUCTION

The objective of the paper is to expose clearly one essential part of standard environmental economics – namely, the externality approach – as well as more recent developments in this field for achieving a solution at the interface between life-support ecosystems and economic processes. The overall concern is about pricing indirect – sometimes unrecognized – environmental goods, services or disservices, which is of paramount importance for economic analysis. Market, public and shadow prices – paid as well as unpaid price-ratios – are supposed to be used in a complementary way as decision criteria for ecologic–economic prospects.

To begin with, we deal with the analysis of spillovers affecting the welfare of individuals and the profit of production firms via the marketplace. This first step constitutes the principle of the theory of externalities; its leads to corrected market prices. Then, as a second externality method, we introduce the problem of environmental externalities, focusing on the physical accounting of materials, which are essentially entering or leaving the economic production process. This more recent paradigm is currently maturing; it leads to accounting prices both as regards extended national, and natural patrimony accounts. Finally, we further explore ecologic–economic regulation with respect to the calibration of environmental goods and services entering economic macro-processes as energy throughputs. Once again, we focus on an externality method, though a more controversial approach than the previous ones, and designed to outleap standard environmental economics. This third step leads to 'eMergy' analysis, energy externalities, and shadow macro-prices.

In order to obtain a synthetic view of the matter in hand, and to make the paper didactic, we present the well-known concept of externality as being the pinpoint of the exposé. In short, we use and develop the very economic concept of externality to deal with environmental facts and figures having a

146

Table 10.1 Externality methods

Concept	Theory	Definition	Diagramming	Modelling	Field, Range	Principles	Application
Economic, market externalities (1950)	Welfare economics; marginalist theory	$\dfrac{1}{\lambda_j}\partial u_i / \partial x_{ik}\begin{cases}<0\\>0\end{cases}$	Graph	Functional interrelationships; perfect competitive market micro-models	Nonmarket measurement of utility	Utility or profit maximization under budgetary constraint; Pareto optimality (compensative variations)	Incidental external effects; nonmarket welfare measurement and policy-making; cost–benefit analysis
Environmental externalities (1972)	Environmental economics; I–O analysis; accounting framework	[R], [W] resource, waste; Welfare function: W[C,R(Z=f[X,C])]	Matrix form	I–O tables with fixed coefficients; (multi-) sectorial models	Extension of the economic paradigm to include environmental links	General equilibrium; constant resource availability; application of physical principles to economics	Pervasive external effects; I–O based materials accounting; taxonomy of materials by economics use
Energy externality (1983)	Ecological economics; Systems ecology; energy analysis procedure (eMergy theory)	F, I → □ → Y, = emJ; F/I energy investment ratio; I/Y energy externality (xE)	Autocatalytic design; circuit language and energy diagrams	Macroscopic minimodels; microcomputer simulations	Interfaced environment and economic systems	Energy laws; maximum power principle; Macro-prices	Structural importance; eMergy analysis of environment economic systems and subsystems; measurement of the external, energy basis of a national economy

Source: Pillet (1986)

variety of form. Table 10.1 has been prepared to be used interchangeably as the analytical sieve of the paper or as the summary table. This brings us to the contents of the paper. Section 2 deals with the first line of Table 1; in Section 3, the second line is the focus; in Section 4, we render the third line explicit. Short case studies are introduced in the course of the paper in order to elucidate the theory.

In sum, this paper is designed as a way towards ecological economics from environmental economics by means of three distinct, though interlinked externality methods. The accent is more on the syntax, or the way in which concepts are put together to form policy instruments, than on the instruments themselves.

Each analytical step includes definitions, an exposé of the method used, a presentation of major results, and a discussion. Suggestions for further reading will be given for each step.

2 MARKET EXTERNALITIES

2.1 Topical background

I hire a carpenter to mend my table. The service which I get from having my table mended and the benefit which he receives from the wage which I pay him are all internal to the transaction which takes place between us. But suppose that his hammering disturbs my neighbour's enjoyment of the snooze which he was taking in his garden. This effect is external to the transaction between the carpenter and myself, who together took the decision to mend the table in my house; the noise has not annoyed the carpenter and myself but someone who was outside or external to our decision and took no part in reaching it.

(Meade 1973: 15)

Perfect market prices are supposed to reflect transparent information and rational choices. Unfortunately, they reflect not only efficiency but also ignorance, both moral and ecological. Market externalities address this ecological ignorance, or at least part of it, in emphasizing those effects that are caused by market decision-making, but that are not included in prices. Noise, smoke, ugliness of waterways and streets are analysed as detrimental external effects, or spillovers; quietness, purity of air, cleanliness of waterways and streets shape into beneficial externalities.

The trouble is that economic production (or consumption) incidentally issues costs (or benefits) that are not internally borne – as are costs of labour, equipment, etc. (or profits) – but externally borne, either by the environment (with adverse effects on the welfare of economic agents) or by the general public. As a result, economic actions give rise to inefficiently large outputs of detrimental spillovers and to inefficiently small amounts of beneficial externalities. In order to resolve these market shortcomings, economists believe that by internalising those 'off-shore' costs (or benefits), according to a polluter-pays-principle (or to a preventing-pollution-pays motto), each manufacturer (or consumer) will be given an incentive to reduce air and water pollution or to implement (buy) new, environment-oriented devices, just as any other incentive to reduce usual production costs, to expand production, or to switch consumption strategies. They believe that in assisting the market, the latter will attain efficiency again, while taking care of market-oriented environmental effects. *Explaining how this works is the task of this section.*

More conventionally, a market externality is defined as the case where a regular action of one economic agent (one consumer or one production firm) incidentally affects the utility level of another consumer or the production possibilities of another firm in a way that is not reflected in the setting up of market equilibrium and in the definition of its optimality (Just *et al.* 1982). In the literature, external effects are classified into effects of consumers on consumers, producers on producers, producers on consumers, and consumers on producers. Correcting these market failures leads to corrected market prices, the policies for dealing with these being discussed in terms of internalization, Pareto optimality, bargaining processes, prohibition, quotas, property rights, social optimality, etc. In the next section, we look into the other instruments.

2.2 Correcting market prices

We first define a market externality according to welfare economics; then, we review the above-mentioned major methodological issues.

Externalities and Pareto optimality

In a perfectly competitive market, a single decision-maker cannot influence market prices. Externalities do not exist, or are ignored. Prices are equating marginal costs; there is neither excess demand nor excess supply. As utility maximizers and profit maximizers, consumers and producers are satisfied in both buys and sales. Therefore, a competitive equilibrium exists, which defines Pareto optimality.

If, however, just one individual is led to improve his or her utility level because of over-the-counter unpriced effects caused by another individual, then an externality is designated which is not reflected in the marketplace. This externality is Pareto relevant if it meets the Pareto criterion; that is, if this individual – let us say A – can be made better off without making any other agent – let us say B – worse off.

More formally, let the utility function of the first individual be given by

(1) $\quad u_A = f(x_1, x_2, ..., x_n, y_1)$

and the utility function of the second agent by

(2) $\quad u_B = f(y_1, y_2, ..., y_n)$

We therefore assume that the utility level of A depends upon the commodities that he or she purchases, and upon y_1 which is imposed upon him or her as an externality by B's consumption decision. A and B are maximizing their utility. Yet the effect of y_1 upon A's utility level is unpriced and results in a marginal gain or loss, as follows:

(3) $\quad \partial u_A / \partial y_1 \neq 0$

With (3) > 0, a beneficial effect occurs; with (3) negative, a detrimental externality results.

When considering the production side, such a situation occurs when an input is imposed upon a firm as an externality by another firm; that is, when the second firm imposes a cost on the first one in its choice of an input that is not reflected in the marketplace and is not considered in its profit maximization.

In general, any externality achieves an economic meaning if and only if it is compensated for or internalized to attain Pareto optimality. This brings us to welfare policies in the presence of externalities.

Internalization

Strictly speaking, internalization is the approach used to determine a new 'social' optimality in the presence of externalities by considering all of the involved economic agents jointly.

Consider a producer–producer case. The externality can be internalized by maximizing the joint profits of the two firms as if they were merged; in other words, by moving to the internalized solution, the first firm will gain more than the second one will lose and thus be able to compensate the second for its losses and still be better off. That is, the solution consists of moving from Pareto suboptimal competitive equilibrium to a new social optimality point which is obtained by equating price and marginal social cost with a new equilibrium production (instead of equating price and marginal private cost and having externally borne social costs). This marginal social cost (MSC) is the sum total of the marginal private cost (MPC) and the marginal external cost (MEC) which defines the externality (Figure 10.3 will help to clarify this scheme). One view of this MEC is as a marginal willingness-to-pay which finally represents the available amount used to compensate the second firm for its losses without making the first one worse off.

Bargaining

A bargaining process can be considered as another way of achieving an internalization and meeting a Pareto optimal competitive equilibrium. Such a situation confirms and matches the approach of assignment of property rights in order to encourage the development of a market for externalities. This means that if a polluter owns the right to pollute, the pollutee may be willing to pay the former to reduce or even to cease pollution. On the contrary, if the pollutee owns the right to no pollution, a polluter may be willing either to buy the right to pollute or to bargain a larger level of pollution with the pollutee. In a theorem known as the Coase theorem, Coase (1960) was the first author to show that Pareto optimality is attained when such bargaining processes (or markets) are developed. The following example may be used for this purpose.

To consider such an example, suppose that a consumer–consumer externality takes place between two agents, i and j. Both of them are utility maximizers in a perfectly competitive market.

Let the utility function and the budgetary constraint of i be given by

$$(4) \qquad u_i = f(x_{i1}, ..., x_{in})$$

and

$$(5) \qquad \sum_{k=1}^{n} p_k \, x_{ik} \leq m_i$$

where p_k is the price of k among n goods which have been already produced, and m_i the income of the ith agent. Assume, further, that the second individual faces a utility function which depends upon the same range of market goods plus upon x_{i1}; that is, x_{i1} is imposed upon him or her as an externality by the ith agent's consumption decision. Thus we have

$$(6) \qquad u_j = f(x_{j1}, ..., x_{jn}, x_{i1})$$

and

$$(7) \qquad \sum_{k=1}^{n} p_k \, x_{jk} \leq m_j$$

Individuals i and j maximize their utility under budgetary constraints as follows:

$$(8) \qquad (1/\lambda_r) \, (\partial u_r / \partial x'_{rk}) - p_k \genfrac{\{}{\}}{0pt}{}{\leq}{=} 0$$
$$\text{iff } x'_{rk} \genfrac{\{}{\}}{0pt}{}{=}{>} 0$$

for k = 1, ..., n;
    r = i, j;
    x' = current x before bargaining begins;
    λ = Lagrangian multipliers.[1]

Equation (8) makes sense as follows: in being measured under budgetary constraint, the marginal utility of good k is lower than or is equal to its price p if that good is not purchased (current $x = 0$), and is equal to its price when purchased (that is, when $x' > 0$).

Yet, if the marginal utility of x_{i1} for j is larger or lower than zero, in the same way as in equation (3), Pareto optimality is not achieved. Obviously, the ith agent imposes a gain or a loss of utility on the jth one in his or her choice of x_{i1} that is not reflected in the marketplace and is not considered in its utility maximization.

Therefore, if

$$(9) \qquad (1/\lambda_j)\,(\partial u_j/\partial x'_{i1}) < 0$$

individual j is facing a detrimental externality.

Suppose that (9) prevails. According to the Pareto criterion, a policy change is socially desirable if by that change everyone can be made better off, or at least one benefits while no one is made worse off. Assume that j wants to do something to make him/herself better off. In this case, he or she will offer to i any amount T per unit of x_1, i will be willing to cut down; that is:

$$(10) \qquad T\,(x'_{i1} - x''_{i1}) \qquad \text{iff } x''_{i1} < x'_{i1}$$

According to this process, j will move forward by trial and error, looking for the best solution. Finally, the best offer will be the one that will match the optimal choice x^*_{i1} at which the amount paid by j will make him/herself better off by depleting the external diseconomy imposed on him or her by i, and compensate i for the cutting down of his or her consumption of x_1.

Now, this process modifies the budgetary constraints of each agent, as follows:

$$(11a) \qquad \sum_{k=1}^{n} p_k\,x_{ik} - T\,(x'_{i1} - x^*_{i1}) \le m_i$$

$$(11b) \qquad \sum_{k=1}^{n} p_k\,x_{jk} + T\,(x'_{i1} - x^*_{i1}) \le m_j$$

At the end of the process, we should have:

$$(12) \qquad (1/\lambda^*_i)\,(\partial u_i/\partial x^*_{i1}) - (p_1 - T^*) = 0$$

and

$$(13) \qquad (1/\lambda^*_j)\,(\partial u_j/\partial x^*_{i1}) - T^* = 0$$

If equations (12) and (13) hold, then a Pareto efficient equilibrium state is obtained.

Prohibition, quotas

Public welfare policies in the presence of externalities can lead to various cases involving compensatory payments.

Going one step further within the former model, one may suppose that the consumption of x_{i1} is prohibited unless i is willing to pay compensation to j. Therefore i will bargain in order to obtain some $x_1 > 0$. Suppose that he or she does this, offering to j an amount T per unit of x^{*}_1 that he or she buys in the marketplace.

As a result, i and j maximize their utility as follows regarding x_1:

(14) $(1/\lambda^{\#}_i) (\partial u_i/\partial x^{\#}_{i1}) - (p_1 + T^{\#}) = 0$

and

(15) $(1/\lambda^{\#}_j) (\partial u_j/\partial x^{\#}_{i1}) + T^{\#} = 0$

Analysing quotas would follow a procedure made of back and forward moves of both j and i.

2.3 Taxes and subsidies

Typical public policies for dealing with externalities generally include Pigovian taxes and subsidies. A Pigovian tax is a per unit tax to be paid for by the generator of a diseconomy. A Pigovian subsidy is a per unit amount of money compensating the generator of a beneficial externality. There are ways and means for dealing with Pigovian taxes and subsidies. That is how Turvey (1963) dealt with this. The concern is the divergencies between social and private costs and one has to suppose that what an agent gains or loses in the presence of externalities is well known in monetary terms, and especially in terms of each individual's revenue. In short, the actions of individual i indirectly affect the revenue of j as well as his or her own. We can therefore figure out Turvey's diagram as it is shown in Figure 10.1.

External costs are represented by the area OLM whereas total benefits are indicated by the area OPM. Social optimality is obtained by equating i's marginal gain (mg) and j's marginal loss (ml) and is given by a consumption level of OQ. At this point, a Pigovian tax is imposed on i on an *ad valorem* or per unit basis, and is represented on the diagram by the area ONQ. At that point, the social product is OPNQ minus ONQ; that is to say, the area OPN. It is maximum and meets the Pareto criterion.

On the one hand, a lower consumption level would make j better off, but i would be made worse off in so far as his or her gain would be lower too (it would no longer be maximized). On the other hand, a larger consumption level would make j worse off and even i because the latter would once again have his or her benefit lower than that at consumption level OQ (individual i would have to pay more and more taxes). Any other i's consumption level is unable to make somebody better off without making someone else worse off.

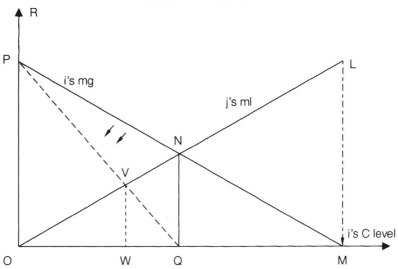

Figure 10.1 Gains and losses in the presence of externalities
Notes: R = revenue, Q = production

Therefore, i's consumption level OQ is the only Pareto efficient level in the presence of negative externalities being corrected by imposing a Pigovian tax.

However, one can imagine that the shape of i's marginal gain curve will not stay the same after taxation and will be shifted downward from PN to PQ as shown in Figure 10.1 (dashed lines). This suggests that a strict Pigovian tax does not attain Pareto optimality and therefore represents a suboptimal policy.[2] According to the new i's curve, j is made better off and i is always maximizing his or her gain (but only after deduction of taxes). Nevertheless, j may be willing to bargain OW or any other new consumption level. The optimal solution lies rather in the redistribution of the tax in order to compensate the victim fully as in the earlier bargaining process.

Let us consider how Just *et al.* (1982) proceed when dealing with social optimality in the presence of both beneficial and detrimental externalities. Once again, it is crucial to distinguish between social and private marginal costs and benefits. A marginal social cost (MSC) is the sum total of a marginal private cost (MPC) and an incidental cost, which is externally borne, and is not reflected in the marketplace. In other words, if a firm increases the smoke that it emits by increasing its output, or if an individual increases the loss that an increased consumption imposes on another individual, then in addition to their private costs as recorded in the company or in the household accounts, expansion of either that production or that consumption imposes external costs on other economic agents (increased laundry bills, ugliness, and so on) which are all part of the economic activity's marginal cost.

Suppose that a firm generates external diseconomies. In this sense, its marginal social cost will be greater than its marginal private cost, or in symbols: MSC > MPC. Since the firm is supposed to be a profit maximizer, it will choose its output by equating its marginal benefits (MPB) and its MPC. But, at this point, those marginal benefits are smaller than the marginal social costs. It follows that society should be willing to tax that output in order to reduce it. In this case, the firm would face the marginal social cost associated with a reduced output. As a result, society is made better off. In other words, if the production firm paid all of the social costs of its output instead of shifting some of them to others (off-shore costs), its marginal cost curve would be no longer that of the marginal private cost but some new marginal social cost curve, as shown in Figure 10.2.

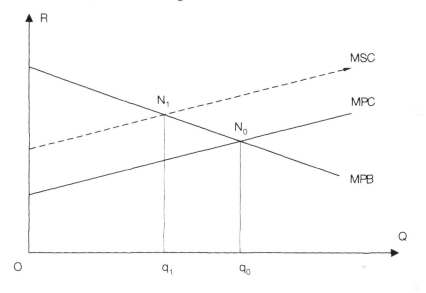

Figure 10.2 Marginal social cost is the sum total of marginal private cost and 'externalities'

Notes: R = revenue, Q = production

In general, one can say that where a firm's profit-maximizing output generates external diseconomies, marginal benefits are less than marginal social costs; therefore, smaller outputs than those given by free markets will be socially desirable (if private companies do not have to pay either the social costs or an equivalent tax, they will produce such output in undesirably large amounts). On the contrary, where a firm's profit-maximizing output generates external economies, free markets will turn out a much too small amount of that product whereas society would be better off with larger quantities; in this case, society may be willing to subsidize that activity.

If we are concerned with prices and compensating variation in terms of applied welfare economics, we can present private equilibrium and social optimality in the presence of both negative and positive externalities as follows. As an example, Figure 10.3 shows MSC and MSB in addition to MPC and MPB associated with a free market output. Private equilibrium must satisfy the required conditions of equating MPC, which represents competitive supply, with MPB, which indicates competitive demand; that is, at price p_0 and quantity q_0. On the other hand, a social optimum is obtained at the intersection of MSC with MSB; that is at price p_1 and quantity q_1. That reveals the presence of detrimental as well as some beneficial externalities.

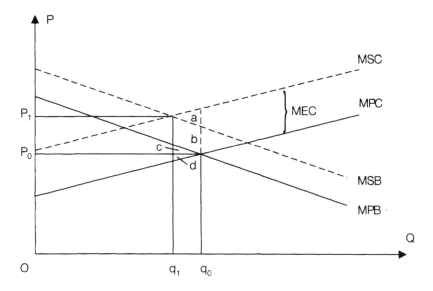

Figure 10.3 Definition of marginal external cost (MEC)
Notes: P = price, Q = production

In Figure 10.3, the vertical distance between MSC and MPC represents marginal external costs, or 'the sum of changes in compensating variation (or equivalent variation as the case may be) with respect to a change in q over all those individuals affected negatively by the increase in q' (Just *et al.* 1982: 273). For precisely analogous reasons, the vertical distance between MSB and MPB indicates marginal external benefits, or the sum of changes in compensating variation or equivalent variation with respect to a change in q over all of those individuals affected positively by an increase in q. As a result, by moving to a social optimum, individual profit-maximizing firms and utility-maximizing consumers jointly lose area c+d; other producers or consumers who suffered external diseconomies gain area a+b+c+d, and those who obtained external economies now lose area b. Therefore, society finally gains area a although it is now facing a decreased quantity q_1 and increased price p_1. But, where marginal external benefits exceed marginal external costs at the

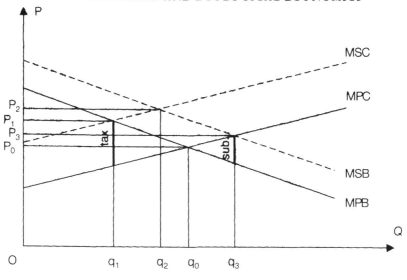

Figure 10.4 Tax-subsidy scheme
Notes: P = price, Q = production

private equilibrium, the social production can be larger than the private one. In the above case, as represented in Figure 10.4, a full Pigovian tax would be determined by the vertical distance between MSC and MPC (i.e. at price p_1 and quantity q_1), and a full Pigovian subsidy would be determined by the vertical distance between MSB and MPB (i.e. at price p_3 and quantity q_3).

In general, the following is important to point out: as the rate of any diseconomy is reduced (and not necessarily completely cut out), marginal external costs are also reduced while marginal market benefits, as represented by the vertical difference between the MSB curve and the MPC curve, outweigh the remaining marginal external costs (i.e. MSC – MPB at price p_2 and quantity q_2 in Figure 10.4).

2.4 Discussion

Unfortunately, all of the above solutions come up against the difficulty of gathering information about any damage function associated with externally affected agents. In addition, it may even be quite difficult to determine all externally affected individuals. Finally, although the theory deals with individual firms or consumers in competitive microeconomics, the external effects may be quite general and hence more public and indivisible effects than private and excludable ones. As argued generally, it is often impossible to allocate private property rights to environmental effects (and *a fortiori* environmental resources), such as the atmosphere, open waters, underground reservoirs covering extensive ground, and so forth. Therefore, correcting market prices often does not allow the market to attain either efficiency or environmental care. A special case is the analysis of externalities as public goods or services, which may be considered as intermediate between the analysis of market externalities and the study of environmental externalities, the task in the next section. For the sake of clarity, we prefer to switch now from incidental externalities to our next analytical step, dealing with pervasive externalities.

3 ENVIRONMENTAL EXTERNALITIES

3.1 Hypotheses

Market external effects are primarily incidental effects indirectly affecting profit and utility functions of economic agents. Theoretically, welfare policies such as compensating variations give rise to meeting Pareto optimality in the presence of such externalities. In reality, if assisted markets may attain both efficiency and environmental care, this is true only where externalities are occasional, market-oriented, and short-term ones. In the very reality, external effects are no longer incidental effects in so far as pervasive social costs and permanent, long-lasting environmental damages (real or monetary) generally occur. Moreover, owing to the scarcity of environmental resources and services, conflicts take place with respect to market/environment allocation of materials and short-range/long-term decision-making. Environmental externalities address, therefore, the cases where actions of economic agents affect the production possibilities of the economy and hence the flow of goods and services that individuals can enjoy in a way which is not reflected in the marketplace but which lies in real, long-range, non-economic terms, with or without accompanying monetary costs. This definition denotes the emerging problem that production and consumption are intimately involved with the real, biophysical dimension of resources, goods, and services.

The 'new' scarcity of environmental resources (Hueting 1980), the permanency of environmental threats, and the accompanying long-term dimension lead to alternative modes of economic and social control in the absence of environmental resource and service markets. In particular, they lead to the production and use of accounting prices (physical units), so as to improve market efficiency and environment management.

The concept of environmental externality has been designed to facilitate the switch to this new paradigm. Its analysis goes to accounting prices of environmental resources entering and leaving the economic process, from abstract models of environmental externalities still belonging to welfare economics. In sum, environmental externalities allow us to deal with the environment as a quasi-economic sector, with extended national, and natural patrimony, accounts, and with accounting, shadow prices.

Nevertheless, we will give less emphasis to this analytical step than to the two others, though this paradigm is really maturing right now. It will first be shown why the environment can be considered, at least to some extent, as a quasi-sector of the economy. Then special attention will be given to the theoretical background of extended accounting models.

3.2 Environment as a quasi-sector of the economy

Individuals are not merely satisfied by buying and selling goods in the marketplace. Their satisfaction also depends upon nonmarket services or disservices associated with the environment. In addition, the emerging view on this subject is that the existence of such externalities is inherent to the economic use of materials and energy – though perhaps once it was not. In other words, unlike market externalities, which were potentially internalized in one way or another, environmental externalities are produced as are economic goods and services. Finally, they have to be integrated into the general economy more than they have to be internalized in a Pareto optimal market equilibrium. This is the very reason why the environment is to be considered as a new, quasi-sector of the economy. First, old concepts such as optimal pollution level, damage function, and welfare function are to be rearranged according to this new paradigm. Next, abstract models of environmental externalities can be built.

Optimal pollution level, damage function, and welfare function revisited

Let us briefly consider the case of a permanent production of pollutants globally affecting the economy. First, pollutants threaten the environment. Next, they generate monetary damage to individuals and production firms as additional expenses and/or diminishing gains, and social costs (e.g. health costs). Finally, environmental protection measures are undertaken in order to counterbalance some damage.

Going straight to the point, one can easily model this figure by plotting on the same diagram (see Figure 10.5) two basic functions:

(a) The monetary cost of environmental damage at different pollution levels.
(b) The cost of environmental improvement according to the above pollution levels; with costs in $ on the y-axis and pollutant concentrations on the x-axis.

As a result of the trade-off between the economic cost of environmental damage (D-curve) and that of environmental improvement (I-curve), a third relation can be diagrammed (I'+D'-curve) whose minimum indicates an optimum pollution level (C^* for $-\Delta D = \Delta I$) from the narrow, economic perspective that we applied here.

A formal statement regarding environmental externalities and welfare economics may be adapted from Forsund (1984), as follows:

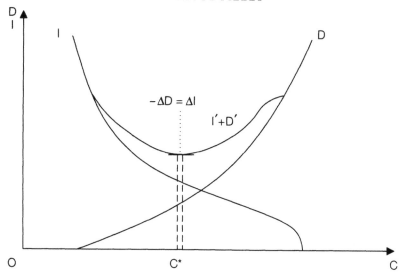

Figure 10.5 Trade-off between economic cost of environmental damage (D-curve) and that of environmental improvement (I-curve)

(16)   W[C, R (Z = f[X, C])]

where   W = welfare function;
        C = consumption;
        R = environmental externalities;
        Z = residuals;
        X = production.

Equation (16) is read as follows: welfare is derived from consumption of human-made goods (C) and environmental goods and services (R). The quality or quantity of R are influenced by residuals (Z) generated by production of goods (X) and consumption (C). This kind of conceptual framework points to abstract models of environmental externalities.

Abstract models of environmental externalities

An early attempt to formulate the whole problem from an economic viewpoint seems to have been that of Daly (1968). According to him, we should consider a simple input–output model figuring some basic relationships between the main economy and the environment (see Figure 10.6). The first sector of this matrix (sector A) represents the economy acting upon the environment, for example in rejecting a huge variety of waste materials. Sector B focuses on the economy as a classical production/consumption ageless machine subject only to effective demand. Money is the 'working fluid' of this

	ECONOMY	ENVIRONMENT
ECONOMY	B	A
ENVIRONMENT	C	D

Figure 10.6   Economy–environment relationships (Daly's model)
Source: Pillet (1986)

machine and any transaction between individuals or firms is subject to an accounting identity between income and expenditures plus any changes in stock (savings, reserves). Sector C deals with the direct supply of natural resources and with waste assimilation from the environment as well as adverse effects upon the economy. Accounting prices are addressed in this sector. Sector D focuses on long-term environmental work. Unlike sector B, which is concerned with economic goods and services, sectors A, C and D deal with environmental goods (zero price on a market basis) or bads (negative price) – though negative prices are non-observable ones referring to damage functions. This model straightforwardly shows the inherent presence of environmental externalities with respect to the general economy.

Other attempts put emphasis on input–output models in which the environment is treated as a distinct sector subject to what has been known as the materials/energy balance principle. This principle states that the laws of conservation of matter and energy 'guarantee that the sum total of all waste flows to the environment from the economy must be equal to the sum total of all resources originally extracted from the environment' (Ayres 1978: 31). This rule makes sense with respect to sectors C plus A of the Daly model (Figure 10.6). The rule applies to the main economy as well as to regions, individual communities, industries, firms, or environmental protection agencies. As an example of a general equilibrium model including environmental externalities, the Ayres–Kneeze model is a production–consumption input–output model associated with an environmental sector, both of them being subject to a balance of all physical flows (see Figure 10.7). Two successive transformation matrices are involved: a resource/commodity matrix ([R] in Table 10.1) and a commodity/product matrix. Goods are distinct from services; the latter involve no physical inputs. Goods produced by the economic sector are priced as are raw materials extracted from the environment. Waste flows from the final consumption carry a negative price (based, for example, on a fee for disposal). In other models (Mäler 1974), they constitute a third matrix ([W] in Table 10.1) and environmental externalities are priced on a physical accounting basis. Recycling is allowed for. This kind of model has been the foundation for the extensions of national accounting (satellite accounts) and welfare measures at issue in the preceding two or three sections.

161

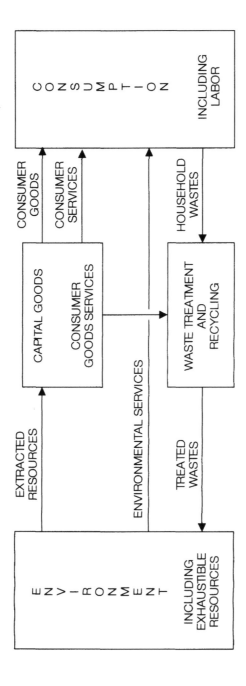

Figure 10.7 Input–output model associated with an environmental sector (Ayess-Kneese model)
Source: Pillet (1986)

3.3 Extension of the economic paradigm to include environmental links: towards extended accounting models

In classical economics, a single stage mapping is required, which goes from factors of production to marketable goods and services. This stage is optimized by means of some functional relationship between economic output and factors of production – land, labour, or capital. From a neoclassical or standard economic perspective, optimization is concerned with a two-stage transformation: from factors of production to goods and services, and then to utility or welfare. In addition, the theory of welfare economics recognizes the possibility of substitution between different goods and services with regard to not only market-allocated but also nonmarket-allocated goods and services (i.e. public goods, as well as economic and/or environmental externalities). This sequence is optimized by reaching competitive equilibrium and Pareto optimality in both production and consumption. Yet, from the viewpoint of a materials/energy balance principle, the economy is a set of unidirectional transformations 'that convert raw materials and natural resources (both renewable and non-renewable) into "final" goods and services' (Ayres 1978). In this sense, economic products are not physically 'consumed', but provide 'services' before being discarded as partially recyclable wastes.

Therefore, it seemed appropriate to develop environmental accounting within a somewhat more elaborate economic paradigm: 'A more complex (but also more realistic) sequence of mappings from exhaustible or renewable natural resources to finished materials and forms of energy; then to material products, and structures (this includes capital goods); then to (abstract) services, and finally utility' (Ayres 1978:67). In this enlarged economic paradigm, the first two stages imply physical transformations (and physical accounting). The latter two transformations are not physical in nature, but their inputs include labour as well as material products generated in the previous stage. However, as regards services, they are non-material ones. It follows that what comes in as material inputs now is converted into wastes. Finally, utility is maximized as in welfare economics. In general, according to the 'physical' principle at work in this revised economic paradigm, each given level of utility or each finished material can be derived from various processes, each different from the other and not from a unique, irreducible set of materials, or energy inputs. Environment appears now as a new, quasi-sector of the economy, becoming part of standard accounting models. Accounting prices – ratios of use of material/energy/pollutant – are derived from a basically similar enlarged paradigm. A special index, namely EROI for Energy Return On Investment, has been developed in some detail by Hall *et al.* (1986). This is the ratio of the energy returned to the energy invested to discover, drill, refine, etc. the energy resources; for example, the ratio of the gross amount of fuel extracted in the energy transformation process to the economic energy required to make that fuel available to society. EROI may be considered as an accounting price in so far as energies put in this ratio are comparable; that is, if they share the same thermodynamic quality. As an example, EROIs have been declining in the US since 1940 for metals, since about 1968 for domestic oil and coal production, and since about 1974 for imported oil. These declines indicate the trend of the trade-off between efficiency, outputs, and costs in the extractive sector ([R] in the above models). Energy as such cannot be used instead of money as a unit of account. However, energy intensities calculated from energies that are comparable give rise to off-shore price-ratios that may be used for possible extensions of national or sectorial accounting. These price-ratios are just unpaid prices; yet, if they reflect some general agreement, they may become public prices. We will say more on energy calibration in the next section.

3.4 Discussion

In this section, it has been suggested how the concept of environmental externality could be used as the common denominator to extended accounting of the quasi-economic sector 'environment' in the absence of resource and service markets. However, we have shortened the exposé to the theoretical background of this so-called enlarged economic paradigm.

This paradigm has not yet passed environmental economics. This leads us to our next analytical step with every confidence that there is still the opportunity to bring constructive controversy to this field.

4 ENERGY EXTERNALITIES

4.1 Prerequisites

The 'eMergy' method, or Odum's conceptualization of embodied energy theory (Odum 1983a, 1986), aims at aggregating energy and building an energy quality hierarchy fitting ecosystems and ecologic–economic bridged systems analyses. As a third analytical step of our externality approach, this section focuses on the eMergy accounting of the external, non-priced contributions to economic processes from ecosystems work (Pillet 1986, 1988, 1993 Pillet and Odum 1984, 1987), using two new units to interlock the study of the latter with that of the former. First, eMergy is used to perform energy calculation for such diverse components as soil, water, fuels, and economic goods and services. Second, monergy is used to price eMergy flows in the macroeconomic domain. Although the matter is still partly controversial, it can be explained and understood inasmuch as it is important to grasp ecological economics on its externality side. We will duly treat the questionable components of this less familiar externality method.

Energy externalities may be defined in the first instance as ecologic–economic oriented external effects; that is, as indirect contributions to economic activity from ecosystems work or services. On the one hand, their real basis is external to market decision-making. On the other hand, they grasp the interface which relates external contributions or limits from the environment to the main economy. We can cite solar energy, including rain and wind, water, topsoil, and natural life-support ecosystems generally as energy externalities.

The analytical idea behind the eMergy method is that at each step of the energy-chain which produces 'higher quality' energy from 'lower quality' resources – such as from sun to plants to animals or from coal to heat to electricity – much of the energy is used in the transformation so that only a small amount is converted into the higher quality energy. This higher quality energy is more concentrated and can accelerate other flows when fed back or used as a catalyst.

The concept of successive energy quality transformations may be illustrated by means of ecological food chains despite the fact that real systems form webs rather than chains; however, the energy changes are similar (see Figure 10.8). At each step, only a small amount of energy is transformed to a higher quality whereas much is used in the transformation. As a consequence, declining energy through the system is accompanied by increasing quality, or transformity – which is a measure of this quality[3] (Figures 10.8a, b). On the one hand, flows of low-quality energy are abundant and widely dispersed, with distinct units small in size (Figure 10.8c). On the other hand, flows of high-quality energy are more concentrated, with individual units larger in size (Figure 10.8c). This drives the arrangement of sources and components of ecosystems as well as of ecologic–economic systems. Finally, the *eMergy* of any (form) energy at any step of such systems or bridged systems is calculated by multiplying their actual energy content by their proper (solar) transformity, which is obtained by the eMergy analysis of that form of energy (see Figure 10.8d). *Em joules* (emJ) are the units of eMergy. The symbol for eMergy is C.

4.2 Establishing shadow macro-prices: energy externalities in ecologic–economic bridged systems

Typical interfaced relationships between environmental work and economic processes can be viewed as shown in Figure 10.9. They relate high-quality information as a processor of lower-quality energy, fitting it with an imported, purchased, higher quality one. Feedback controls are of prime importance in order to optimize the efficiency of the whole bridged system.

For coupling economic activity to environment, a feedback loop is thus needed. Indeed, a simple connection means that the economic activity drains some of the product, as fisheries or forestry do (Figure 10.9a). Continuing interfaced relationships require a feedback loop (Figures 10.9b, c) which amplifies the chain that is used inasmuch as the use is a drain. Traditional

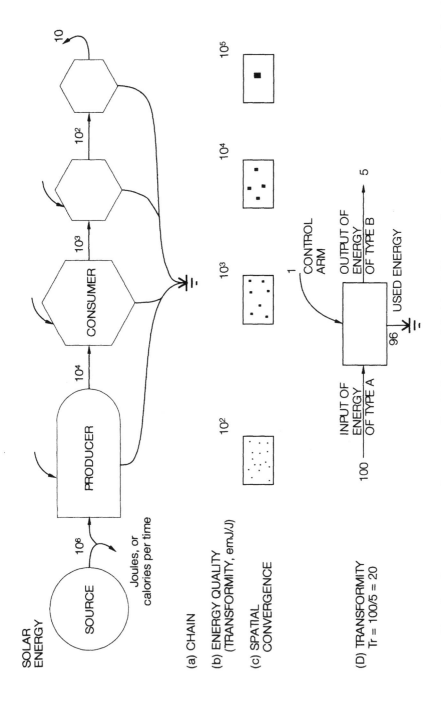

Figure 10.8 Transformity as a measure of energy quality (Odum's model): (a) and (b) successive transformations of work in a food chain; (c) spatial convergence; (d) definition of transformity
Source: Pillet (1986)

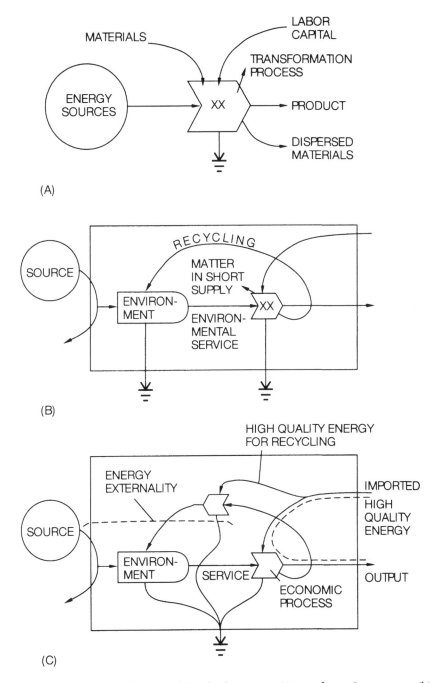

Figure 10.9 Ecological–economic bridged systems: (a) transformation process; (b) recycling; (c) energy externality
Source: Pillet (1986)

agriculture and oyster cultures generally satisfy such feedbacks that keep the chain of use competitive, and economic activity stable. On a larger scale, the *suido* control feedback realizing this relationship between forests, rivers and human society during Japan's seventeenth century is another example (Murota 1984).

Going back to Figure 10.9, it is obvious that we can emphasize the control of the economy from an energy system viewpoint and not merely as a drain from environment to economy. In addition, economic prices do not provide sufficient information in an ecologic–economic model. They are often as far away from the realities of any environmental control of the economy as they are from controls from the economy into the environment. Regarding a whole system and human labour involving energy-work, any work/labour interface is intimately involved in transformations from low-quality to higher-quality energy. So the real basis of economies is outside the money measure but participates in the economic output; that is, as energy externality. Thus, maximization must consider the whole system and not only the resulting dollar. One can thus consider as energy externalities the ultimate non-priced sources that generate economic processes through pathways of action that are indirect and even unrecognized, as shown in Figure 10.9c. They deal with the evaluation, by means of indices and ratios, of external environmental contributions to the main economy or of economic miscontributions regarding the environment. Finally, the question is: how much do energy externalities contribute to an economic outcome?

4.3 Case study of the role of environment as energy externality in ecologic–economic systems

Using the eMergy method, let us consider the analysis of the role of environment and the measurement of energy externalities based on three case studies, as follows:

(a) Vineyard cultivation and wine production in Switzerland.
(b) Sugar-cane and sugar processing to ethanol in Louisiana.
(c) Rice-growing and saké-making in Japan.

EMergy calibration

Swiss vineyards cover around 14,000 ha (less than one per cent of the farming area in Switzerland). The must (or unfermented wine) yield is over one million hl per year. Seventy-five per cent of the area and 85 per cent of the yield are located in western, French-speaking Switzerland. Data used are for 1983, mainly from the Lake Geneva Region (Pillet 1987, 1988).

The eMergy method is used first to evaluate all of the flows of the system in eMergy units, then in dollar flows using the monergy of the country (Pillet and Odum 1984). Energy flows of the Swiss vineyard production and the

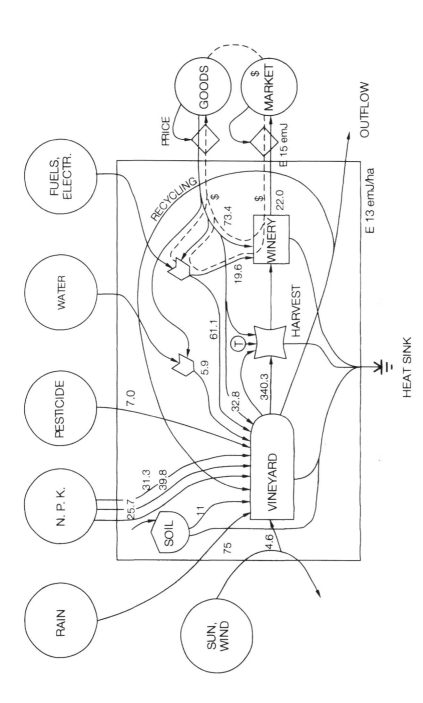

Figure 10.10 Energy diagram of Geneva vineyard cultivation and wine production
Source: Pillet (1988)

processing of wine are portrayed in Figure 10.10. Each pathway is calibrated in E13 emJ/ha/yr. Calculations appear in Table 10.2.

Table 10.2   EMergy flows in vineyard and wine production, Switzerland, Lake Geneva Region

Form of energy	Actual energy (J/yr)	Solar transformity (emJ/J)	EMergy (1 E 13 emJ/J)
1 Direct solar energy	4.6 E 13	1.0	4.6
2 Rain	5.0 E 10	1.5 E 4	75.0
3 Soil used up	17.6 E 8	6.24 E 4	11.0
4 Organic matter	1.02 E 9	6.24 E 4	6.4
5 Irrigation water	3.95 E 8	1.5 E 5	5.9
6 Nitrogen	15.2 E 7	1.69 E 6	25.7
7 Potassium	11.93 E 7	2.62 E 6	31.3
8 Phosphate	9.61 E 6	4.14 E 7	39.8
9 Pesticide	10.64 E 8	6.6 E 4	7.0
10 Direct fuels	9.25 E 9	6.6 E 4	61.1
11 Steel in machinery	1.81 E 7	1.01 E 7	18.3
12 Iron wires, stakes	3.25 E 7	1.01 E 7	32.8
13 Services	$ 6 E 3/ha/yr	0.12 E 12 emJ/$	72.0
14 Vine grapes yield	4.74 E 10	7.18 E 4	340.3
15 Sugar added	6.66 E 9	8.39 E 4	55.9
16 Electricity	1.23 E 9	15.9 E 4	19.6
17 Water	1.11 E 8	1.5 E 5	1.7
18 Capital	$ 2.3 E 4/ha/yr	0.72 E 12 emJ/$	16.6 E 15
19 Services	$ 1.02 E 3/ha/yr	0.72 E 12 emJ/$	73.4
20 Wines yield	3.3 E 10	6.7 E 5	22.0 E 15

Source: Pillet (1988)

First, actual energy of each form of input is measured in joules (column 2). Second, the transformity (emJ/J) of each category of flow is reported (or calculated; e.g. grapes and wine in this study) in column 3. Finally, the eMergy (emJ per surface unit per time) of each energy flow is obtained as the product of the actual energy multiplied by the transformity attributed to the category under consideration (column 4). EMergy values are ultimately reported on the vineyard and wine energy diagram (Figure 10.10). An aggregated three-armed diagram, calculated from data in Table 10.2, is used to help inter-pretation (Figure 10.11).

The same procedure has been applied to case studies of the production of ethanol in Louisiana (Odum and Odum 1983) and of Japanese saké (Pillet and Murota 1990).

In particular, energy externalities clearly appear on the energy diagram in Figure 10.11 (I). They are expressed in E13 emJ/ha/year; in this case, I represents the rain and the soil. It is calibrated to 86 E13 emJ/ha/year. The used soil is evaluated as 11.1 E13 emJ/ha/year. Along with water and wind, its work is indispensable for the continuity of the economic product.

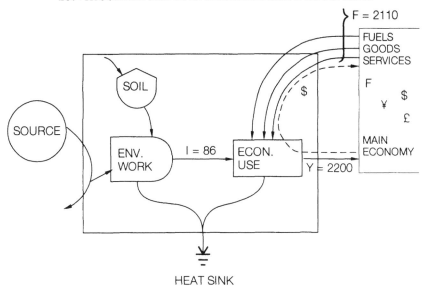

Figure 10.11 Aggregate view of Geneva vineyard cultivation and wine production
Source: Pillet (1988)

Ratios and indices

The aggregated diagram on Figure 10.11 allows us to define various ratios:

Net eMergy yield ratio Y/F   This ratio is the energy of the output (Y) divided by the eMergy of the market inputs (F). The results are as follows:

Swiss wine	1.04
Louisiana ethanol	1.05
Japanese saké	1.06

This measured the high percentage of purchased inputs with respect to the product yield.

EMergy investment ratio F/I   This ratio is the ratio of the eMergy fed back from the economy (F) to the eMergy inputs from the natural ecological system (I). The results are as follows:

Swiss wine	24
Louisiana ethanol	21
Japanese saké	17
US agriculture	25

This confirms the intensive use of high-quality energy inputs in comparison with the use of low-quality, environmental ones.

Energy externality ratio I/(I+F)   This ratio (xE) is the ratio of use that is free. It is given as follows: xE = I/(I+F). The results are:

Swiss wine	0.04
Louisiana ethanol	0.04
Japanese saké	0.06

This is much more than economists usually attribute to environmental services, arguing that economic macro-products are composed of as much as 99 per cent of embodied human labour.

Macro-prices Environmental macro-prices (i.e. shadow prices) can be calculated for typical energy externalities according to the monergy of the country.

With respect to the production of Swiss wine, energy externalities can be evaluated at SFr 18,000/ha/year using the monergy of the free environmental contribution to the Swiss economy (0.12 E12 emJ/$ 1983) (Pillet and Odum 1984). This is less than the annual capital flow/ha (about SFr 58,000) but more than the flow of services spent per hectare and per year in vineyard and harvest (SFr 15,000). To sum up, one could say that water and soil, considered as energy externalities, contribute approximately 15 per cent of the economic expenses in vineyard cultivation and wine production per hectare per year, in Switzerland.

4.4 Discussion

One main point to be discussed here could be about the uncertainty of the quantitative information used for or issued by shadow-pricing energy externalities. Thus, let us try to qualify these numbers in a more responsible manner for their effective use in policy-making, and let us do that according to the Funtowicz–Ravetz NUSAP notational system (see Funtowicz and Ravetz 1988). By means of this account (NUSAP stands for Numeral, Unit, Spread, Assessment and Pedigree), the state of the art of the production of a quantity can be well characterized.

We have, for N:U:S:A:[P], with A (confidence) being codified from 4 (total) down to 0 (none):

Energy	1: E6 J: E3: 4: [established theory/field data/total acceptance/consensus]
Transformity	1.5: E4: E2: 3: [theoretically based model/field data/ medium acceptance/competing schools]
EMergy	75: E13 emJ: E6: 3: [theoretically based model/calculated data/medium acceptance/ competing schools]
Energy externality	4: %: ±0.5: 1: [theoretically based model/calculated data/medium acceptance/embryonic field].

This is given in order not to isolate quantified expressions from their partly controversial, though still relevant and qualifying properties.

4.5 Conclusion

Ecologic–economic systems interlock natural life-support ecosystems providing free services for humans, and human-made subsidies of materials and energy provided via the marketplace. We have shown that it was possible to derive from the neo-classical externality approach quite different approaches for achieving a solution at this interface. First, environmental externalities have been thought of as being a common denominator for possible extensions of national accounting. Second, the energy externality method has been used to calibrate the flow of free services provided by natural life-support ecosystems to economic processes.

How much does the environment contribute in free services for humans through economic processing? This was the task of shadow-pricing indirect, environmental contributions to an economic process. In short, pricing market externalities led to corrected market prices, pricing environmental externalities to accounting and public prices, and pricing energy externalities to macro shadow-prices. Thus, we have in a complementary way:

(a) Micro-prices, for market consideration, so as to act at the market-economy level.
(b) Public prices, resulting from collective agreements, so as to act at a political level.
(c) Macro-prices, having sense in environmental economics and in ecological economics, so as to act at the interface between the main economy and life-support ecosystems.

To sum up, starting from a basic, neo-classical relationship, the subject has branched out into quite diverse applications. None the less, once this basic principle was well understood, the common problems of these different applications and methods could be seen through the externality sieve. This is what we did. In this way, we made the externality approach an interdisciplinary one. We believe that this can help economists and environmentalists to have a common interest in ecological economics.

5 NOTES

1 The Lagrangian multiplier is the dual variable associated to the constraint. It equals the incremental change in value from an incremental change in the constraint parameter: λ represents the marginal value of relaxing the kth constraint.
2 This statement is made according to our interpretation of Giffen's paradox which suggests a potential downward shift of i's marginal gain curve after taxation (Pillet 1984).
3 The term 'transformity' names a ratio measuring the quantity of A required for a quality of B (see Figure 10.8d). It has been hypothesized that transformity was a measure of both efficiency and maximum power according to natural selection, but this has never been measured.

6 REFERENCES

Ayres, R.U. (1978) *Resources, Environment and Economics*, New York: Wiley.

Coase, R.H. (1960) 'The problem of social cost', *Journal of Law and Economics* 3:1–44.

Daly, H.E. (1968) 'On economics as a life science', *Journal of Political Economy*, May/June: 392–406.

Forsund, F.R. (1984) 'The introduction of environmental considerations into national planning models', in A.-M. Jansson (ed.) *Integration of Economy and Ecology – An Outlook for the Eighties*, Stockholm: Askö Laboratory, University of Stockholm.

Funtowicz, S.D. and Ravetz, J.R. (1988) 'Managing the uncertainties of statistical-information', in J. Brown (ed.) *Environmental Threats: Social Science Studies, Risk Perception and Risk Management*, London: Pinter.

Hall, C.A.S., Cleveland, C.J. and Kaufmann, R. (1986) *Energy and Resource Quality – The Ecology of the Economic Process*, New York: Wiley.

Hueting, R. (1980) *New Scarcity and Economic Growth – More Welfare Through Less Production?*, Amsterdam: North-Holland.

Just, R.E., Hueth, D.L. and Schmitz, A. (1982) *Applied Welfare Economics and Public Policy*, Englewood Cliffs, NJ: Prentice Hall.

Maeler, K.-G. (1974) *Environmental Economics – A Theoretical Inquiry*, Baltimore, MD: Johns Hopkins University Press.

Meade, J.E. (1973) *The Theory of Economic Externalities*, Genève: Institut Universitaire des Hautes Etudes Internationales.

Murota, T. (1984) 'Heat economy of the water planet Earth: an entropic analysis and the water–soil matrix theory', *Hitotsubashi Journal of Economics* 25 (2): 161–72.

Odum, H.T. (1983a) *Systems Ecology – An Introduction*, New York: Wiley.

Odum, H.T. (1986) 'Enmergy in ecosystem', in N. Polunin (ed.) *Ecosystem Theory and Application*, New York: Wiley.

Odum, H.T. and Odum, E.C. (1983) *Energy Analysis Overview of Nations*, Laxenburg, Austria: WP, IIASA.

Pillet, G. (1984) *New Proposals in the Theory of Economic Externalities and Applied Environmental Economics – A Report*, Gainesville, FL: Environmental Engineering Sciences and Center for Wetlands, University of Florida.

Pillet, G. (1986) 'From external effects to energy externality: new proposals in environmental economics', *Hitotsubashi Journal of Economics*, 27 (1): 77–97.

Pillet, G. (1987) 'Case study of the role of environment as an energy externality in Genève vineyard cultivation and wine production', *Environmental Conservation* 14 (1): 53–8.

Pillet, G. (1988) 'Water, wind and soil: hidden keys to the water planet Earth and to economic macroprocesses', *The Energy Journal* 9 (1): 43–52.

Pillet, G. (1993) *Economic Ecologique*, Gèneve: Georg Editeur.

Pillet, G. and Murota, T. (1990) 'Shadow pricing the role of environment as an energy externality in Geneva wine, Louisiana sugar-cane-alcohol, and Japanese saké', unpublished paper, Genève: Centre of Human Ecology and Environmental Sciences, University of Geneva.

Pillet, G. and Odum, H.T. (1984) 'Energy externality and the economy of Switzerland', *Schweiz. Zeitschrift für Volkswirtschaft und Statistik / Swiss Journal of Economics and Statistics* 120 (3–4): 409–35.

Pillet, G. and Odum, H.T. (1987) E^3 – *Energie, écologie, économie*, Genève: Georg Editeur.

Turvey, R. (1963) 'On divergencies between social cost and private cost', *Economica* 309–13.

7 FURTHER READING

Section 2

Baumol, W.J. and Blinder, A.S. (1979/1983) *Economics: Principles and Policy* (ch. 31, 'The economics of environmental protection'), New York: Harcourt Brace Jovanovich.

Pillet, G. (1980) 'Joint production of external diseconomies', *Economie appliquée*, 33 (3–4): 651–2.

Section 3

Amir, S. (1987) 'Energy pricing, biomass accumulation, and project appraisal: a thermodynamic approach to the economics of ecosystem management', in G. Pillet and T. Murota (eds) *Environmental Economics – The Analysis of a Major Interface*, Genève: Leimgruber.

Gaudard, G. (1987) 'Regional economic development and the future of environment', in G. Pillet and T. Murota (eds) *Environmental Economics – The Analysis of a Major Interface*, Genève: Leimgruber.

Paruelo, J.M., Aphalo, P.J., Hall, C.A.S. and Gibson, D. (1987) 'Energy use and economic output for Argentina', in G. Pillet and T. Murota (eds) *Environmental Economics – The Analysis of a Major Interface*, Genève: Leimgruber.

Section 4

Lavine, M.J. and Butler, T.J. (1982) *Use of Embodied Energy Values to Price Environmental Factors: Examining the Embodied Energy/Dollar Relationship*, Ithaca, NY: Center for Environmental Research and Department of Environmental Engineering, Cornell University.

Lotka, A.J. (1922) 'Contribution to the energetics of evolution' and 'Natural selection as a physical principle', *Proceedings of the National Academy of Sciences* 8 (6): 147–54.

Odum, H.T. (1983b) 'Maximum power and efficiency: a rebuttal', *Ecological Modelling* 20: 71–82.

Odum, H.T. (1984) 'Energy analysis of the environmental role in agriculture', *Advanced Series in Agriculture* 14, Berlin: Springer.

Odum, H.T. and Pinkerton, R.C. (1955) 'Time's speed regulator: the optimum efficiency for maximum power output in physical and biological systems', *American Scientist* 43: 331–43.

Pillet, G. and Murota, T. (1987) *Environmental Economics – The Analysis of a Major Interface*, Genève: Leimgruber.

Scienceman, D.M. (1987) 'Energy and eMergy', in G. Pillet and T. Murota (eds) *Environmental Economics – The Analysis of a Major Interface*, Genève: Leimgruber.

11

THE SELF-ORGANIZATION OF ECOLOGICALLY SOUND BEHAVIOUR

Hans-Joachim Mosler

1 SEEKING A THIRD SOLUTION

It can be assumed that the environmental problems so often cited do not refer to problems just between people and the environment, but much more to those between members of a social system who have an impact on the environment. Ecology-oriented behaviour is always social behaviour, because ecological resources are a common good, that is, they are meant for the use and the enjoyment of all. Each time a common resource is used its condition, and hence the possibility for others to use it, is altered. Similarly, environmental problems arise in connection with a social conflict of interests: every individual is interested in getting maximum use from a resource, while the diminished usefulness or damage thus incurred must be borne by all (cf. Spada and Opwis 1985).

In the field of social research, these socially based environmental issues are approached in two ways: using surveys and using experimental conflict situations. The surveys are generally aimed at finding decisive factors for environmental awareness, such as socialization within the family, schooling, the effect of mass media, the influence of personality variables and of personal attitudes and values (cf. Langeheine and Lehmann 1986; Schahn *et al.* 1988; Urban 1986). Experimental conflict situations seek to investigate the conditions under which cooperative and resources-appropriate behaviour is revealed, such as the influence of the social characteristics of a group, the effect of possibilities for communication, the influence of social orientations and of social institutions (cf. Dawes 1980; Edney 1980; Wilke *et al.* 1986).

As it is not expedient here to describe in detail the results of surveys and of research in conflict situations, I would like instead briefly to take a closer look at the two sociological solutions to the problem of promoting ecologically sound behaviour, as often cited in political discussions in the public realm:

1 Government measures and ordinances aimed at influencing behaviour toward the environment.

2 Changes in the individual's values, attitudes, and so on, aimed at calling forth ecologically sound behaviour.

Both solutions have accompanying disadvantages which call into question their effectiveness. If government measures cannot be decreed in a dictatorial fashion, the decision-makers are required to use democratic means. The almost inevitable result is that solutions which would definitely benefit the environment cannot be employed because the decision-makers have to satisfy everyone (e.g. when speed limits were introduced in Switzerland, it was not possible to enforce a dramatically low one). The wish to bring about changes in the individual's values, attitudes, and so on presupposes an image of a human being who is easily swayed. It would appear that politicians in responsible positions, judging by their readiness to commission PR firms with environmental protection campaigns, have just such an image of the average individual. In this context it must be borne in mind that there is a definite difference between trying to sell people a product and trying to influence them to behave in an ecologically responsible way. This is because ecology-oriented behaviour is certainly guided more by attitudes and values than are purchasing habits. However, even if it were possible to change people's values, this is still a long way from ensuring a change in their behaviour. Other behavioural factors would also have to be considered in this case (e.g. the hierarchy of values and attitudes, willingness to act, living situations, habits).

In addition to these two potential solutions, that is, government measures and changes in individual attitudes and values, a third solution is desirable whereby, without regulations and attempts to change people's values, the individuals on their own would adopt a common, that is, collective, ecologically sound mode of behaviour.

2 THE PROCESS OF SELF-ORGANIZATION IN SYNERGETICS

There are a number of spontaneously occurring, collective behaviours that can be observed in everyday life, for example, the way in which a large number of people move along in a narrow street devoid of cars. If at the outset there are only a few people on the street, they each move in whatever way they choose. As more and more people enter the scene, a variety of modes of forward motion emerges, e.g. walking in a zig-zag, walking in single file. However, there is almost always an abrupt change to walking together in the same direction on one side, that is, a crowd of people flowing along. The stronger such a moving stream of people becomes, the fewer the number of people who do not move as part of it. Without being ordered to do so, or agreeing upon it among themselves, these moving streams of people organize themselves quite automatically.

177

There is a new branch of science concerned with such phenomena of self-organization: synergetics, the study of interactive processes. Synergetics is viewed as the science of ordered, self-organized collective behaviour (Haken 1986: 21), within which according to Probst (1987: 14) self-organization is a meta-concept for the understanding of the emergence, maintenance and development of order patterns. Synergetics seeks laws which describe how well-ordered structures emerge from the interaction of very many individual parts. Concepts from the science of synergetics can be adapted profitably in a variety of natural and social sciences, for example in physics, chemistry, biology, sociology, or economics (cf. Haken 1983). Initial research into self-organization can be traced to various authors, especially Prigogine, Eigen and Winkler, Maturana and Varela, as well as Haken (a more detailed listing can be found in Probst 1987, chap. 2). As it was Haken more than any other who most consistently investigated the application of synergetics in the social realm, the following remarks are based on publications by him and his collaborators (Haken 1983, 1984, 1986; Haken and Wunderlin 1986; Wunderlin and Haken 1984).

Generally speaking, within this type of self-organization process the following phases can be distinguished.

Phase 1: Destabilization of the existing condition

Self-organized systems are made up of homogeneous elements (units, subsystems, micro-structures) whose condition is described by means of micro-variables. The decisive feature of the elements of such a system is that they constantly influence one another in some way or other.

Self-organized systems are always open systems, that is, they exist in an exchange of energy and matter with their environment. In order for an existing system state, for example, the resting state, to become instable, the system must be fed with energy from the outside or the number of individual elements must be increased. In both cases, each element of the system experiences an activation, because the number of interactions is increased.

Phase 2: Phase transition with fluctuations

If the system reaches a so-called point of instability it is subject to fluctuating energies, which lead to strong fluctuations in the state of the system. At this point a variety of ordered states come into being. These fluctuations allow the system to 'try out' various conditions, until a new, stable condition can take over. By means of fluctuations the system 'recognizes' whether, in view of its energy condition, it really is situated in a minimum, or whether even deeper minima exist. In many cases the newly emerging condition is not clearly determined a priori. A symmetry can exist between different, equivalent states, that is, there exist equivalent minima. In this realm of phase transition, it is often the randomly occurring micro-fluctuations which decide which new state of stability the system takes on in the end.

Phase 3: Slaving and the emergence of macro-structures

In Phase 2 various order patterns are 'competing for domination' in the system. With their behaviour the elements join on to the order pattern which under the given conditions is the most appropriate one for them.

Each individual element that joins itself and its behaviour with an order pattern contributes to the maintenance of this pattern, as well as strengthening it and exciting other elements to behave in just such a way. If more and more elements join on to an order pattern, this pattern will be massively strengthened and drive out all other models: it will become the macro-structure of the system. However, there are also systems in which various patterns co-exist and are mutually stabilizing. It is even possible for order patterns to cooperate, thereby building up increasingly complex systems.

In the context of microvariables and order parameters the slaving principle becomes clear. (The irritating term 'slaving' has been retained purposely by Haken [1984] because he believes that it best expresses what is meant in the social context. Perhaps 'unification' [of behaviour] would be a more propitious choice of term.) The behaviour of a subsystem is described as well as prescribed by means of the order pattern and thus via the order parameter, however, the order pattern is only made possible through the collective behaviour of the subsystems. Each element that joins the collective behaviour contributes to the maintenance of the order pattern, whereby it is reinforced and, for its part, prompts the individual elements to behave in a collective manner. The process of self-organization can only be initiated at all by means of the retroaction of the collective behaviour on the behaviour of the individual elements. For each element, the only real possible behaviour is the dominant one, which the macro-structure acquires, with the result that the order pattern unifies the behaviour of the elements.

The three phases of system organization can be distinguished in the example of the surging crowds. The destabilization of the 'people on the street' system results from a steady increase in the number of people. At first a number of different order patterns can emerge, for example, walking in a zig-zag; however, the most suitable form of forward motion, walking in a throng, tends to assert itself because under the given conditions (narrowness of the street) the act of meeting head-on is minimized by the necessity of making way for others. Each person who enters into the throng becomes part of it and strengthens it, thereby making it more difficult for other people to move forward in any other way.

The extent to which models from the scientific field of self-organization can be useful in the social sciences will be discussed first in a theoretic sense in the next section, and then further elucidated in the following one with the aid of examples.

3 ADAPTING SELF-ORGANIZATION MODELS TO SOCIOLOGICAL QUESTIONS

In an article dealing with scientific theory, Druwe (1988) investigates the possibility of transferring self-organization models to sociological questions. He developed transferability criteria for quantitative and qualitative models.

In order to transfer the formal quantitative self-organization model, structural isomorphs must exist between 'natural' (in the sense of natural sciences) and 'social' self-organized phenomena, in such a way that the 'natural' non-linear stochastic equilibrium systems are able to describe social phenomena. In order to transfer qualitative models it is necessary to proceed according to the principles of linguistic precision and the adequate inter-pretation of definitions, that is, the notionalities which stem from the natural sciences must be filled with empirical sociological contents. To avoid working with pure conceptual shells, Druwe first requires an (empirical) search for macroscopic variables, order parameters, etc., followed by a mathematical analysis with the aid of self-organization principles.

Weidlich and Haag have fulfilled the requirements for the quantitative description of social processes. The starting point for their mathematical analyses was provided by the following synergetic concepts (1983: 10 ff):

1 Human society is an open system because it exists in the interaction between its external environment and its technological surroundings.
2 The individuals in a society can be considered as elements. These elements are not all alike, of course; however, their attitudes (behaviour modes, etc.) classified in certain subgroups, can be considered alike, that is, comparable with one another, even though they may be based on extremely complex mental processes. In this sense, the attitude toward a political party can be the result of the various mental considerations, emotional preferences and motives of one person, while remaining at the same time a measurable and comparable attitude. This means that it is useful to investigate the con-ditions under which the average attitude of a group of persons evolves as a macrovariable, even if the details of how the individual attitudes evolved are not known.
3 The concept of a socio-configuration is introduced, which describes the distribution of attitudes of subpopulations of the society, making it possible to draw up a useful set of macrovariables.
4 It can be assumed that attitude changes by the members of a society provoke changes affecting the entire society. The dynamics of a social system arises from individual, interrelated changes.
5 Social self-organization systems have a self-consistent (self-preserving)

180

structure, which is dependent on a combining, a cyclical retroaction, of causes and effects in the society: by means of their cultural and economic activities the individual members of a society contribute to the setting up of a general 'field' of civilization. This collective field determines the socio-political atmosphere and the cultural and economic standard and can be considered as the order parameter that characterizes the phase in which the system finds itself. The collective field in turn strongly influences the individuals by directing their activities, activating or deactivating their latent qualities and abilities and by expanding or reducing the scope of their thoughts and actions.

6 From the feedback mechanism between collective field and individuals emerge self-accelerating and self-saturating processes, which normally lead to stable system states. However, on the basis of inner or outer interactions the order parameters can absorb certain values which lead the system to a critical area where often very differing yet equivalent paths are open to the social system. In this situation of phase transition the microfluctuations – e.g. the actions of a few influential people – determine the direction in which the behaviour of the society evolves.

It can be seen from these conceptual considerations that it is possible to assign sociological meanings to synergetics terminology borrowed from the natural sciences. A number of possible applications of the formation of synergetic models in the social sciences will be presented briefly in the following section.

4 EXAMPLES OF THE FORMATION OF SYNERGETIC MODELS FOR SOCIOLOGICAL QUESTIONS

Weidlich and Haag drew up various quantitative models, for example, a stochastic model for the formation of public opinion (1983, chap. 2; abridged version also in Haken 1983, chap. 11). They take as a starting point the probability distribution of two different opinions, such as the opinion for and against the death penalty. They determine the transition probability from one opinion to the other by inserting into a mathematical model of the ferro-magnet the following microvariables, which could play a role in the changing of an opinion:

(a) The adaptation of one's own opinion to that of one's neighbour, as a measure of the individual's tendency to take on the opinions of others.
(b) The individual preference for an opinion, as a measure of indifference to a topic.
(c) The dominant social climate (e.g. tolerance), which favours or hinders the changing of opinions.

The probability distribution function can now be described with a differential equation, whose solution can be investigated using various forms of the

microvariables. This leads, for example, to the result of a polarization of the society into two groups with differing opinions when the personal opinion is strongly adapted to that of one's neighbour, or when a change in opinion is hindered by the social climate, for instance, great intolerance (as in the case of totalitarian societies). However, in the case of a high degree of intolerance, two equally common opinions occur when for the microvariable of 'individual preference for one opinion' there are no differences between the persons, that is, when no one in a society specially favours one opinion. As soon as there is even a very slight preference for one opinion, however, the frequency distribution is heavily displaced for a long period in favour of the preferred opinion. If the adaptation to a dominant opinion is slight, that is, the dominant social climate favours the changing of opinions – as in liberal societies – a balanced distribution of opinions in the population occurs.

Haken (1986, chap. 13) provides examples of qualitative models by investigating various social phenomena on the basis of synergetic terms. Only his comments about the formation of public opinion are presented here.

He uses as his starting point a measurable opinion structure, for example, a frequency distribution of the opinion for or against the current government. He explains the occurrence of such an opinion structure in the following way. As revealed by conformity research, people often personally adopt the views of the majority. Thus individual views can be influenced, and the mutual capacity to be influenced is, in the synergetic sense, the root of all collective effects, and hence is operative in the formation of public opinion. Haken and also Noelle-Neumann (1982) name as the source of this capacity to be influenced the individual's fear of isolation (rejection), that is, the fear of being cut off from fellow human beings and ignored by them. A publicly expressed opinion that is generally accepted will, according to Noelle-Neumann, influence other people to accept this opinion, because of their fear of rejection. Thus, this mutual susceptibility to being influenced results in a type of reinforcement, that is, a growing number of people will adopt this opinion, which will continue to snowball, until a dominant public opinion emerges. This dominant public opinion now plays the role of regulator (order pattern), which enslaves the opinions of individuals and is in this way itself able to survive.

The influence of the mass media and of the government also play roles, of course, in the formation of public opinion. The printed mass media are regulators under pressure to select. They influence the opinions of the readers, but are themselves dependent on the collective purchasing behaviour of these readers. For reasons of competition and because of the limited resources available they are forced to make a preselection, in such a way as to guarantee their own continued existence. Even a government can be seen as a regulator which influences the opinions of the people, but which for its part is also influenced by public opinion.

In response to the question of how a change in opinion occurs, Haken offers the following explanation: changes in opinion have to be prepared for by external circumstances, that is, the system must be destabilized from the outside, for example, by changes in the economic situation. People finding themselves in a state of instability are of the opinion that something new has to happen, whereby the 'what' is often unclear. In such a situation the objectives of a small group of determined people can often indicate a direction, or unpredictable, local events can acquire a strong emanating force, enabling them to bring about a new ordered structure.

A related question, namely, how a firmly established, environmentally harmful behaviour structure can be transformed into an environmentally respectful behaviour structure, will be pursued in the next section. First, however, the question should be posed as to what is to be gained from the synergetic viewpoint on sociological situations. The advantages are the following:

1 The most widely differing social phenomena can be explained with the use of a standardized principle. Such differing phenomena as the formation of a public opinion, the functioning of a newspaper or a government, in the role of a regulator under pressure to select, can be understood using the 'slaving principle'. The principle abstracts from the concrete example and points out the common element of the examples.
2 Different conditions and developments of a social system can be explained by varying only a few variables. For example, the distribution of two opinions can be explained on the basis of the variables.
3 A concrete heuristic can be drawn up now for self-organization phenomena that are widespread in the social realm. This means that precise data regarding the factors to be considered now exist to aid in the investigation of social self-organization phenomena. These factors include, for example, the interplay between individuals as well as the existing social climate.
4 It is now possible to derive more widely reaching implications for self-organization phenomena. One implication is that in order to effect a change in behaviour, measures must be taken which trigger all the phases of the synergetic change in the system. This explains why measures carried out in isolation do not lead to success.

5 SYNERGETIC CONCEPTS FOR ENVIRONMENTAL PROBLEMS

It will now be set out how, by applying the synergetic self-organization model to the ecological behaviour of individuals in a society, new light is shed on environmental problems and possibilities thereby opened up for new problem-solving methods.

Let it first clearly be stated that order parameters in the realm of ecological behaviour can be estimated:

(a) In many cases scales have been developed for ecology-oriented values, attitudes, willingness and actions, which make it possible to name mutually comparable average values as macrovariables for the population (cf. Kley and Fietkau 1979; Langeheine and Lehmann 1986; Schahn et al. 1988; Urban 1986).

(b) It is also possible to measure directly macrovariables of ecological behaviour, that is, without having to depend on personal statements. For example, it is possible to weigh how many tonnes of waste are separated by a community.

The central concept of synergetics – the standardization of the behaviour of the individual elements by the order pattern – now applies exactly to the connection between individual and collective use of an environmental resource. The behaviour pattern of the collective in the use of a resource greatly influences the behaviour of the individual, because this determines the cost–benefit ratio for the individual resource use; if many people over-use a resource, the individual must also over-use it, otherwise, in addition to the damage caused by over-use, which affects everyone, he or she also has a smaller individual benefit. The over-using collective behaviour permits (purely in the material sense) only over-using behaviour by the individual. This means, for example, that the individual must also drive a car when others are driving, otherwise in addition to bearing the damage caused by noise and polluted air, he or she also benefits less in the form of diminished mobility. Through this effect of the collective behaviour on the behaviour of the individual the collective behaviour pattern maintains itself on its own, at least for the time being, because it is made up of the behaviours of individuals. The members of an ecologically over-using society 'trap' one another in this over-using behaviour in that each person hinders the other from behaving in a worthwhile, alternative way.

In order to break out of such a vicious cycle, the following steps, to be integrated in a solution strategy, are required by the synergetic self-organization model:

(a) Reinforce destabilization in the system.
(b) Increase fluctuations in the system.
(c) Promote the spread of a new form of behaviour.

5.1 Reinforcing destabilization in the system

A destabilization of the system, 'ecology-oriented behaviour of the members of a society', seems to occur through 'external effects', that is, changes in the environment brought about by the system, which in turn affect the system (for example, air pollution). This is ascertained again and again in surveys, in which the majority of those questioned claim to hold 'green views' ('people are thinking green'). A change in opinion seems to have taken place; what is missing is a 'change in behaviour'. It is also possible to destabilize the dominant behaviour structure if the individual loses confidence in his or her preferred mode of behaviour. The worsening environmental conditions (for the influence of individual affectedness, see Hippler 1986) as well as a social climate in which hardly anyone admits openly and directly to environmentally harmful behaviour are already having an effect in this direction.

Changing the boundary conditions of individual behaviour possibilities represents a further way to destabilize the established behaviour structure. Government measures, in the form of behaviour incentives, prohibitions or related laws, etc., would certainly be an appropriate means for making the current preferred behaviour pattern seem unattractive. I have already described the difficulties in introducing government measures. Haken (1986: 112) too described this situation using the synergetic self-organization model: although in a democracy the government, in the role of regulator, can influence the behaviour of the individual, it in turn is dependent on the votes of these individuals. The political parties as well, in competition for the votes of the electorate, will adapt their proposed environmental policies to suit a wide spectrum of the population (cf. Haken 1986: 173), thereby watering them down. This means that government measures, although necessary for the destabilization of an existing behaviour, are hardly likely to suffice.

5.2 Increasing fluctuations in the system

There exist the following indications of a certain instability in the dominant environmentally over-using modes of behaviour, that is, there are areas in which alternative forms of behaviour also exist:

(a) There is an alternative living culture.
(b) Demonstrations are held for ecology-oriented living styles.
(c) In some areas of activity (e.g. used-paper collection) behaviour is environmentally sound, in others not (e.g. car driving).

Fluctuations can thus be ascertained, yet the system, 'ecology-oriented behaviour of the members of a society', tends to remain stable, which in the sense of Haken (1983: 219) means that the fluctuations are too slight to effect a full transition of the system to another state. The variables responsible for this are known from various surveys: higher income and laziness are the main elements which make environmentally harmful behaviour seem attractive.

If the fluctuations are to be reinforced, it would surely be reasonable to comply with the demands of a few ecological psychologists (e.g. Fietkau and Kessel 1981), namely to offer good alternative behaviours, so that each person has the opportunity to try out environmentally sound modes of behaviour.

On a collective societal level this would mean evolving into an 'experimenting society' as required by Fietkau (1985), thus enabling one, in the synergetics sense, to become acquainted with the advantages of other system states.

5.3 Promoting the spread of a new mode of behaviour

It was shown in the two preceding sections that the system, 'ecology-oriented behaviour of the members of a society', is already unstable and subject to certain fluctuations. As the development of a system cannot be predicted on the basis of the instability points, what is needed here is groups of people who behave in an ecologically sound way, regardless of what the majority is doing, and thereby are able to serve as a focal point of ecologically sound behaviour for the others. In order for an alternative, ecologically sound behaviour to become widespread as well as dominant, two conditions must be fulfilled, according to the synergetic model of self-organization:

1 Ecologically sound behaviour must, at least in the longer term, be the appropriate behaviour, so that the individual's preferences will influence his or her selection of this behaviour rather than of another. This selection can, via the mutual reinforcing effect, trigger a mushrooming of ecologically sound behaviour. Which behaviour is appropriate in the short term will be determined by conditions named in the previous sections (e.g. government measures, choice of alternative behaviours).

2 Ecologically sound behaviour ought to influence others (that is, inspire them) to behave likewise. This reciprocal influencing of elements of a system is, according to Haken (1986: 178), the fundamental prerequisite of self-organization.

Members of society already do influence one another when it comes to environmental issues; otherwise there definitely would be no changes in opinion in favour of ecology-oriented thinking, as has been established positively in surveys. However, this influencing seems to be limited to opinions only and does not include ecology-oriented behaviour. It seems that in societies in which thousands of people live anonymously next to one another, the ability to influence each another is rather weak in some areas of activity critical to the environment. Which individual is able to know, or even better see, beyond the limits of a small neighbourhood, who is behaving how, in terms of the environment?

Ecologically sound behaviour is only worthwhile for the individual when the vast majority of a collective also behaves in this way. In a city whose air is polluted primarily by car exhaust, the individual who forgoes the use of his or her car benefits in the form of clean air only if the majority of car drivers join him or her in not driving. This means that individuals would have to carry out this transition to a collective ecologically sound behaviour together. Only when many individuals together change their behaviour do they achieve a benefit through their changed behaviour. If only a few people change their behaviour, they do not benefit at all. When the transition to a collective ecologically sound behaviour has to be carried out together it means that each individual is very much dependent on the others, as he or she relies on them to 'participate' also.

For situations in which there is great dependency of individuals on unknown other persons, trust in the behavioural strategy of these anonymous other persons is indispensable. Trust engenders solidarity. Trust can give rise to the willingness to bridge the gap of uncertainty concerning the actions of others (cf. Luhmann 1968, chap. 5; Lewis and Weigert 1985).

It will be demonstrated in the section that follows how an operating mechanism for the formation of trust can achieve a mutual influencing between individuals in an anonymous society toward ecologically sound behaviour. Mutual reinforcement could lead to the mushrooming of eco-logically sound behaviour, from which an ecologically sound collective behavioural pattern could emerge.

6 VERIFIABLE PUBLIC COMMITMENT AS THE TRUST BASIS FOR ECOLOGICALLY SOUND BEHAVIOUR

6.1 Sociological theories

Various sociological theories concerning trust are drawn from in order to create an operating mechanism for the formation of trust, which is meant to bring about the mutual influencing between anonymous individuals of a social system. These, briefly described, are as follows.

Deutsch's concept of 'social trust'

The term 'trust' encompasses for Deutsch (1958; 1973, chap. 7) both the motivational relevancy of a trust-oriented mode of behaviour for a person, and the predictability of the behaviour of other persons. In his view, trust within the framework of interpersonal relations represents a special case of the general attitude toward the environment. In the interpersonal situa-tion, he emphasizes that a person is aware of the trust of others, and that the trusting person feels that others are obliged, to a certain extent, to fulfil this trust. This is the origin of mutual trust, whereby each party assumes that others are aware of the intentions and the trust of those involved in the relationship, and behave accordingly. He concludes from his investigations of the trust relationship that statements of one's views on the following topics are important for the building of trust: one's own inten-tions, one's own expectations, and one's way of reacting if the trust is betrayed.

187

The concept of trust as a generalized expectation, according to Rotter

Proceeding from the theory of social learning, Rotter (1967, 1971, 1980) states that individuals tend to generalize their experiences of whether promises are kept or not on to their behaviour toward others. This means that a person is more likely to trust others if he or she has had the experience that promises generally are kept. Thus, Rotter defines trust as the expectation that one can place one's trust in the word of another. Based on this definition, Rotter has developed the so-called 'Interpersonal Trust Scale'.

Luhmann's personal 'procedural' concept of trust

Luhmann (1968, chap. 6) names the following prerequisites for the building of trust:

(a) Human behaviour must be seen as the behaviour of individual persons, because only in this way can it be linked to a person.
(b) The relationship must be initiated by means of a number of small acts, which may not be precisely measured and reciprocated immediately.
(c) The opportunity for breach of trust must be at hand, but may not be used. Trust cannot develop in a risk-free relationship.
(d) Familiarity is essential, or, when this is missing, the motivational structure of the other must be clear, that is, one person reveals what benefits the other could obtain from the relationship, or what sanctions the other would experience if he or she were to violate the trust accorded to him or her.
(e) It is beneficial to the relationship if one knows that it will extend over a longer period of time.
(f) Alternating dependency of the parties on one another benefits the formation of trust.
(g) A certain unpredictability of the advantages of situations is advantageous to building trust.

In spite of all of these possibilities to establish criteria for placing one's trust in another, trust will always remain an 'overdrawing of the information that one possesses from the past, and on the basis of trust one takes the risk of determining the future' (Luhmann 1968: 18).

The sociologically oriented concept of Luhmann, Barber and Lewis and Weigert

These authors regard trust not so much as an attitude of individuals, but more as a type of relationship between people. For them trust is more social and normative than individual and calculating. Trust exists in a social system when its members act in such a way that they expect an assured future, which arises from the presence or symbolic representation of each other person

(Barber 1983). This means that, according to this definition of trust, the feeling of having a secure place in a social system is at the forefront. This becomes clear within the framework of the three dimensions which Lewis and Weigert (1985) name as being the basis of trust:

1 Cognitive processes influence trust by deciding what and whom is trust-worthy, drawing upon familiarity and experience with others. Trust is regarded as a collective, cognitive reality, whereby each individual can overextend the information that he or she has about others, that is, the individual can abandon the basis of assured experience, because he or she may assume that other people also have trust and want to trust.
2 Every trusting relationship has an emotional basis, which is why the relationship is deeply affected if the trust is misused.
3 The fact that one person trusts another must find expression in his or her personal behaviour, because this helps to create or reinforce the emotional feeling of trust.

According to Luhmann (1968, chap. 4), trust serves to reduce the complexity of a social system. Trust strengthens the tolerance for ambiguity.

An operating mechanism for the formation of trust, which can function in anonymous relationships and refers only to behaviour relevant to society and not to personal acquaintanceship, must, in keeping with all of the concepts named, include the following determining elements.

Verification

In the case of a lack of familiarity with the anonymous other, trustworthiness must be established through means independent of the relationship. Legal foundations, possibilities for sanctioning or checking up on others seem, for someone who trusts, more likely to assure that the other will behave in a corresponding manner (Luhmann 1968, chap. 5). If one has no experience with the other, the possibility to check up on this person can decisively improve his or her credibility and reliability in behaving appropriately. Trust is only possible where truth is possible (Luhmann 1968: 48).

Publicness

Human behaviour must become visible to others as behaviour specific to the individual so that trust can be placed in an individual person (Luhmann 1968: 39). Trust-oriented actions help to establish or reinforce the emotional basis for trust (Lewis and Weigert 1985). 'Revealing oneself to be trustworthy' (Luhmann 1968: 62) by being open and showing that one has consideration for the expectations of others promotes an intensification and generalization of social relationships.

In addition to this trust-promoting effect of publicness, a display of (ecologically sound) behaviour has the effect of influencing others not merely at the time when the behaviour is carried out. This is very beneficial for the spread of such behaviour in a large, anonymous society.

Commitment

A trustworthy person is one whose actions are consistent with what he or she has communicated about him/herself. Self-revelation gives others criteria for the formation of trust and for the standardization of continuity expectations (Luhmann 1968: 37–8). This means that the person who wants to be trusted by others must consistently fulfil the expectations that he or she has aroused in them.

Trust also includes expectations of others (Rotter 1980). Actions which express trust include the expectation of others not to betray the trust placed in them. If, when interacting with another, a person undertakes a commitment whose outcome is dependent on the behaviour of others, he or she expresses through this commitment his or her trust in others. A one-sided commitment has the characteristic of appealing to others not to take advantage of the commitment and to behave in such a way that all can benefit from the interaction. If the benefit from an interaction is dependent on the goodwill of all, then a one-sided commitment expresses the expectation of reciprocity (Deutsch 1958), which is based on the need for social justice (cf. Equity Theory, Walster and Walster 1975).

6.2 Discussion

In summary, it was postulated in this section that a verifiable, public commitment can be assumed as the operating mechanism for the formation of trust. This mechanism could cause a mutual influencing between anonymous individuals of a society, which, together with a preference for ecologically sound behaviour, has a snowball effect, giving rise to a collective, ecologically sound behaviour pattern.

In the preceding sections it was shown how, with the aid of a synergetic viewpoint, it is possible to provide a new outlook on environmental problems. An important statement emerged, namely that the members of a society which exploits the environment 'trap' each other in an environmentally exploitative behaviour, in that each person, by his or her own behaviour, prevents the others from pursuing a worthwhile alternative behaviour. The promotion of the spread of ecologically sound behaviour was worked out as a significant approach to a solution. In order to achieve this, a mutual influencing of the society members toward ecologically sound behaviour, proceeding from the focal point of those who do behave in an ecologically sound way, was considered necessary. It was postulated that, by establishing mutual trust between the anonymous individuals of a society, it would be possible to bring about mutual influencing toward ecologically sound behaviour, and 'verifiable, public commitment' was derived from the pertinent social theories as the operating mechanism for this process.

The last section deals with the problem of how to realize these theoretically derived ideas in a concrete way in a real society, and which supportive measures are likely to be needed to transform the existing collective, environment-damaging behaviour pattern into an ecologically sound one.

7 A STRATEGY FOR INTRODUCING A COLLECTIVE, ECOLOGICALLY SOUND MODE OF BEHAVIOUR

The general principles of synergetics offer an answer to the frequently posed question of 'Why is it that so many people think green, but so few actually behave that way?' The response is that ecologically sound behaviour is meaningless for the individual (at least in a material sense, perhaps in the philosophical sense) as long as most other people do not also behave in this way, and, since the same is also true for the others, people trap one another in this environmentally harmful mode of behaviour. 'Yes, but someone has to be the first,' is the remark that follows. This is true, there needs to be a group of people willing to behave in an ecologically sound way, regardless of what the majority is doing, thereby serving as a sort of crystallization point in the synergetic sense. However, such a group unfortunately fails to have much effect on the actions of others, is not able to influence the behaviour of the anonymous multitude.

This is where the most important strategy for the solution, based on the synergetic view of the issue, of (socially caused) environmental problems comes in: the spread of ecologically sound behaviour must be fostered by making it visible to the public, and making it a part of a commitment that includes an unspoken expectation to behave likewise.

This could be accomplished by means of a very simple and already known mechanism, that is, by identifying oneself by means of badges, stickers, etc. There exist already examples of the open demonstration of certain types of ecologically sound thought and behaviour, such as 'Nuclear power – no thanks!' or 'I drive 100 km/h maximum'. However, what is missing here is the crucial element of verification possibilities, as mentioned in an earlier section. The veracity of a public commitment must be guaranteed in order to be credible. Lack of credibility is reflected in counter-stickers, e.g. 'Nuclear power – no thanks. My electricity comes from a socket.' For this reason it is essential that the commitment expressed on the car sticker, 'I drive only every second Sunday', can be verified by means of a recording device or a similar and less complicated apparatus already available.

When, how and by whom these checks would be carried out could be decided upon by the participants themselves. Aversion to such inspections could be answered to by pointing out that one subjects voluntarily to inspection for the sake of upholding one's credibility. The existence of cheaters, free riders and opponents of such a campaign will not in themselves prevent the individual from taking part, as long as enough people participate to make it worthwhile for each one. There will, however, definitely be a need for shoring-up measures to support the individual in his or her ecologically sound behaviour during a transition phase when the majority of the members of society are not participating and there is as yet no evident benefit for the individual.

This leads to the two further components of a solution strategy based on the synergetic view of the problem, namely the reinforcement of destabilization and the increasing of fluctuations. When carried out together with the initial components, the potential, though not very effective, solutions mentioned in the introduction can be applied well: destabilizing the system means, among other things, the perplexing of the individual regarding his or her current mode of behaviour. This probably can be achieved through advertising campaigns if characteristics of the 'microsystem' human being are taken into consideration. Here it is a matter of the subtle interplay of attitudes and willingness to act within certain hierarchies of values. Similarly, the peripheral conditions for individual behaviour could be changed through government measures in such a way as to make the current modes of behaviour seem unattractive.

Fluctuations can be caused by enabling people to pursue a range of good alternative behaviours, so that it becomes easier for the individual to try out ecologically sound ways of acting. Certainly the best way to be supportive during this period of change would be to offer a combination of rewards and a range of positive alternative behaviours.

A total strategy could look something like this (greatly simplified): a core group of community members organizes the issuing and verification of a car sticker that states, 'I drive only every second Sunday'. Anyone who commits him/herself to do this will receive a bonus coupon for use on a greatly expanded public transportation service.

This all appears quite simple but no doubt, in addition to problems of feasibility, difficulties would arise, unknown to us at present, relating to the human element. It would not be tenable, in an ethical or in a material sense, to attempt to put such a theory into practice without broadening our scientific knowledge of the social mechanisms of self-organization.

There are still many 'ifs' and 'buts' connected with this idea; however, it appears worthwhile to continue to think about and to research self-organization mechanisms, not least in order to prevent a commitment to environmental matters from degenerating into an imposing of rules and regulations, and to promote the seeking of ways in which the members of a society can achieve a mode of interaction which is meaningful and beneficial to all.

8 REFERENCES

Barber, B. (1983) *The Logic and Limits of Trust*, New Jersey: Rutgers University Press.

Dawes, R.M. (1980) 'Social dilemmas', *Annual Review of Psychology* 31: 169–93.

Deutsch, M. (1958) 'Trust and suspicion', *Conflict Resolution* 2: 265–79.

Deutsch, M. (1973) *'The resolution of conflict*, New Haven and London: Yale University Press.

Druwe, U. (1988) '"Selbstorganisation" in den Sozialwissenschaften: Wissenschaftstheoretische Anmerkungen zur Uebertragung der naturwissenschaftlichen Selbstorganisationsmodelle auf sozialwissenschaftliche Fragestellungen', *Kölner Zeitschrift für Soziologie und Sozialpsychologie* 40: 762–75.

Edney, J.J. (1980) 'The commons problem: alternative perspectives', *American Psychology* 35: 131–50.

Fietkau, H.-J. (1985) 'Psychologische Aspekte umweltpolitischen Handelns', in P. Day, U. Fuhrer and U. Laucken (eds) *Umwelt und Handeln*, Tübingen: Attempto.

Fietkau, H.-J. and Kessel, H. (1981) 'Handlungsleitende Konsequenzen', in H.-J. Fietkau and H. Kessel (eds) *Umweltlernen*, Königstein: Anton Hain.

Haken, H. (1983) *Synergetik* (2nd edn), Berlin: Springer (English edition: *Synergetics*, Stuttgart: Teubner, 1973).

Haken, H. (1984) 'Can synergetics be of use to management theory?', in H. Ulrich and G.J.B. Probst (eds) *Self-Organization and Management of Social Systems*, Berlin: Springer.

Haken, H. (1986) *Erfolgsgeheimnisse der Natur*, Stuttgart: Deutsche Verlags-Anstalt (English edition: *The Science of Structure: Synergetics*, New York: Van Nostrand Reinhold, 1986).

Haken, H. and Wunderlin, A. (1986) 'Synergetik: Prozesse der Selbstorganisation in der belebten und unbelebten Natur', in A. Dress, H. Hendrichs and G. Küppers (eds) *Selbstorganisation*, München: Piper.

Hippler, H.-J. (1986) 'Determinanten des Umweltbewusstseins in einer belasteten Grossstadt', in R. Günther and G. Winter (eds) *Umweltbewusstsein und persönliches Handeln*, Weinheim: Beltz.

Kley, J. and Fietkau, H.-J. (1979) 'Verhaltenswirksame Variablen des Umweltbewusstseins', *Psychologie und Praxis* 1: 13–22.

Langeheine, R. and Lehmann, J. (1986) *Die Bedeutung der Erziehung für das Umweltbewusstsein*, Kiel: Institut für die Pädagogik der Naturwissenschaften an der Universität Kiel.

Lewis, J.D. and Weigert, A. (1985) 'Trust as a social reality', *Social Forces* 63: 967–85.

Luhmann, N. (1968) *Vertrauen: Ein Mechanismus der Reduktion sozialer Komplexitäten*, Stuttgart: Enke.

Noelle-Neumann, E. (1982) *Die Schweigespirale*, München: Piper.

Petermann, F. (1985) *Psychologie des Vertrauens*, Salzburg: Müller.

Probst, G.J.B. (1987) *Selbst-Organisation*, Berlin: Parey.

Rotter, J.B. (1967) 'A new scale for the measurement of interpersonal trust', *Journal of Personality* 35: 661–5.

Rotter, J.B. (1971) 'Generalized expectancies for interpersonal trust', *American Psychologist* 26: 443–52.

Rotter, J.B. (1980) 'Interpersonal trust, trustworthiness, and gullibility', *American Psychologist* 35 (1): 1–7.

Schahn, J., Holzer, E. and Amelang, M. (1988) 'Psychologische Beiträge zur Ermittlung und Beeinflussung des Umweltbewusstseins bei Erwachsenen', in F. V. Cube and V. Storch (eds) *Umweltpädagogik*, Heidelberg: Edition Schindele.

Spada, H. and Opwis, K. (1985) 'Ökologisches Handeln im Konflikt: Die Allmende

Klemme', in P. Day, U. Fuhrer and U. Laucken (eds) *Umwelt und Handeln*, Tübingen: Attempto.

Urban, D. (1986) 'Was ist Umweltbewusstsein? Exploration eines mehrdimensionalen Einstellungskonstruktes', *Zeitschrift für Soziologie* 15: 363–77.

Walster, E. and Walster, G.W. (1975) 'Equity and social justice', *Journal of Social Issues* 31: 21–43.

Weidlich, W. and Haag, G. (1983) *Concepts and Models of a Quantitative Sociology*, Berlin: Springer.

Wilke, H.A.M., Messick, D.V. and Rutte, C.G. (eds) (1986) *Experimental Social Dilemmas*, Frankfurt: Lang.

Wunderlin, A. and Haken, H. (1984) 'Some applications of basic ideas and models of synergetics to sociology', in E. Frehland (ed.) *Synergetics: from Microscopic to Macroscopic Order*, Berlin: Springer.

12

SUBJECTION OBJECTED

Dagmar Reichert

We did not do very much today. Down by the river, lying flat on our faces on the smoothly polished boulders, we watched the currents. I can do this for a long time. Wild water smashing against the rocks, currents whirling, swirling around a stone, white, green, blue, not one colour really, flowing, rushing water, released force, endlessly beating against its banks... 'Down at the end of the valley', K. said, 'it is dammed.' Damned! Why do they have to catch it in a dam, take away all its life, the splashing, the currents... all smoothened to slack water. Tamed water. Water caught in concrete pipes. Taken. Funnelled into a network of wires. Reliable conduct. Standardized units. Energy, accessible for everyone. How would our society work without it? But this river here. Its wild flow... the way it changes with the light, with the weather, the way how its sound rises in the night...

subject vs. object...[1] can man's destruction of the environment be reduced as long as we define ourselves as the other of nature... is there a more humane society possible as long as we see ourselves as that which the others are not... dominating forms of human self-understanding... written down in the grammar of our language the idea of our sciences the logic of our

The relationship between the subject and the object is that of negation. In its background stands the classical metaphysical identity of thinking and being. Classical logic is based on the idea of relating logical forms (which correspond to forms of being) according to rules which correspond to the relations between the forms of being. Hence the axioms of logic are the basic laws of being, first of all: there exists what is identical to itself.[2] Our contemporary thinking of the human being as 'subject' and even the very idea of being able to capture human being in thought are still part of this tradition.

laws the relations between men and women the functioning of the economy ... dominating form of thinking that defines the subject by its difference from the other which in the course of this becomes the object... breathless object... the one lying below... or at the edges ... in the periphery... and then... fucking morality of helpfulness... subject looking after... down at the object... charitably avoiding its own emptiness... subject I... Iye...

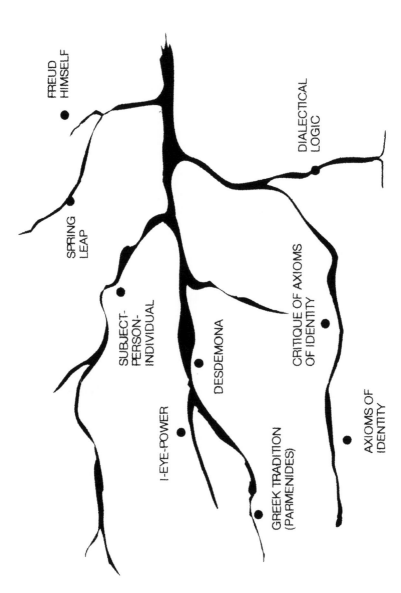

Figure 12.1 Streams of thought

Bentham's Panopticon is the model of modern power: 'Its principles are well known: in the peripery there is a circular building and in the middle there is a tower with large windows opening to the inner side of the circle. The circular building is separated into little cells. ... Each of these cages is like a small theater in which the actor is all by himself, completely individualized and always visible. ... Those subjected to visibility and knowing of this subjection take over the sanctions of power and direct them against themselves; they internalize a relation of power in which they play both roles at the same time; they become the originators of their own subjection.

(Foucault 1977: 256, 260)

In the western tradition, confession has become the most important technique for the production of truth. ... One confesses one's crimes, one's sins, one's thoughts and desires, one's illnesses and troubles. ... One confesses in public and privately, to one's parents, teachers, doctor and those one loves; one admits to oneself, in pleasure and in pain, things it would be impossible to tell anyone else.... . Western man has become a confessing animal.

(Foucault 1983: 76)

and the eye you cannot see ... panopticon of absolute control ... absolution ... visual logic abstracting aspects towards the general... towards the trinitary generality of author-mediumreader ... or producercommodityconsumer or origintraditiondestiny or scienceknowledgepolitics ... to guarantee unbroken view ... penetrating line... way of truth ... topo-logic of clear and distinctive thinking ... identifying...

what a woman is... am I that name... why ask me... how could I know what I am... how could I say it... clearly... you are asking for a good purpose you say... fighting for equal rights ... 'we women' ... 'women are' ... not like the others ... definitely ... no ... sorry ... I cannot ... I cannot think therefore I am ...

Desdemona: Am I that name, Iago?
Iago: What name, fair lady?
Desdemona: Such as she says my lord did say I was.

(Shakespeare, *Othello*, Act 4, Scene 2)

topo-logical visual objectifying thinking if this is 'thinking' how can we think without degrading it isolating it binding it domesticating it strai(gh)tening it canalizing it normalizing it ... it ... the world wild life chora Id human being ... prior to all bad intentions ... out of the laws of form ... the upright form of the 'I' ... the transcendental signifier if you like ...

I will call the subject the general-, the person the specific-, and the individual the particular (singular) selfconscious being. (Frank 1986: 9, Frank *et al.* 1988: 25; trans. D.R.)

how did it come about ... this idea of identifying the human being ... identifying it in general and in particular ... as the I... as a subject... as a person... how

did it come about this idea that what a human being is could be thought, defined, understood... I remember one day my father told me a fairy-tale about this...

The Story of how the human being was turned into an 'I'

Once upon a time there was a rebel.[3] He happened to ask a question which provoked a powerful answer: the peculiar bond between epistemology and ontology. It holds that there exists what can be known, and that 'knowing something' is ... yes, is what only some people can claim to do. And these people said: 'Knowing something is seeing the truth. And truth is not something that

197

is simply there or comes to you all by itself. What do you think? You have to see it. Naturally not everybody can see it. It takes some distance, because truth is something very peculiar, it is "an idea". But don't worry, if you study carefully, you can learn how to see it. We can teach you, just come and have dinner with us.' And so it happened. Once upon a time in Greece.

It took quite a while, but then, on a stormy night in the Netherlands, the subject was born. It was an eye. It could have been an ear, but it was an eye. And the women said: 'What poor remains of a human being, a single, lonesome eye!' But the eye did not mind the loneliness at all. Rather be lonely than accept my doubtful origin, it said to itself. (Understandable. The rumour went that its father was an Italian archbishop and its mother a French lady-in-waiting.) Then something important happened. I don't quite know why, but it was the fault of these bourgeois who baptized it: 'Do you know that you are something special?' they asked the eye. 'You are a subject, and a subject can see the truth. Not everbody can see it, but you are not everybody. You are not immoral. There is nothing that interferes. Look around, there is the eternal order of nature in perfect clarity. You need not feel insecure, even if your parents do not take care of you. You are part of this wonderful order, and representing it clearly and distinctively is your role in it.' Then the eye was very proud. It did not believe anything it could not see itself. And it saw many things.

One day it saw something very immoral. 'What are you doing?' it asked the people. 'Dear eye,' they laughed, 'this is nothing for you. If you do this, you will turn short-sighted.' Then the subject became very angry. 'What damned freedom is that supposed to be? I can see the true order of nature, but I cannot see the disorder?'

The subject really was very angry. It decided that it wanted to grow up and organize everything itself. That was when it turned into a big 'I'. 'I am the I,' it said with a deep voice,' and I am the one who can see the truth. My truth. I can be as immoral as I like, but I also can be moral. If I am, it is because I will it. I am free.' Then the I went out. After a while it realized that being an I was quite exhausting. It was always running behind itself, it was lost and could not find where it had started, and worst of all, it did not know who it was any more.

quiet ... listen ... something is cracking ... *ding ding* ... as if the strands of a cable were bursting ... one after the other ... *ding* ... a cable yes ... the powerful bond between ontology and epistemology ... thinkingbeingbeingthinking ... must be that ... identity ... key concept in all the different versions of this bond ... ideas of identity cracking ... *ding* ...

The original truths of reason are what I call identical truths ...: 'Everything is what it is; and in whatever case, is A=A and B=B.' ... In general the principle of contradiction says: 'A sentence is either true or false'; this includes two true assertions: first, that ... a sentence cannot be true and false at the same time; second, ... that there is no intermediate position between the true and the false one, in other words, that it is impossible for a sentence to be neither true nor false at the same time. ... What is A cannot be Not-A. ... All this can be accepted, independently of any experiment or any indirect proof.'   (Leibniz 1915: 420ff, trans. D.R.)

The propositions thus arising have been stated as universal laws of thought. Thus the first of them, the maxim of Identity, reads: 'Everything is identical with itself, A=A; and, negatively, A cannot at the same time be A and not A. – This maxim, instead of being a true law of thought, is nothing but the law of abstract understanding. The prepositional form itself contradicts it, ... It is asserted that the maxim of Identity, though it cannot be proved, regulates the procedure of every consciousness. ... To this alleged experience of the logic-books may be opposed the universal experience that no mind thinks or forms conceptions or speaks in accordance with this law, and that no existence of any kind whatever conforms to it. (Hegel 1929/69: 136–7)

here ... there exists what has an identity ... identity is the basis of logical thinking to have an identity is being the same with itself and different from the other ... *ding* ... cracks in the monotheistic monuments of
science
patriarchy
cartesian
I ...

but here too ... in spite of its critical capacity ... there exists what moves reflectively in-itself and passes through moments of identity ... identity is what is not the same with itself but the same with the other and still different from it ... it is what is driven by dia-lectical contradiction... *ding* ... cracks in the course of history ... victim of its own hope-ful teleology ... but the old god is dead ...

Essence is mere Identity and reflection in itself only as it is self-relating negativity, and in that way self-repulsion. It contains therefore essentially the characteristic of difference. ...

Difference is (1) immediate difference, i.e. Diversity or variety. In Diversity the different things are each individually what they are, and unaffected by the relations in which they stand to each other. This relation is therefore external to them. In consequence of the various things being thus indifferent to the difference between them, it falls outside them into a third thing, the agent of comparison. This external difference, as an identity of the object related, is Likeness; as the non-identity of them is Unlikeness. ...

(2) [Everything] has an existence of its own in proportion as it is not the other. The one is made visible in the other, and is only in so far as that other is. Essential Difference is therefore Opposition; according to which the different is not confronted by any other but by its other. That is, [everything] is stamped with a characteristic of its own only in its relation to the other: the one is only reflected into itself as it is reflected into the other. Either in this way is the other's one other.
(Hegel 1929: 139–40)

when the bond between ontology and epistemology is breaking and identity cracking language obtains a different function ... writing as well as reading become different activities ... 'For the same is thinking as well as being' ... reinterpreting Parmenides is kneading language between the finger tips ... persu-

For the same are thinking as well as being' (Parmenides). The question of the meaning of this Same is the question of the active nature of identity. The doctrine of metaphysics represents identity as fundamental characteristic of Being. Now it becomes clear that Being belongs with thinking to an identity whose active essence stems from that letting belong together which we call the appropriation. The essence of identity is a property of the event of appropriation.

ading to trust and wait for the event... the time to stop remembering and begin to be alife ... ironically Heidegger himself was not patient enough with world history ... driven by the believe in one

We must experience simply this owing in which man and Being are delivered over to each other, that is, we must enter into what we call the event of appropriation.

How can such an entry come about? By our moving away from the attitude of representational thinking. This move is a leap in the sense of a spring. The spring leaps away, away from the habitual idea of man as the rational animal who in modern times has become a subject for its objects.

[This...] would bring the appropriate recovery – appropriate hence never to be produced by man alone – of the world of technology from its dominance back to servitude in the realm by which man reaches more truly into the event of appropriation.

(Heidegger 1969: 39, 36, 32, 37)

god ... the god of the promised land ... and still there is war ... how difficult to be patient ... how difficult to trust in those who do not trust themselves in those who do not trust without difficulty ... admitting the apparent loss of certainty and autonomy ... naïvity is helpful ... and I will not repair the bond between being and thinking... but neither will I wait... and shut up

and accept that there is something 'whereof one cannot speak' ... rather try alternatives ... giving language this different function to be the event it only used to lead to ... giving up the distance ... understanding differently ... writing in a different way ... one that the Wien School boys could only dream about... and not fear trouble when language is celebrating ... but celebrate language ...

processes to which the dream-thoughts, previously constructed on rational lines, are subjected to in the course of dream-work:

1 The intensities of the individual ideas become capable of discharge *en bloc* and pass over from one idea to the other, so that certain ideas are formed which are endowed with great intensity (compression or condensation).

2 'intermediate ideas', resembling compromises, are constructed under the sway of condensation. ... Composite structures and compromises occur with remarkable frequency when we try to express preconscious thought in speech. They are then regarded as species of 'slips of the tongue'.

3 The ideas which transfer their intensities to each other stand in the loosest mutual relations. They are linked by associations of a kind that is scorned by our normal thinking and relegated to the use of jokes (in particular associations based on homonyms and verbal similarities).

4 Thoughts which are mutually contradictory make no attempt to do away with each other, but persist side by side. They often combine to form condensations just as if there were no contradictions between them, or arrive at compromises such as our conscious thoughts would never tolerate but such as are often admitted in our actions.

(Freud 1976: 753)

1 NOTES

1 Other articles in this book have pointed to the inadequacy of dualistic thinking for a human ecology and to the importance of overcoming anything that presupposes a clear separation between subject and object. It is for these reasons, for example, that P. Weichhart (this volume) discusses the notion of 'transactional thinking'. In this chapter I think about the self-understanding of the human being as a subject (i.e. something that is identical to itself and different from the other,

the object it faces). In the way I think about it, however, I try – as much as possible – to refrain from reproducing subject–object relations myself. The traditional notion of an author writing about something and thereby communicating a definite message to a passive reader is rooted in such relations. Hence I experiment with writing techniques taken from literary traditions such as stream of consciousness writing, intertextual writing and surrealist writing and emphasize the metonymic movement.

2 After G. Günther 1976: 45, 143, 147.

3 Story inspired by Frank *et al.* 1988: 9ff.

2 REFERENCES

Foucault, M. (1977) *Uberwachen und Strafen,* Frankfurt a/M: Suhrkamp.

Foucault, M. (1983) *Sexualität und Wahrheit ,* Frankfurt a/M: Suhrkamp.

Frank, M. (1986) *Die Unhintergehbarkeit der Individualität,* Frankfurt a/M: Suhrkamp.

Frank, M., Raulet, G. and Van Reijen, W. (1988) *Die Frage nach dem Subjekt,* Frankfurt a/M: Suhrkamp.

Freud, S. (1976) *Interpretation of Dreams,* London: Penguin (German original: *Traumdeutung,* Frankfurt a/M: Fischer, 1961).

Günther, G. (1976) 'Die aristotelische Logik des Seins und die nicht-aristotelische Logik der Reflexion', *Beiträge zur Grundlegung eineroperationsfähigen Dialektik* 1: 141–88, Hamburg: Meiner.

Hegel, G.W.F. (1929) *Hegel, Selections,* ed. J. Loewenberg, New York: Scribner's Son.

Heidegger, M. (1969) *Identity and Difference ,* New York: Harper & Row (German original: *Identität und Differenz,* Pfullingen: Neske, 1957).

Leibniz, G.W. (1915) *Neue Abhandlungen über den menschlichen Verstand,* Leipzig: Meiner.

Shakespeare, W. (1960) *Complete works,* London: Spring Books.

3 FURTHER READING

Acker, C. (1990) *In Memoriam to Identity,* New York: Grove Weidenfeld.

Baudrillard, J. (1986) *Subjekt und Objekt,* Bern: Fraktal, Benteli.

Benjamin, J. (1989) 'Herrschaft – Knechtschaft. Die Phantasie von der erotischen Unterwerfung', in E. List and H. Studer (eds) *Denkverhältnisse,* Frankfurt a/M: Suhrkamp.

Bruno, G. (1980) *Eroici furori,* Darmstadt: Wissenschaftliche Buchgesellschaft.

Descartes, R. (1971) *Meditationen über die Erste Philosophie,* Stuttgart: Reclam.

Descombes, V. (1981) *Das Selbe und das Andere,* Frankfurt a/M: Suhrkamp.

Elias, N. (1988) *Die Gesellschaft der Individuen,* Frankfurt a/M: Suhrkamp.

Elster, J. (1986) *The Multiple Self,* Cambridge: Cambridge University Press.

Foucault, M. (1974) *Die Ordnung der Dinge. Eine Archäologie der Humanwissenschaften,* Frankfurt a/M: Suhrkamp.

Frank, M. (1990) *Das Sagbare und das Unsagbare,* Frankfurt a/M: Suhrkamp.

Freud, S. (1982) *Psychologie des Unbewußten* 3, Frankfurt a/M: Fischer.

Hegel, G.W.F. (1975) *Wissenschaft der Logik,* Hamburg: Meiner.

Heidegger, M. (1967) 'Was heißt denken?', *Vorträge und Aufsätze* 2: 3–51, Pfullingen: Neske.

Hrachovec, H. (1989) *Vermessen. Studien über Subjektivität,* Basel: Stroemfeld/ Roter Stern.

Kristeva, J. (1978) *Die Revolution der poetischen Sprache*, Frankfurt a/M: Suhrkamp.

Mauss, M. (1989) 'Der Begriff der Person und des "Ich", in M. Mauss (ed.) *Soziologie und Anthropologie 2*, Frankfurt a/M: Fischer.

Meyer, E. (1983) *Zählen und Erzählen. Für eine Semiotik des Weiblichen*, Wien: Medusa.

Riley, D. (1988) *Am I that name? Feminism and the category of 'women' in history*, Minneapolis, MN: University of Minnesota Press.

Staten, H. (1985) *Wittgenstein and Derrida*, Oxford: Blackwell.

Stein, G. (1988) *Die geographische Geschichte von Amerika oder die Beziehung zwischen der menschlichen Natur und dem Geist des Menschen*, Frankfurt a/M: Suhrkamp.

Taylor, C. (1989) *Sources of the Self*, Cambridge, MA: Harvard University Press.

Tyler, S. (1987) *The Unspeakable. Discourse, Dialogue, and Rhetoric in the Postmodern World*, Madison, WI: University Press of Wisconsin.

Weigert, A., Smith Teitge J. and Teitge, D. (1986) *Society and Identity*, Cambridge: Cambridge University Press.

Wittgenstein, L. (1982) *Tractatus logico-philosophicus*, Frankfurt a/M: Suhrkamp.

Part III
STRUCTURATION

INTRODUCTION
TO PART III

As indicated earlier[1] it is a central notion of a general human ecology that the so-called 'ecological crisis' really is a 'human crisis', in particular of our modern Western society, and that, as a consequence, the way in which the human sciences see the problem should be given a prominent place. Indeed there is a need for social theory as discussed in the Introduction to Part IV and it is there that this need is stated in the context of geography, which finds itself in a position where it can no longer explain the rapid transformation of features of the environment without exploring the role of social forces and individual actions behind such transformations. As a consequence we should investigate the usefulness of existing social theories in an environmental context. In Part IV some aspects of the theories by Luhmann and by Habermas will be discussed. Here the focus is on the theory of structuration of society by Giddens (1984). The Geography Working Group mentioned in the Preface and also in the General Introduction, whose work led to the Appenberg conference and eventually to the present volume, chose this theory as a core feature for a general human ecology. Obviously this was not a 'once-and-for-all' kind of decision, but rather a signal to indicate that we wanted to take a serious look at its potential value as an important building block.

It is appropriate, therefore, that a separate part of the present volume is devoted to a discussion of this theory.[2] Two of the three papers contained in this part will address it specifically. They give a description of the theory and thereby, depending on the course of argumentation, stress different features of it. The contribution by Roderick Lawrence provides a general overview of the potentialities as well as some shortcomings of the theory. Markus Nauser in his contribution criticizes the usual assumptions made in socio-psychological research about environmental concern and points to insights which the theory of structuration can provide in this respect. The third paper by Alfred Lang does not refer to Giddens at all; the reason for its inclusion in this part of the book should become clear through the further argumentation below. Suffice it to mention at this point that we see its theme as dealing with a possible extension of the theory of structuration as it is commonly understood. It is for this reason that Part III has been entitled simply 'Structuration' and not 'The Theory of Structuration'. We will have a look at these points in the sequel in a way that allows us again to see connections to the remainder of the book.

1 REASON FOR HOPE, REASON FOR DESPAIR: CAN SOCIETY BE TRANSFORMED OR DOES IT TRANSFORM ITSELF?

The foremost reason that the structuration theory is of interest for a general human ecology is that it has a relational character which would seem to fit perfectly into the framework of a new evolutionary or transactional world-view as discussed by Steiner and by Weichhart (both this volume). Such a worldview abstains from reductionisms of any kind, that is, neither a reduction from the level of a whole to the level of its parts nor the converse reduction is admissible. In the context of the person–society question the former would correspond to an individualist conception, that is, to the claim that what we call society is simply the product of individual decisions and actions, the latter to a structuralist conception, that is, to the opposing position that what individuals do is entirely determined by pregiven social structures. The notion that, as in Giddens' theory, both one-sided perspectives should and can be combined into a circularly connected conception of the duality of persons and structures is well described by Bhaskar (1978):[3] 'Men do not create society. For it always pre-exists them. Rather it is an ensemble of structures, practices and conventions that individuals reproduce or transform. But which would not exist unless they did so. Society does not exist independently of conscious human activity (the error of reification). But it is not the product of the latter (error of voluntarism). . . . What is the counterpart . . . to the relationship of reproduction/transformation in which individuals stand to society? . . . The process whereby the stock of skills and competences appropriate to given social contexts are acquired could be generically referred to as "socialization" (Bhaskar 1978:12).'

Thus the theory of structuration has a circular rather than a linear notion of the relationship between individuals and society. Of what importance is this in a human ecological context? Reichert (1992) mentions three aspects: (1) the theory can provide a sense for the self-dynamic inherent in human societies; (2) at the same time it can make obvious that society is transformable, and (3) it can sharpen one's consciousness of the fact that human actions always have unintended consequences.

Indeed the circularity between persons and structures is a source of self-dynamic, that is, a human society may change itself as a result of not only outside influences but also endogenous factors. In this situation individuals may not readily experience themselves as agents of change; rather they may have a feeling of helplessness. Surveys about the existence of environmental concern show that people often are aware of ecological problems, but may suffer under the impression that their own actions do not have any bearing on whether or not these problems exist, or, in cases where they do see connections between their own doings and these problems, may find themselves in a straitjacket which seemingly prevents them from acting differently. This presumably is particularly true for our modern economic society. Giddens (1984: 171) points out that 'the greater the time–space distanciation – the more their institutions bite into time and space – the more resistant they are to manipulation or change by any individual agent.'

On the other hand, actions by a person may not simply reproduce given structures perfectly, but they always also carry the potential of transformation with them. Bhaskar (1978: 14) puts it thus: 'Because social structures are themselves social products, they are themselves possible objects of transformation and so may be only relatively enduring.' It is for this reason that he talks about 'the transformational model of social activity' (Bhaskar 1978: 12). Obviously, it is to be expected that the potential of transformation may be greater in contexts involving institutions of a lesser scope.[4] All the more important would seem to be such notions as the 'radical decentralization' advocated by Dryzek (1987).[5] At any rate this model 'can sustain a genuine concept of *change*, and hence of *history*' (Bhaskar 1978: 13). It is an aspect which, as criticized in the contribution by Lawrence, does not get Giddens' full attention, at least not in the context of crisis.[6]

This recognition of the transformability of society may alleviate any feelings of helplessness that one might have. Nevertheless, the question arises of whether this can be a source of hope in the face of the ecological crisis. How far does the possibility of change imply that a society can be transformed consciously? If it can, what are the chances that the resulting transformation matches the intentions, that we can avoid becoming involved in another 'autonomous process'[7] which may lead us astray for good?[8] How far does the notion of change in a model of structuration not simply refer to history after the fact? The question of the steerability of society away from the abyss moves into a very central position if we set ourselves a goal such as Milbrath's 'sustainable society' (Milbrath 1989).[9]

One of the problems we may have to face with respect to any notion of 'environmental concern' is the degree of human consciousness which attaches to social practices. Here the reader should remember the three levels of consciousness discussed earlier:[10] discursive consciousness, practical consciousness, and the unconscious. As mentioned by Nauser in his contribution, Giddens makes a point of describing day-to-day human agency as happening largely in a routinized fashion in a state of practical consciousness. To be able to act routinely, that is, without much explicit reflection, actually is a fundamental prerequisite for a feeling of ontological security.[11] It is a fallacy to assume that most of what people do happens in the way of conscious decisions between alternatives. However, this is, as pointed out by Nauser, exactly the assumption made by many social and environmental psychologists who try to pinpoint determining factors for individual environmental consciousness or concern and to meet the hopes of policy-makers who would like to know how people can be influenced in this respect. It cannot come as a surprise, therefore, that the results of these studies are disappointing, that is, not in any way conclusive.

It is the role of practical consciousness in everyday life with which we should be concerned if we wish to entertain any hope of gaining a better understanding of how the emergence of environmental consciousness is fostered or hampered. Steiner (this volume) mentions the importance of tacit knowledge of traditional farmers in ecologically unstable areas; this is an example of human actions grounded in practical consciousness that are environment-friendly. The farmers themselves may not be aware of it, but they act in the way that they do out of necessity: the rightness of actions has a survival quality. In contrast, for most people in modern Western society to follow a course of routinized agency means to adhere to some generally accepted social rules. These rules may have no bearing on the biophysical environment at all. In fact, in the course of cultural evolution humans have moved from acting within an originally heavily environment-related practical consciousness to that which is much more purely socially related. This explains part of the destruction that follows, even if it is unintentional. If practical consciousness is the source of environmentally responsible action on the one hand, it becomes an obstacle to it on the other.

2 THE MISSING LINK: THE CONNECTION TO THE ENVIRONMENT

The theory of structuration by Giddens is strictly a social theory and, as such, its usefulness to human ecology is limited as long as it is not linked up to the biophysical world (cf. Reichert 1992). True, it does contain references to environmental features, such as 'locales' and 'allocative resources' (cf. the contribution by Lawrence). A 'locale' is a physically concrete place such as an office building, a market square or a sports field, on which social interactions can unfold repeatedly. This is possible because such places are not simply a backdrop for these interactions. Instead the physical aspects of locales, including their spatial arrangement, have a constitutive character for them.[12] The concept of 'allocative resources' refers to the availability of material resources and the ensuing constitution of 'structures of domination'. Indeed, if we have mentioned before that each human action carries the potential for transformation in principle, then we have to modify this statement by saying that there is an unequal distribution of such a potential among the members of a society which mirrors an unequal distribution of access to resources. The problem with both locales and allocative resources in a human ecological context is that, in the theory of structuration, their existence is assumed to be given. The question of interest is how they enable or constrain social practices and interactions, and not how they themselves come into existence and become transformed.[13]

Indeed, as suggested by Reichert (1992), one might think of developing a model of structuration applying to the concrete world of the biophysical environment in parallel to the one dealing with the abstract world of social rules. In an archaic setting people live in a predominantly natural landscape which is not drastically transformed by their activities. However, they already make consistent and repeated use of particular places, very much in the sense of the 'affordances' referred to by Carello in her concept of ecological psychology (this volume). At any rate, pre-existing, naturally provided structures are used again and again and perhaps maintained and slowly transformed. Conversely, as long as these places exist, the activities related to them can be repeated. From here there is a gradual development to the totally transformed landscape of our times, consisting largely of artefacts or at least anthropogenically influenced natural components. Nevertheless, the basic situation is the same: these concrete structures are used, maintained and/or transformed. They have the function of locales and allocative resources in the sense of Giddens, where social practices are enabled or constrained not only by the social rules underlying those practices but equally well by the physical properties of those places. These properties can thus be seen as material (as opposed to immaterial) structures.

It is Lang who in his contribution develops a concept of this kind from a psychological perspective. He sees a process of structure formation resulting in what he calls an 'external memory' or a 'concretization of the mind', whereby he is interested in the circularities functioning between mind and environment in a way that is akin to Carello's notion of a continuous perception–action cycle (this volume). Lang's concept, therefore, would seem to fit again into an evolutionary or transactional worldview. He points out that in the past objects and spaces usually have been conceived of as opposites of human subjects, whereas the present kind of thinking would rather see them as extensions of persons. Despite the psychological standpoint on which Lang's paper is based it also addresses the role of the social background. Of course, the eventually decisive question in a human ecological context is how the social rules guiding the activities of the members of a society translate, via these activities, into concrete spatial–environmental structures and, conversely, how these structures lead, via the activities of people again, to the persistence or transformation of social structures. Take the example of the relationships between the advent of motor traffic and the development of settlements structures mentioned earlier.[14] This example also serves as an illustration for possible conflicts arising from the discrepancy in temporal existence between persons and external concretizations (referred to by Lang). Many people may be aware of the ecological consequences of driving a car but, finding themselves living far from their workplace and having no access to a reasonable public transportation system, they are practically forced to continue driving it.

209

Another point of possible human ecological interest is Giddens' notion of 'contradiction', which he sees as an inherent feature of human existence and human societies. Actually, he makes a distinction between 'existential contradiction' and 'structural contradiction'. 'By existential contradiction I refer to an elemental aspect of human existence in relation to nature or the material world. There is . . . an antagonism of opposites at the very heart of the human condition, in the sense that life is predicated upon nature, yet is not of nature and is set off against it. . . . Structural contradiction refers to the constitutive features of human societies. I suggest that structural principles operate in contradiction. What I mean by this is that structural principles operate in terms of one another but yet also contravene each other' (Giddens 1984: 193). Giddens then applies this concept to a characterization of aspects of the different types of society, archaic, political and economic (cf. Steiner, this volume).[15] In archaic societies we find, in the absence of a state, a preeminence of existential contradiction that is mediated cognitively by myths. In political societies, while the importance of existential contradiction is still substantial, the emergence of states in the form of city–countryside relations also gives rise to structural contradiction: 'The city is a *milieu* alien to that of nature. . . . The city wall may symbolically and materially seal off the urban *milieu* from the outside. But traditional cities could exist only through their transactions with their agrarian hinterlands' (Giddens 1984: 195). Finally, the economic society is characterized by a pre-eminence of structural contradiction with respect to the relation between the political and the economic system: 'The capitalist state, as a "socializing" centre representing the powers of the community at large, is dependent upon mechanisms of production and reproduction which it helps to bring into being but which are set off and antagonistic to it' (Giddens 1984: 197).

It would seem that the concept of contradiction can be seen as an evolutionary principle, and it is fairly evident that it can be brought into connection with the notion of recursive systems being embedded within each other, whereby new structures emerge out of old ones, emancipate themselves from them, but can do so only with the supporting basis of the old structures (cf. Steiner, this volume).[16] In the context of an evolutionary perspective one might argue that the distinction between existential and structural contradiction is artificial, that in fact what Giddens calls 'existential contradiction' is a structural contradiction resulting from the process of hominization, involving a contradictory interplay between purely bioecological principles as old structures and psychological and social principles typical for humans as new structures. It would be intriguing to explore the human ecological potential of the concept of structural contradiction. Not only may it find its spatial expression in the biophysical world (such as in the opposition of city and countryside in political societies, which Giddens calls the 'dominant locale organization' of those societies), but also the question arises as to the role of structural contradictions in the development of ecological problems.[17]

3 NOTES

1 Cf. Introduction to Part I, Section 2, and also Steiner (this volume).

2 Further remarks concerning the theory by Giddens can be found in Steiner, Weichhart, Werlen and Söderstrom (all this volume).

3 However, Bhaskar provides his description without reference to Giddens. Bhaskar is a representative of the philosophy of science known as 'transcendental realism' to which another reference in a different context is made in the Introduction to Part IV. Its notions seem to be very compatible with an evolutionary worldview.

4 As argued by Steiner et al. (this volume) such small-scale institutions as families, schools and (small) private firms may become the nuclei of change in a bottom-up transformation of society.

5 Cf. Introduction to Part IV, Section 3.

6 According to Lawrence, 'an explicit historical dimension is lacking'. However, Giddens (1984) in ch. 4 gives a kind of a historical breakdown, characterizing the types of society that have evolved in terms of existing structural contradictions, for example. Also, in ch. 5 he discusses theories about the formation of states rather extensively. It seems that it would be more correct to say that Giddens explores 'social change' in connection with past developments at some length, while he seems to be reluctant to discuss the possibilities of change in the context of contemporary problems, particularly in the context of the ecological crisis.

7 Cf. Introduction to Part I, Section 2.

8 Hoyningen-Huene (1983) mentions the story of the decline of the Roman Republic: measures were taken consciously with the intention of maintaining the status quo; instead these measures contributed to the radical dissolution of previously existing social structures. Presumably one can imagine the opposite happening, namely that efforts taken to induce change actually support the maintenance of the status quo.

9 Tentative answers to this question and references to components which could possibly contribute to a change in a desired direction are distributed throughout this book. For example, Marler in his contribution refers to the possible triggering effect of social trust in a self-organizing process. The reader is also reminded of the attempt mentioned earlier (see Introduction to Part I, Section 4) by Dryzek (1987) to come up with a comparative evaluation of various social choice mechanisms in terms of 'ecological rationality'. It is obvious that one single and clear-cut recipe simply does not exist. It is equally obvious that the issue of steerability is linked to the question of who can achieve what. In other words, the distribution of power or, in Giddens' terminology, the 'authoritative resources' available within persisting 'structures of domination' become decisive. It goes without saying that the feminist critique of the still prevalent male domination of society features prominently in this context.

10 Cf. Steiner (this volume) and the Introduction to Part II.

11 There is a link here to the acquisition of social identity discussed in the Introduction to Part IV and by Werlen (this volume). The identity that a person acquires is the result of social interactions repeated time and again with the same fellow human beings.

12 For that matter, not only space but also time plays a constitutive role in the theory of structuration. For examples see Steiner et al. (this volume).

13 Perhaps we should differentiate between the theory of structuration and its author, Giddens. The latter is definitely open-minded with respect to questions with an ecological bearing. In a seminar held at the Department of Geography of the ETH, Zürich on 18 February 1988, he offered the suggestion that perhaps we should be treating the environment like a person.

14 Introduction to Part I, Section 2.
15 For the three types of society Giddens (1984) employs the terms 'tribal society', 'class-divided society' and 'class society'.
16 To see structural contradiction as a possible evolutionary principle does not mean that all contradictions actually emerging in the course of cultural evolution are unavoidable. Take the example of the contradiction between political and economic structures referred to above. Historically speaking the latter evolved out of the former, yet today, as described by Giddens, the political system, for its continued existence, depends on the well-being of the economic system. This fact may be interpreted as still another aspect of the overemancipation of the economic system diagnosed in Steiner (this volume). Parallel to the 'moral inversion' mentioned there one can see the present state of affairs as an inversion in the contradiction of structural principles. A political system and an economic system may always in some respects contradict each other, but, in an evolutionary sense, the economic should depend on the political system, not the other way round (cf. the discussion of a hierarchy of rationalities, Section 5, Introduction to Part I).
17 This question stands in a relation to the issue of identity acquisition referred to in the Introduction to Part IV and discussed by Werlen (this volume).

4 REFERENCES

Bhaskar, R. (1978) 'On the possibility of social scientific knowledge and the limits of naturalism', *Journal for the Theory of Social Behaviour* 8: 1–28.
Dryzek, J.S. (1987) *Rational Ecology. Environment and Political Economy*, Oxford: Blackwell.
Giddens, A. (1984) *The Constitution of Society. Outline of the Theory of Structuration*, Berkeley and Los Angeles: University of California Press.
Hoyningen-Huene, P. (1983) 'Autonome historische Prozesse – kybernetisch betrachtet', *Geschichte und Gesellschaft* 9 (1): 119–23.
Milbrath, L.W. (1989) *Envisioning a Sustainable Society. Learning Our Way Out*, Albany, NY: State University of New York Press.
Reichert, D. (1992) 'Wie das Tun verstanden wird und was das Tun bewirkt', Section 1.3.2 in D. Reichert and W. Zierhofer, *Umwelt zur Sprache bringen*, Research Report, Zürich: Department of Geography, ETH (publication in book form forthcoming).

14

CAN HUMAN ECOLOGY PROVIDE AN INTEGRATIVE FRAMEWORK?

The contribution of structuration theory to contemporary debate

Roderick J. Lawrence

1 INTRODUCTION

Human ecology is a term that has been, and still is, characterized by confusion and a lack of consensus about what it means (Young 1983). According to Bruhn (1974: 105), human ecology 'has been proposed as a science, a separate discipline, a philosophy, a point of view, and an approach for studying a given problem'. In general, human ecology has focused on the relations between people and their immediate surroundings. This subject has been studied by academics and professional practitioners with training in established social science disciplines. Bruhn (1974) has presented a brief yet instructive overview of the development of human ecology studies in the disciplines of anthropology, geography, psychology and sociology. He also endeavours to identify whether the contributions by each of these disciplines can enable the formulation of an interdisciplinary approach for people–environment studies. His conclusion is not an optimistic one.

Bruhn and others have shown that social scientists have frequently (implicitly or explicitly) borrowed analogies from the natural sciences, especially biology. Overviews of ecological studies in diverse disciplines (e.g. ecological anthropology [Hardesty 1977]; ecological geography [MacArthur 1972]; ecological psychology [Barker 1968, Wicker 1979]; and ecological sociology [Park *et al.* 1925]) show, in general, that concepts, principles and methods from the natural sciences have been repeatedly used analogically at the expense of those in the social sciences. For example, many environmental and economic analyses of cities have frequently adopted a biological analogy by treating human settlements as metabolisms, and by examining flows of energy and materials at the expense of other social processes, especially human knowledge, communication and information. Consequently, the term 'environment' has been interpreted and studied restrictively, according to academic concepts and methods that often emphasize human products and processes, whereas many inorganic, biological and symbolic constituents of the environment have commonly been overlooked. In this respect, the theory of structuration presented by Giddens does not explicitly challenge the status quo, yet it does provide unexplored cues for future developments.

Moreover, many contemporary people–environment studies do not identify the impacts or consequences of human activities on either the human-made or the inorganic and biological constituents of the environment. Consequently, it seems reasonable to suggest that, in general, these contributions are not ecological but spatial or environmental anthropology, geography, psychology and sociology, because they usually do not account for both the natural and human ecosystems, their constituent parts, and the interrelations between them. This shortcoming warrants further comment.

Although the majority of social scientists, and notably geographers, have been divided between those who describe 'nature' and those who examine 'culture' as separate entities, it is instructive to recall those significant integrative contributions dating from the early nineteenth century; indeed many decades before the term human ecology was widely used. According to Bruhn (1974), some geographers (and others) need reminding that Alexander von Humboldt made significant contributions to regional geography, the distribution of plants, and human demography, especially during his studies in Mexico and South America; von Humboldt examined the interrelations between the natural constituents of the environment and human activities. Moreover, Karl Ritter argued for the mutual dependence of human and natural constituents of the world in his holistic view of the earth. The ideas of von Humboldt and Ritter were developed by Ratzel, who examined the distribution and growth of human resources. Ratzel interpreted land as a basic resource of human life which has both an ecological and a political stake. In sum, these seminal contributions showed that it is possible to integrate, at least conceptually, the inorganic, biological and human sciences in order to construct an integrative framework. Bearing these and other important contributions in mind, it is appropriate to state the meaning of human ecology that will be used in the remainder of this paper.

2 WHAT IS HUMAN ECOLOGY?

Human ecology is an holistic, integrative interpretation of those processes, products, orders and mediating factors that regulate natural and human ecosystems at all scales of the earth's surface and atmosphere. It implies a systemic framework for the analysis and comprehension of three logics and the interrelations between their constituents using a temporal perspective. These three logics are:

1 A bio-logic, or the orders of biological organisms.

2 An eco-logic, or the order of inorganic constituents (e.g. water, air, soil and sun).

3 A human-logic, or the ordering of cultural, societal and individual human factors.

It is suggested that this macro-system of three logics regulates the world with respect to energy flows, material resources, human labour, and knowledge, communication and information. Consequently, it is inappropriate to emphasize one set of constituents to the detriment of others. Moreover, it is erroneous to distinguish between the 'physical' and the 'social' constituents of environments. This definition implies that a contextual approach is pertinent. This kind of integrative approach would examine specific situations in terms of the reciprocal relations between the constituents of these three logics, both at one point in time, and over an extended period of time.

This interpretation challenges the 'human–environment paradigm', which has consistently been used since antiquity to distinguish human beings from their 'natural habitat', and to claim that the transformation of the material constituents of that habitat by people is an 'underlying force' that has guided human history. Giddens, quite correctly, challenges this interpretation. He is critical of the dualism between people and their habitat. Such chasms are bridged if it is accepted that it is misleading to study the inorganic, biological or human constituents of the environment, because they are mutually defined by, and defining components of, one ecosystem in which people are but one constituent. Human attitudes, motives and values influence what people perceive and construe, how they use precise settings, and how they modify them over time. Moreover, the location, composition and organization of a setting has some bearing on how it is perceived and used. In sum, it is not 'the people' or 'the environment' which should be given priority, or become the methodological unit of study, as Bruhn (1974: 112) has claimed. Rather, the interrelations between energy flows, material resources, human labour, and knowledge, communication and information should be examined over an extended period of time in the context in which they occur.

The purpose of the remainder of this paper is to examine some specific concepts and principles that Giddens has presented with respect to his theory of structuration, and to appraise whether they are heuristically pertinent for the elaboration of a conceptual framework for studies in human ecology.

3 A SELECTIVE OVERVIEW OF THE THEORY OF STRUCTURATION

In elaborating the concept of structuration theory, I do not intend to put forward a potentially new orthodoxy to replace the old one. But structuration theory is sensitive to the shortcomings of the orthodox consensus.

(Giddens 1986: xvi)

This section will examine some concepts and ideas developed by Giddens in his writings on the theory of structuration. In general, all references in this section pertain to his most recent book (Giddens 1984/1986), although his earlier publications during the formulation of this theory are extensive.

The concepts and ideas related to the theory of structuration that will be briefly examined here include:

(a) Structure and structuration.
(b) Locale.
(c) Purposive behaviour.
(d) Unintended consequences.
(e) Power, regulation and control.
(f) Allocative and authoritative resources.

These concepts and ideas will be examined in order to establish if and how they can be used to comprehend and explain the interrelations between human individuals, societies, and their built and 'natural' environments. Where deemed necessary, these concepts and ideas will be complemented by, or related to, others presented by some authors in this volume, as well as others who have published in the vast field of people–environment studies. Collectively, if both sets of concepts and ideas are considered germane, then they may be adopted by scholars who wish to formulate an integrative framework for human ecology.

3.1 Structure and structuration

In the introduction to his book, Giddens (1986) states that many contemporary interpretations of structural sociology have maintained that the structural properties of societies constrain human activities. Giddens challenges these recurrent interpretations of structure and structuralism, which he confronts with the theory of structuration.

In elaborating the theory of structuration, Giddens endeavours to rework established, shared conceptions of 'human being and human doing, social reproduction and social transformation'. In order to achieve this goal Giddens states that it is necessary to replace common 'either/or' dualisms by what he terms 'the duality of structure'. He notes that the term 'structure' has frequently been interpreted as 'the more enduring aspects of social systems' (p. 23), which he adopts and explicitly relates to 'rules and resources recursively involved in social reproduction'. According to Giddens, 'the duality of structure' means that 'the structural properties of social systems are both medium and outcome of the practices they recursively organize' (p. 25). From this perspective Giddens elaborates the 'theory of structuration' which is also founded on the principle that structure is both enabling and constraining, owing to the active and recursive role of societal and individual human factors in daily life. Consequently, 'structuration' is defined as 'the structuring of social relations across time and space, in virtue of the duality of structure' (p. 376).

This interpretation challenges those contemporary, structuralist conceptualizations of people–environment relations, that commonly uphold absolute, lawlike mechanisms (cf. Lawrence 1989 for an overview). In contrast, Giddens suggests that structural parameters, whether constraints, catalysts or enablers, do not operate independently of the motives and reasoning of people. These motives and reasons are associated with values, which are context dependent and susceptible to change over the course of time. From this perspective, 'the epistemological and ontological chasm' dividing objectivism and subjectivism in the guise of 'the imperialism of society' (the social object) and 'the imperialism of the individual agent' (the subject) can be bridged. This integrative perspective is necessary and important, because Giddens has shown in many of his anterior writings that dualisms have commonly divided functionalism and structural social laws, on the one hand, from humanism, hermeneutics and methodological individualism, on the other hand. Giddens argues that 'a decentring of the subject' is basic to structuration theory, but he also adds that this does not imply 'the evaporation of subjectivity into an empty universe of signs'. On the contrary, according to Giddens, social practices are considered to be the foundation of 'the constitution of both subject and social object' (p. xxii). In other words, according to the theory of structuration, the basic domain of study of the social sciences 'is neither the experience of the individual actor, nor the existence of any form of societal totality, but social practices ordered across space and time' (p. 2).

Given the rich, theoretical contribution formulated by Giddens, which is meant to provide an integrative perspective for people–environment studies, it is instructive to examine whether those concepts, ideas and principles that are components of the theory of structuration are pertinent for the formulation of an integrative framework for studies in human ecology.

3.2 Locales

According to Giddens a locale is an enriched interpretation of 'place' that encompasses contextual meaning, not merely geographical location. In this respect, context defines how the physical and/or architectural setting is used by social actors to organize the reproduction of their social lives across time and space.

Giddens is critical of many contemporary contributions by social scientists concerned with human behaviour 'in' space and time. Accordingly, he states that 'time-geography has some very distinct shortcomings', which he relates to Hägerstrand's (1975) interpretation of the human agent, the dualism of action and structure, the constraining nature of human action, and an undeveloped conceptualization of power.

Giddens also challenges those contemporary interpretations of 'place', which he states cannot be used 'to designate "point in space", any more than we speak of points in time as a succession of "nows"' (p. 118). This critique is timely, given the continued formulation and application of the concept of 'behaviour setting' by 'ecological psychologists' including Barker (1968) and Wicker (1979), and the corpus of works in the fields of 'architectural morphology' and 'space syntax' by Hillier and Hanson (1984) and Steadman (1983). Surprisingly, Giddens makes no reference to this large contribution, but focuses on the works of a few other geographers, sociologists, and psychologists, to show quite convincingly that time and space are not merely settings for human activities, because they are constitutive of what people do.

Giddens elaborates the concept of 'locale', which is a setting of human activity that, in turn, defines 'contextuality'. He states that locales cannot be defined solely in terms of their physical characteristics – an error made by many architects, ecological psychologists and time-geographers. Although the features of a locale have some influence on the meaningful content of human activities, these activities cannot be detached and studied apart from the societal and cultural characteristics of daily life – the constituents of meaning – in this setting, and at much broader scales. In this respect Giddens (1986: 29) states:

> The communication of meaning, as with all aspects of the contextuality of action, does not have to be seen merely as happening 'in' time–space. Agents routinely incorporate temporal and spatial features of encounters in processes of meaning constitution.

From this perspective, the concepts of 'positioning' and 'regionalization' are formulated by Giddens, who emphasizes the settings, and the socio-psychological nature of encounters, routines and patterns of daily life. Although Giddens is indebted to Goffman, it is noteworthy that the latter considered the settings of human activities, especially encounters, using dramaturgical metaphors and analogies. None the less, some rituals, routines and practices, including the use of many biological and inorganic constituents of the earth's surface and atmosphere, cannot be interpreted in this way. This shortcoming is not primarily a question of temporal–spatial scale, which Giddens does account for in his formulation of regionalization and 'time-space distanciation'. Rather, it highlights the orthodox, restrictive interpretation of 'environment' by many social scientists that has been mentioned in the introduction to this paper. This shortcoming presents an obstacle to the application of concepts and ideas of the theory of structuration for the elaboration of an integrative, conceptual framework for human ecology. It is suggested that the definition of environment must be re-examined in order to achieve such a framework.

3.3 Purposive behaviour and unintended consequences

Giddens maintains that the 'recursive' nature of human activities and the 'reiterative' nature of social systems are crucial for an integrative approach to study people–environment relations. According to Giddens, 'human agents or actors ... have, as an inherent aspect of what they do, the capacity to understand what they do while they do it' (p. xxii). However, Giddens notes that 'reflexivity operates only partly on a discursive level'. 'What agents know about what they do, and why they do it – their knowledgeability as agents – is largely carried in practical consciousness' (p. xxiii). In stark contrast to the role of the human individual in orthodox structuralist theories of people–environment relations (cf. Lawrence 1989) this formulation of 'practical consciousness' is important for the development of the theory of structuration. Concurrently, this interpretation does not challenge other concepts of consciousness ('discursive consciousness') and the unconscious: Giddens uses each of these concepts to formulate his conceptualization of 'routinization' which is a fundamental concept of structuration theory. Routines are interpreted in terms of not only psychological mechanisms and meaning but also societal and cultural ones.

Unfortunately, an explicit historical dimension is lacking from this discussion, and this oversight underlines the fact that structuration theory does not provide an explanation of mechanisms of societal change. The contributions of Goffman (1959, 1963, 1972, 1981) and Hall (1959, 1966, 1983) confirm that it is meaningless to distinguish between individual and group, as well as between space and time. Moreover, these contributions underline the pertinence of concepts such as 'purposive human behaviour', and its 'intended and unintended consequences'. According to Giddens 'it is hard to exaggerate the importance of the unintended consequences of intentional conduct' (p. 12). Giddens presents a critical review of Merton's seminal contribution. Merton (1936) stated that a human activity can have insignificant or significant consequences that may be of single or multiple magnitude. The definition of what is 'significant' is context dependent (both temporally and spatially), and will depend on the nature of the human conduct and its enquiry. Giddens then challenges Merton's claim that the analysis of unintended consequences 'can make sense of seemingly rational forms or patterns of social conduct'. To recall, Merton equated intentional activity with manifest functions, which he then contrasted with its unintended consequences (latent functions). As Giddens stresses, latent functions are not forcibly unintentional or irrational. Moreover, institutionalized societal practices and routinized group and individual human activities may be irrational! This principle can be illustrated by studies of the interrelations between the components of the three logics – the bio-logic, the eco-logic, and the human-logic. For example, an earthquake can destroy a human settlement without the residents having any means of intervention, whereas the flooding of a village by the collapse of a human-made water dam should be considered in relation to the motives and values associated with the construction of a reservoir on that site, and the diverse impacts of the reservoir on the human ecosystem over the course of time.

This discussion and example illustrates that the discussion of purposive behaviour and unintended consequences could benefit from a complementary account of the concept of the 'impacts' of human products and processes on all constituents of the environment, as well as a discussion about the formulation and application of 'scenarios' (especially in the sixth chapter of Giddens [1986], which addresses empirical research and social critique). Although environmental impact assessment is a growing field of enquiry, it has largely followed a technocratic approach, producing a reiteration of the status quo rather than the formulation of a range of alternatives grounded on the elaboration of scenarios.

3.4 Power, regulation and control

The concepts of power and social control have been elaborated by social scientists but rarely in the context of people–environment studies. Giddens notes that power is not necessarily linked with conflict and it is not inherently oppressive, as Marx and many of his disciples have commonly maintained. According to Giddens (1986: 157–8):

> Power is the capacity to achieve outcomes; whether or not these are connected to purely sectional interests is not germane to its definition. Power is not, as such, an obstacle to freedom or emancipation but is their very medium – although it would be foolish, of course, to ignore its constraining properties. . . . Power is generated in and through the reproduction of structures of domination.

While Giddens is critical of the interpretation of power presented by Marx and Durkheim, he is also dissatisfied with the analysis of power subsequently presented by Parsons and Foucault. According to Giddens, the latter do not adequately consider the role of individuals or their knowledgeability, which he explicitly incorporates in the concept of 'authoritative resources' (see below). None the less, the roles of individuals and small groups could also have been discussed by Giddens with respect to implicit norms and rules that regulate information and symbolic exchange. These social regulating mechanisms warrant further consideration, prior to the formulation and application of an integrative framework.

The interpretation of power as both enabling and constraining human activities does not contradict the notion of 'social control' elaborated by Ross (1901/1970), who discussed the concepts of social order, social cooperation, and social hierarchical organization. Ross noted that, in general, the members of communities or groups do not explicitly try to aggress one another, and hat individuals usually make adjustments to escape conflicts. Successful cooperation implies social order, which is founded on implicit rules and customs – 'tacit knowledge' – that regulate human activities, including the sharing of information and symbolic exchange. Both of these human processes define and are mutually defined by hierarchically ordered powers that regulate communication between individuals and groups in human societies. These implicit regulatory mechanisms are not static or absolute but change according to the contextual conditions in which they operate. Traditional methods of food production – cultivation, harvesting and storage – food processing and packaging prior to consumption are but one example (cf. Goody 1982). When these implicit regulatory mechanisms related to food are replaced by imported foreign practices, as illustrated by Boyden *et al.* (1981) in Hong Kong, the consequences are manifold: changes occur in agricultural productivity, the use of energy, the accumulation of toxic and non-toxic waste products, the importation and exportation of materials, and the diet of the local population. Apart from a few studies such as that by Boyden *et al.* (1981), many studies in the field of human ecology do not identify or study the scope and consequences of the interrelated impacts of this kind on human ecosystems. This subject area is beyond the range of concepts and ideas addressed by Giddens.

Hierarchical organization is a complex indicator of social order, particularly in societies with strong economic differentiation. The modern city, for example, is not only the locus of human products and processes defined by economic, legal, political, professional and religious differentiation but also the seat of administrative and political power. It is from this locus that information is shared or withheld, thus exacerbating or reducing those differentiations just mentioned. As Ross (1901/1970: 19) states:

It is in the urban group, in fact, that social order finally parts company with the sociable impulse. For one reason why this impulse tends to harmonious conduct is that bad actions cause forfeiture of companionship, exclusion from social pleasure, ostracism. But in the city the variety of social grouping is so great that the anti-social disturber need not forego festal enjoyment.

In a word we have spun out the web until the social relations of the individual are far too many to elicit any response from this jaded social instinct. We have spun out amazingly the relations of persons to things, i.e. property rights.

Ross notes that, in contemporary societies, private property is 'a great transforming force which acts almost independently of the human will'. Although we are far removed from the 'knowledgeability of human agent', it is noteworthy that private property is the pretence, motive and goal for the acquisition, use and misuse of the biological and inorganic constituents of human ecosystems, by individuals, groups and/or whole societies.

Until recently, traditional economic theories did not include the use of natural resources (especially air, water and wind) when accounting for the production and use of human artefacts. Furthermore, they have usually overlooked the production, treatment and disposal of waste products, which have usually not been considered as a kind of reusable resource; and the cost of treating them has commonly been overlooked, at least until recent decades. As these natural resources have not been accounted for, there have been no explicit regulating mechanisms for their monitoring and use, in relation to the energetic laws of the planetary ecosystem. Such misleading theories and practices, coupled with biased economic calculations, have exacerbated economic differentiation in human societies, as well as the continual harvesting of natural resources and the accumulation of toxic and non-toxic wastes. These impacts have not been related to ecological knowledge, but are products of the quest for private possessions. At the beginning of this century Ross (1901: 53) noted:

> As [private property] warps society further and further from the pristine equality that brings out the best in human nature, there is a need for artificial frames and webs that may hold the social mass together in spite of the rifts and seams that appear in it. Property is, therefore, the thing that calls into being rigid structures. It is in the composite society, then, where the need of control is most imperative and unremitting, that the various instruments of regulation receive their highest form and finish.

Such instruments of regulation, which encompass Giddens' concept of authoritative resources, are fundamentally different from those implicit customs, norms and rules found in human societies, and transmitted orally from generation to generation. Whereas implicit instruments of regulation are both conditionally and contextually defined and applied, explicit instruments of regulation in many contemporary societies are absolute, fixed and prescribed by written deeds and codes of conduct. Moreover, whereas implicit instruments include tacit regulatory knowledge and are self-administered, those which are explicit are imposed by administrators, politicians, or property owners. Explicit regulators usurp many implicit customs, norms and rules; they may challenge 'the knowledgeability' of some people, and they become a resource of abstract power and social control in human groups and societies. One consequence is that face-to-face conduct is gradually replaced by technical forms of communication and administrative controls which increasingly regulate daily affairs.

The suppression of implicit instruments of regulation and the growth of explicit ones has been related to long-term processes of urbanization by Mumford (1961) in his classical analysis of the history of human settlements. Clearly, both these sets of instruments are important constituents of the human logic and they could have been addressed more explicitly by Giddens.

As Raffestin (1980) states, although they provide complementary kinds of information, the tacit regulatory knowledge of implicit instruments of regulation have increasingly been usurped by explicit functionalist regulations. Raffestin argues that more regulatory knowledge is needed in contemporary societies.

3.5 Resources

According to Giddens, those resources which constitute structures of domination are either 'allocative' or 'authoritative', and 'any coordination of social systems across time and space, necessarily involves a definite combination of these two types of resources' (p. 158). These two types of resources are classified by Giddens in the way shown in Table 14.1.

Giddens states that both sets of resources are not static, and that together they form the media of the expandable character of power in different types of society' (p. 258). Giddens quite rightly notes that evolutionary theories have usually given priority to allocative resources. He could have added economic theories and geographical studies too. In general, however, following the example of many orthodox economic theories, these studies have not explicitly accounted for the use of aeolian, solar and water energy and resources, as well as waste products generated by human activities. Moreover, many social scientists who have published studies in human ecology have followed suit. Although authoritative resources are an integral component of the human-logic defined earlier in this paper, it is curious that many recent urban studies that have adopted a human ecology perspective have commonly overlooked them.

Table 14.1 Classification of resources according to Giddens

Allocative resources	*Authoritative resources*
Material features of the environment (raw-materials, material power resources)	1 Organization of social time–space (temporal–spatial constitution of paths and regions)
Means of material production/reproduction (instruments of production, technology)	2 Production/reproduction of the body (organization and relation of human beings in mutual association)
Produced goods (artefacts created by the interaction of 1 and 2)	3 Organization of life chances (constitutions of chances of self-development and self-expression)

Whereas architectural, economic, social and urban historians have commonly upheld that the foundation of human settlements from hunter-gatherer societies to contemporary metropolitan agglomerations and nation-states corresponds with the accumulation of material possessions and the

223

production and storage of food supplies, it is noteworthy that Ross, Mumford and Giddens argue that the composition and application of explicit authoritative resources developed in tandem with human settlements. Giddens (1986: 160) states that it is quite misleading to separate allocative resources from authoritative resources:

> described in such a manner, human history would sound (and has often been made to sound) like a sequence of enlargements of the 'forces of production'. The augmenting of material resources is fundamental to the expansion of power, but allocative resources cannot be developed with the transmutation of authoritative resources, and the latter are undoubtedly at least as important in providing 'levers' of social change as the former.

Although the interrelations between allocative and authoritative resources have been elaborated by Giddens, he underplays the role of information and communication. None the less, his integrative interpretation of these two types of resources is an important contribution for studies of human ecology, provided that both implicit and explicit instruments of regulation are taken into account, as Raffestin (1980) has argued.

4 SYNTHESIS AND DISCUSSION

This paper has briefly presented and appraised some of the concepts and principles elaborated by Anthony Giddens for the theory of structuration. This paper also indicates that if academics and professionals are to broaden their concern for people–environment studies, then they need to account for a range of ideas, concepts and principles that define, and are mutually defined by, a bio-logic, an eco-logic and a human-logic, and the interrelations between the constituents of these three logics, using a temporal perspective. To our knowledge, such a tripartite integrative framework has not been elaborated for studies in human ecology, and yet this set of papers will help to meet this goal. In this respect, Söderström (this volume) argues that if human ecology is to provide an integrative perspective for the study of human ecosystems, then it should overcome the following limitations:

1 The lack of articulation of contemporary interpretations of human ecology to a coherent social theory.
2 The continued application of a biological analogy (largely with respect to the flow of energy, material resources and waste products) at the expense of social processes including human communication and information.
3 The restricted technocratic study of environmental problems, and the explicit administrative, economic and juridic regulation of them, largely from monetary and energetics perspectives.
4 The popularization of 'the environmental crisis' without due recognition of the underlying foundations of contemporary problems.

This concise assessment by Söderström is useful in the context of this collection of papers because it helps to identify the strengths and weaknesses of structuration theory as a means for achieving an integrative perspective. Provided the limitations of structuration theory outlined earlier in this paper are borne in mind, Giddens has undoubtedly helped to overcome the first and second limitations. Furthermore, with respect to the third and fourth limitations, it is noteworthy that although structuration theory does not give explicit priorities to particular methods of study, it implicitly underlines the pertinence of ethnography and diverse survey methods. Also, although an historical or temporal perspective is implicit, it is not underlined. In this respect, both Giddens and Boyden (this volume) quite rightly examine the dynamic interrelations between constituents of human ecosystems, but their approach is basically analytical and descriptive, rather than explanatory: neither Boyden nor Giddens account for the foundations of societal change, or explanations of past events. In Boyden's own terms, the biohistorical approach does not encompass the tacit characteristics of human society. None the less, it is precisely these characteristics that ought to be examined in tandem with manifest, physical characteristics if an integrative approach is to be applied to comprehend the most mundane events in daily life (Lawrence 1989). Only an understanding of both of these sets of factors, examined historically, can explain, for example, why a housewife in Australia washes her children's clothes in a room (the laundry) demarcated from the kitchen, whilst her counterpart in England commonly undertakes the same activity in the kitchen, and usually in the kitchen sink (Lawrence 1982)!

In the field of environmental economics the externality approach is one means of accounting for the indirect pricing of natural resources, and especially flows of energy and materials, as Pillet (this volume) discusses. Externalities can account for both the direct and indirect effects that result from decisions related to human products and processes, but which are not included in the market price of goods and services. This approach is one application of the biological analogy mentioned above. From this perspective, those impacts such as air pollution, noise and solid wastes can be calculated as negative external effects, whereas the purity of air, quietness and reusable waste products are calculated as beneficial or positive externalities. This kind of assessment enables traditional interpretations of economic production and consumption to encompass many ecological costs and benefits that are borne either internally or externally.

The definition and implementation of market externalities can be enlarged and enriched, as discussed by Pillet (this volume) in his examination of the interrelations between human activities and their ecological impacts. However, the application of such externalities to serve as explicit regulators of economic activity (e.g. the-polluter-pays-principle) are limited in as much as they are explicitly tied to economic affordability rather than market efficiency, social consensus, or ecological sustenance. Moreover, although they can account for intended and unintended consequences of human products and processes, they do not encompass one of the fundamental constituents of human ecosystems, namely human knowledge, communication and information (e.g. both symbolic exchange and the technological transmission of information). These crucial characteristics cannot only be analysed by flows of energy and materials.

Beyond the limitations of contemporary economic and energetic theories and applications it is appropriate to mention other approaches for calculating the costs and benefits of human products and processes. One approach, for example, suggests that the impacts of human products and processes can be evaluated and monitored in relation to the health and well-being of people. Epidemiological and socio-psychological studies can establish the morbidity and mortality of populations, be they animal, human or plant species. Research on 'sick building syndromes', including recent studies of the propagation and effects of legionnaires' disease, is but one application of this kind of approach with respect to the design and use of the built environment (Raffestin and Lawrence 1990).

The economic, energetics and health approaches and applications briefly presented in this paper indicate that there are distinctly different ways of evaluating and monitoring the costs and benefits of human products and processes. This non-exhaustive range of approaches illustrates that these approaches are not exclusive but complementary, and they can be implemented simultaneously. Unfortunately this kind of integrative assessment has rarely been undertaken.

5 CONCLUSION

This paper shows that many concepts presented by Giddens are theoretically rich and pertinent for the elaboration of an integrative framework for human ecology. However, it is also clear that some definitions and concepts need to be redefined and reconceptualized if they are to be useful for the elaboration of the three logics that constitute an integrative conceptual framework. In this respect, first, it is necessary to enlarge the definition of 'environment' used by Giddens (and most social scientists) in order to account for the biotic and inorganic constituents of human habitats. Second, it is necessary to examine the impacts of human processes and products on these constituents of the environment that exist across all levels or scales of the earth's surface and its atmosphere. The larger scales of the biosphere are not explicitly accounted for in Giddens' definition of locales or allocative resources. Third, the concept of 'authoritative resources', coupled with those of power and social control, should consider both implicit and explicit instruments of regulation, which have been outlined earlier in this paper. Both of these fundamental constituents of the human-logic are not explicitly examined by Giddens, yet this paper also shows that they do not contradict the rich analysis of authoritative resources

226

that he has presented. Last but not least, the concept of 'allocative resources' should be enlarged to include the administration and use of biological and inorganic constituents of the environment; the study of energy flow, including the production and consumption of energy to entropic laws of the planetary ecosystem, and the use of non-renewable resources; and the production, treatment and disposal of toxic and non-toxic wastes and by-products.

In sum, although Giddens has provided a rich, integrative contribution, some of the concepts and principles deemed necessary for an integrative framework for human ecology are not included in the theory of structuration. Yet to be fair, this was not the purpose of formulating this theory. However, as indicated in this paper, it is necessary and feasible to complement or redefine some of these concepts and principles in order to identify and examine explicitly the constituents of extant eco-logics, bio-logics and human-logics that exists at all levels or scales of human ecosystems, both at one point in time, and over an extended time period. Suggestions have been presented briefly in this paper in order to achieve this difficult goal.

6 REFERENCES

Barker, R. (1968) *Ecological Psychology: Concepts and Methods for Studying the Environment of Human Behavior*, Stanford, CA: Stanford University Press.

Boyden, S., Millar, S., Newcombe, K. and O'Neill, B. (1981) *The Ecology of a City and Its People: The Case of Hong Kong*, Canberra: Australian National University Press.

Bruhn, J. (1974) 'Human ecology: a unifying science?', *Human Ecology* 2: 105–25.

Giddens, A. (1986) *The Constitution of Society: Outline of the Theory of Structuration* (1st edn 1984), Cambridge: Polity Press.

Goody, J. (1982) *Cooking, Cuisine and Class: A Study in Comparative Sociology*, Cambridge: Cambridge University Press.

Goffman, E. (1959) *The Presentation of Self in Everyday Life*, New York: Doubleday.

Goffman, E. (1963) *Behavior in Public Places*, New York: Free Press.

Goffman, E. (1972) *Interaction Ritual*, London: Allen Lane.

Goffman, E. (1981) *Forms of Talk*, Oxford: Blackwell.

Hägerstrand, T. (1975) 'Space, time and human conditions', in A. Karlquist (ed.) *Dynamic Association of Urban Space*, Farnborough: Saxon House.

Hall, E. (1959) *The Silent Language*, New York: Doubleday.

Hall, E. (1966) *The Hidden Dimension*, London: Bodley Head.

Hall, E. (1983) *The Dance of Life: The Other Dimension of Time*, New York: Anchor Press.

Hardesty, D. (1977) *Ecological Anthropology*, New York: Alfred Knopf.

Hillier, W. and Hanson, J. (1984) *The Social Logic of Space*, Cambridge: Cambridge University Press.

Lawrence, R. (1982) 'Domestic space and society: a cross-cultural study', *Comparative Studies in Society and History* 24 (1): 104–30.

Lawrence, (1989) 'Structuralist theories in environment–behavior–design research: Applications for analyses of people and the built environment', in E. Zube and G. Moore (eds) *Advances in Environment, Behavior and Design* 2, New York: Plenum Press.

MacArthur, R. (1972) *Geographical Ecology: Patterns in the Distribution of Species* (2nd edn 1984), New York: Harper & Row.

Merton, R. (1936) 'The unanticipated consequences of purposive social action', *American Sociological Review* 1: 894–904.

Mumford, L. (1961) *The City in History: Its Origins, Its Transformations, and Its Prospects*, London: Secker & Warberg.

Park, R., Burgess, E. and McKenzie, R. (1925) *The City*, Chicago, Ill.: Chicago University Press.

Raffestin, C. (1980) 'Plaidoyer pour une écologie humaine', *Archives suisses d'anthropologie générale* 44: 123–9.

Raffestin, C. and Lawrence, R. (1990) 'An ecological perspective on housing, health and well-being', *Journal of Sociology and Social Welfare* 17 (1): 143–60.

Ross, E. (1901) *Social Control: A Survey of the Foundations of Order* (reprint edn 1970), New York: Macmillan.

Steadman, P. (1983) *Architectural Morphology*, London: Pion.

Wicker, A. (1979) *An Introduction to Ecological Psychology*, Monterey, CA: Brooks/Cole.

Young, G. (ed.) (1983) *Origins of Human Ecology*, Stroudsberg, PA: Hutchinson Ross.

15

ENVIRONMENTAL CONCERN AND THE THEORY OF STRUCTURATION

Steps towards a better understanding of environmentally harmful agency

Markus Nauser

1 INTRODUCTION

Amongst the many effects of the globally growing concern about environmental problems are the rising expectations placed in contemporary science. Once the inherently social nature of environmental problems is acknowledged it is evident that they present a great challenge to social science, too. However, owing to its hitherto modest presence in public debate, the question may arise whether social science actually is in the position to offer policy-relevant contributions.

In this paper a social scientific approach is presented for discussion which – besides its political relevance – deserves attention for the interest that its object meets in the public realm: the perception and assessment of the environmental problems, in confrontation with individual agency, have for over twenty years been the focus of research related to the concept of 'environmental concern' (or 'environmental awareness'). Many studies conducted in the field of environmental concern are motivated by the topicality connected to it. Yet, as Stern and Oskamp (1987: 1070) have pointed out, 'researchers [in environmental psychology] often overstate the importance of their work for solving major world problems'. Notwithstanding the intended or hoped-for effects, the way an object is conceptualized in science bears implications relevant to social practice. As Giddens (1984: xxxv) puts it: 'Theories and findings in the social sciences are likely to have practical (and political) consequences regardless of whether or not the sociological observer or policy-maker decides that they can be "applied" to a given practical issue.'

On this background, the subject of this paper is a critical appraisal of applied research on environmental concern and an attempt to show a possible way to improve its value for social practice. First, in a generalizing manner, a description of the conceptual foundations of environmental concern research is given. Problems and shortcomings with respect to its descriptive and interpretive capacities in a practical–political context are elaborated. In view of overcoming these deficiencies, a selective presentation of Giddens' (1984) theory of structuration is given, followed by the application of some of its concepts to the topic of environmentally responsible agency. The paper concludes with a number of arguments regarding the presuppositions of applied social scientific enquiry in the field of environmental concern. These may contribute to a more pertinent debate on responsible agency on the academic as well as on the public and political level, giving way to more adequate, credible and thus successfully implementable policy approaches.

2 THE ENVIRONMENTAL CONCERN APPROACH AND ITS DEFICIENCIES

'Environmental concern' is a term used in a non-standard way in research to denote the causal relationship between individual bodies of knowledge, attitudinal positions and behavioural characteristics with respect to environmental problems. Research on this topic is rooted in the theoretical concepts and widely varied behavioural models from the field of social psychology. Even though environmental psychology (focusing primarily on mental processes) and environmental sociology (biased more towards socio-cultural determinants of conduct) are its heartland, related studies have been carried out extending far beyond the fields of sociology and psychology.

The main areas of concern in this research tradition are, besides the documenting of the individual valuation of environmental problems and environmental policy, the elucidation of the cognitive structure of environmental attitudes and the investigation of the attitude–behaviour–relation.[1] A further, policy-related goal of research is to contribute to an assessment and improvement of strategies employed to foster environmentally sound attitudes and practices. With respect to contemporary political issues the main attention is focused on the domains of waste and recycling, energy conservation, transportation and environmental education. Studies in the German-speaking realm – to which most of the discourse in this contribution refers – are mainly based on sample survey methodology while in American research experimental methods are of considerably greater importance.

The claim to political relevance of the environmental concern approach derives from the fundamental assumption (implicit or explicit) that changes in individual attitude and behaviour are essential for coping with environmental problems. Research is likely to foster this process if it contributes to a better understanding of individual problem handling.

This assumption is based on the fact that most environmental problems are in a relatively easily demonstrable causal relation to individual modes of conduct. It is in contrast to the empirical finding that there are scarcely any changes to be noted in individual agency, in spite of the general recognition of the seriousness of environmental problems and a widespread grasp of the ecological consequences of the modern consumer lifestyle. Thus, a social science which does not stand aloof from the practical and political challenge presented by environmental problems is necessarily confronted with the question of how to interpret this apparently contradictory situation. What are the consequences to be taken regarding the role of the general population in environmental policy-making?

In the public forum, the term 'environmental concern' is an emotionally laden one. It provides the basis – though seldom one with a clear content – of numerous discussions concerning the assessment of environmental problems and the willingness of a broad segment of the population to make a contribution toward protection of the environment. In Switzerland, a country known for its well-developed system of participatory democracy, such discussions are carried on perhaps with a somewhat greater intensity than in other places.[2]

From the fundamental assumption presented above, often the conclusion is drawn that, in the interest of combating the problems at their roots, one must begin with the individuals, with their everyday practices and the motivation behind them. Establishing and propagating an adequate understanding of the problem and an associated change in attitudes ought to be the first step in building the basis for more environmentally respectful lifestyles and for a change in environmentally harmful structures at all levels of society.

Owing to its plausibility (at least on the surface) this mode of argumentation has become widespread. It is the basis for many activities of environmental organizations in their attempt to sensitize and mobilize the public in favour of environmentally responsible agency. This argument is also frequently cited by public officials and politicians, when the responsibility of individuals in coping with environmental problems is being stressed. And finally, the manufacturers and distributors of goods and services have reacted to criticism from the ecological front by emphasizing the consumer-dependent nature of their activities.

2.1 Conceptual features of the research on environmental concern

The background for the presentation and discussion of the environmental concern approach which follows is provided by numerous empirical and conceptional studies conducted in particular in Germany and Switzerland in the 1980s (Grob 1990, Spada 1990, Dierkes and Fietkau 1988, Urban 1986 and 1987, Balderjahn 1986, Langeheine and Lehmann 1986a, Tampe-Oloff 1985 and 1986, Fietkau 1984, Kessel and Tischler 1984 [cf. also Milbrath 1984 for the corresponding English report], Nöldner 1984, Fietkau et al. 1982, Fietkau and Glaeser 1981, Fietkau and Kessel 1981, Winter 1981, van Raaij 1979). In many cases these are built upon American studies from the 1970s (written up in detail in Langeheine and Lehmann 1986a, chap. 2). Additionally, a selection

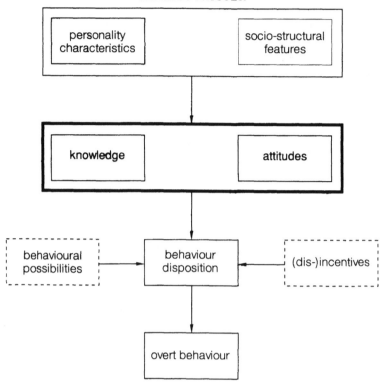

Figure 15.1 Behaviour model of environmental concern research

of recent American contributions have been considered (Buttel 1987, Stern and Oskamp 1987, Gray 1985).

Figure 15.1 depicts the most important elements and relationships of the behavioural model relevant to these investigations. Numerous differentiations and modifications of the model arise from the various theoretical concepts, used in theory-based studies for the clarification of modes of action (e.g. norm-activation, dissonance, self-perception, decision theory).

Attitudes are the most intensively investigated element of research on environmental concern. 'Attitude', in the context of this paper, is used as a collective term to denote beliefs, values, opinions, intentions, motives and feelings. These can be of a rather general nature (e.g. to developments in society or to individual life goals), but they can also relate to the evaluation of very specific, real or hypothetical situations, circumstances or outcomes. In investigations of the cognitive structure of environmental orientations, attitudes are commonly treated as dependent variables. Independent variables are, on the one hand, socio-structural features which, in addition to the classic demographic characteristics such as age, gender, education, and professional status, may also include conditions of socialization, use of media, extent affected by environmental ills. On the other hand, environment-related attitudes are also brought in relation to psychological personality characteristics such as neuroticism, extraversion, authoritarianism and optimism.

'Knowledge' relates to the understanding of ecological relationships, the influence of human actions on natural ecosystems and/or to knowledge as behaviour-relevant know-how in the narrower sense. Knowledge as it interacts with attitudes is also investigated occasionally.

Of primary importance is the assumption that attitudes (together with knowledge) are the central determinants of environmentally relevant behaviour.[3] In the investigations of the relationship of attitudes (in the broad sense of the word, as described above) to behaviour, the latter becomes the dependent variable. The empirical clarification of the attitude–behaviour relationship is a central and controversial problem of environmental concern research. The conceptual extension of the behaviour model by the behaviour disposition factor represents the attempt at structuring the diversity of the potential behaviour determinants in a hierarchal way. The 'behaviour disposition' corresponds to the emphasis placed on the variables 'expectations about the outcome of alternative acts' and 'expectations of others' (adopted from a general behaviour model by Fishbein and Ajzen 1975).

In applied environmental concern studies, only marginal attention is paid to externally set behaviour possibilities and incentives and their relationship to behaviour. They are of significance in studies which concentrate on evaluation research as part of governmental environment programmes (cf. Stern and Oskamp 1987).

The following representation of the behavioural concept can be derived from the applied research on environmental concern.

Knowledge data, attitudes in various forms and behaviour dispositions are causally related to behaviour and are its foremost determinants. Socio-structural factors and personality traits have an indirect effect, conveyed by means of attitudes and behaviour dispositions, on behaviour. Behaviour itself as a dependent variable is sufficiently determined from the other elements of the model and can be interpreted in their light.

· The concept presumes the freedom of individuals to act consciously *vis-à-vis* the environment in a harmful or careful way. The execution of an act is understood as the product of cognitive, emotional and norm-related information processing that flows into actions via processes of assessment of behaviour alternatives. Preferences for certain methods and contents, as well as conventions in socio-psychological research, seem to be responsible for the lack of regard for the features of the actual context of agency, which are recognized theoretically but scarcely set out in an empirical way.

The foregoing is a rough sketch of the conceptual and interpretive framework of applied research on environmental concern. Supported by the field-specific discussion, this framework will be critically evaluated in the next two sections: how useful is it for the description and interpretation of environmentally relevant conduct and what is its relevance as a contribution to the solution of environmental problems?

2.2 Discussion

A survey of the attempts to achieve a social-scientific understanding of environment-related cognition, evaluation and behaviour yields sobering results. The reciprocal relationships and dependencies of the investigated variables are generally considered to be insufficiently clarified. Langeheine and Lehmann (1986b) agree with the opinion of Dunlap and van Liere (1978) that little is known about the determinants of environmental concern even after a tradition of research spanning fifteen years. Dunlap (1985: xiii) writes: 'hundreds of surveys have been conducted to assess attitudes toward environmental issues, while scores of experiments have been conducted to understand and modify environmental behaviors. Yet, despite this large amount of research, our knowledge of what are now called "ecological" attitudes and behaviors remains limited.'

There are a number of possible explanations for this situation. Kruse *et al.* (1990: 8) maintain that the following are typical features of ecological psychology:

(a) The heterogeneity of theoretical concepts and topics.
(b) The non-relatedness of a great number of the contributions.
(c) The general absence of interdisciplinary research in this field.
(d) The non-standardness of research strategies.

No doubt this statement applies equally well to the subdiscipline of environmental psychology.[4] Dunlap's statement is also to be seen in this light. A recapitulation of some of the most dominant conceptual and methodological problems discussed within the research tradition of the environmental concern approach is given by Weigel (1985: 60–74).

Notwithstanding the limited comprehensiveness of knowledge gathered to date, there is definitely some merit in research aimed at identifying and assessing psychological and sociological factors which are relevant to the investigation of how individuals deal with environmental problems. However, in the light of the global threat which these problems represent today, it seems not only justified, but indeed urgently necessary, to demand a widening as well as a focusing of the scientific research agenda. Needed are contributions which do justice to a complex reality and which counterbalance the numerous unworldly approaches that accumulate masses of detailed knowledge at the cost of excessive reductionism. If scientists acknowledge their being part of a more encompassing socio-political milieu, existing criteria for guiding and valuing scientific work seem to be – at least partially – in need of questioning and reorientation. A critical assessment of environmental psychology from this perspective is given by Stern and Oskamp (1987) and by Sommer (1987: 1494–6).

The subsequent discourse aims at outlining some of the problems and shortcomings of the environmental concern approach with a view to attaining a better understanding of the factors relevant to anti-environmental behaviour. These on the one hand are drawn from the literature and on the other are based on the confrontation of standard concepts and the defining of issues, with observations and experiences taken from everyday life. Two aspects in particular merit closer consideration: the conceptualization of agency and the way in which individual conduct is related to social reality.

2.3 Problems and shortcomings of environmental concern research with regard to policy

The conceptualization of agency

In studies on environmental concern, actions are normally interpreted as a conscious choice between perceived alternatives and competing goals. Various authors modify this assumption by pointing to the relative position of habitual behaviour (e.g. Spada 1990, Fietkau 1984). On the basis of an extensive treatment of approaches used in psychology to explain environmentally relevant behaviour, Nöldner (1984: 515) concludes that the models concerned implicitly go on the assumption that everyday behaviour occurs in a relatively conscious form. However, much of the lack of environmental responsibility in human behaviour can be explained by the fact that behaviour takes place automatically. Thus, it does not have an explicit relation to social norms, values and attitudes.

According to the concept of behaviour embraced by most empirical studies on environmental concern, the negative impact that practices have on the environment is – by means of the evaluation of alternative options – either consciously avoided or consciously accepted. If consideration is also to be given to unconscious, habitual behaviour, there must follow a further differentiation between habitual behaviour which harms the environment unintentionally, and that which, also unintentionally, does not harm it (which is the case for many types of behaviour expressing traditional values such as simplicity or frugality, for example).

When a causal relationship 'attitude – willingness-to-act – behaviour' can no longer be assumed automatically, the relevance of the findings on knowledge and attitudes is diminished, both for the interpretation of certain practices and for recommendations concerning how to promote pro-environmental behaviour through changing attitudes (cf. Fietkau 1984: 110–13). This has important practical implications, for example, for the consideration of environmental education: what contents to convey, what methods to choose, what effects to expect. The question arises anew of what conditions habitual or specifically motivated behaviour is dependent on, and what are the prerequisites for habitual conduct to become intentional conduct (and vice versa).

MARKUS NAUSER

The significance of the situative and structural context of agency in society

When the environmental concern research includes in its studies the social context in which individuals act, it normally limits itself to the normative milieu and the formative influences of upbringing and socialization. The – usually implicit – idea that individuals are free to decide how to act, within this framework, tends to be dominant. Consequently, the term 'environmental concern' refers to the idea of a generalized, consistent attitude which corresponds to a particular, generalized type of behaviour. However, on an everyday level, it is easy to ascertain that pro-environment behaviour as a general criterion for distinguishing between individuals has very limited usefulness. Situation-specific inconsistencies of acting, the extent to which ecological aspects are considered in different domains (e.g. waste reduction, energy saving, use of public transportation), and the very varied success of government programmes in these domains indicate that, in addition to the determinants investigated as part of the research on environmental concern, there are other important (context-related) factors which influence modes of actions.

These factors comprise the concrete behavioural context of a person, that is, the ways and means which influence the scope and the available alternatives for action, as well as various socio-structural preconditions for individual behaviour. From a policy-related point of view, it is precisely these factors which are of most importance for the promotion and facilitation of environmentally sound modes of behaviour. The significance of these has been hinted at in conceptual studies, too (cf. Dierkes and Fietkau 1988, Fietkau 1984, Fietkau and Kessel 1981). The first-mentioned also emphasize the socio-structural conditions for individual action: the ecological relevance of technical, social, political, economic and legal arrangements is often far from being immediately recognizable.

Buttel (1987) points out this negligence at the structural level. Basing his comments on an overview of the development and the status of research in environmental sociology, he reaches the conclusion that there is a great need for theoretical innovation which goes beyond the refinement of existing concepts and methods: 'Much of the empirical literature in environmental sociology receives theoretical guidance that is exclusively subjectivist and microsociological and . . . could benefit from a more macrostructural orientation' (Buttel 1987: 484).

In the following section a sociological approach will be considered which has a number of interesting points for the description and interpretation of environmentally relevant conduct. Anthony Giddens' theory of structuration is especially worthy of note, with respect to Buttel's statement cited above: the attempt is made to link the individual (micro-)level of behaviour in an innovative manner with the structural (macro-)level of the social context. In this way greater justice could be done to the requirements of a realistic integrative perspective which focuses primarily on agency while still not losing sight of the overall social relationships.

3   THE CONCEPTION OF AGENCY IN THE THEORY OF STRUCTURATION

In this second section a brief and highly selective presentation of the theory of structuration as elaborated by Giddens in his book, *The Constitution of Society* (1984), will be given. It serves as a frame of reference for the subsequent discussion of environmentally harmful agency.

As Giddens (in Held and Thompson 1989: 294) points out, it was not his intention to formulate a theory that offers a ready-made analytical scheme for empirical exploration. He considers it a 'sensitizing device', and in this sense it will be used in the remainder of this contribution.

According to Giddens (1984), from which the subsequent quotations are taken (unless otherwise noted), individual agency can be characterized and interpreted along the following lines.

'Routine . . . is the predominant form of day-to-day social activity. In the enactment of routines agents sustain a sense of ontological security' (p. 282). Human agency, seen as the competent and skilful mastering of daily life in a continuous flow of conduct, is fundamental to Giddens' conception of society. In German, the term 'routine' has a double meaning which seems to express very well the semantic content referred to by Giddens. The first meaning stresses the repetitive, taken-for-granted, habitual character, containing an element of passiveness. However, as emphasized by Giddens, habitually performed practices do not imply mindlessness: 'Routine is founded in tradition, custom or habit, but it is a major error to suppose . . . that these phenomena are simply repetitive forms of behaviour carried out "mindlessly". . . . the routinized character of most social activity is something that has to be "worked at" continually' (p. 86).

The second meaning of 'routine' in German concerns its use to attribute experience to a person's way of acting. A person may 'have routine', 'be routinized', or do something 'with routine', meaning that he or she has attained high skills by virtue of long practice. Thus, according to this second use, interpreting most of human agency as routine accounts for the familiarity of individuals with many practices and for the great extent to which these have proven their suitability and appropriateness in a given context. This perspective, of course, stands in obvious contrast to the 'inside' perspective assumed by research approaches that are interested in agency under a very specific motivational aspect, for example, the concern with environmental degradation.

The outlined understanding of routine is closely connected to the notion of 'knowledgeability' with reference to which Giddens develops his conception of the cognitive organization of human consciousness: individuals are knowledgeable agents, that is, they 'know a great deal about the conditions and consequences of what they do in their day-to-day lives' (p. 281). Yet, routinized action does not require a permanent reflection on motives, intentions and available means. Although individuals continually monitor actions (their own and those of others) and contexts (social and physical) in a reflexive manner, this

> reflexivity operates only partly on a discursive level. What agents know about what they do, and why they do it – their knowledgeability as agents – is largely carried in practical consciousness. Practical conscious-ness consists of all the things which actors know tacitly about how to 'go on' in the contexts of social life without being able to give them direct discursive expression.
>
> (p. xxii-xxiii)

Accordingly, most of the day-to-day conduct of individuals is to be under-stood as diffusely motivated. Instead of attributing specific motives to specific acts, Giddens suggests 'a generalized motivational commitment to the inte-gration of habitual practices across time and space' (p. 64). If, for whatever purpose, reasons for specific forms of conduct must be given, these are subject to the limitations of discursive consciousness, that is, the level of awareness on which human beings express themselves verbally.[5]

Individual knowledgeability itself is limited by motivations and repressions hidden in the unconscious, and by 'false theories, descriptions or accounts both of the contexts of [a person's] action and of the characteristics of more encompassing social systems' (p. 92). Consequently, the discursive inter-pretation of social reality as expressed in conversations need not always be in agreement with the routine management of everyday tasks consistent with practical consciousness.

Finally, the relation of human agency to social institutions and structures bears mention: it is conceptualized in the core concept of the theory of structuration, the duality of structure. Giddens defines structure as 'the medium and outcome of the conduct it recursively organizes' (p. 374). Through their routine actions individuals unintentionally, and usually without being aware of this, reproduce the structural properties of the social systems of which they are part. On an everyday level, this implies an interdependence of change in conduct with change of structure: patterns of agency will not change independently of structural arrangements but only in a process of mutual adaptation. As it is in the nature of habitually established practices that they – similarly to a thread of a spider's web – are part of their context beyond the narrowly definable purpose they may serve at first sight, campaigns calling for a change in habits are futile unless there is sufficient flexibility to integrate this change in the given structural frame of reference.

Structures, as context-related rules of conduct and (unevenly distributed) resources which individuals habitually dispose of, are enabling on the one hand, while they impose, on the other, certain, only partially acknowledged, boundaries. The enabling and constraining properties of social structure, as well as the limited knowledgeability that agents have of themselves and the social system in which they live, are contained in the encounters and settings of day-to-day life.

Three conclusions will be drawn in regard to the subsequent discussion of environmentally harmful agency:

1   Understanding actions presumes considering their embeddedness in diffusely motivated flows of conduct which relate to and depend on specific social contexts.

2   The pre-eminence of routine in human conduct calls for particular attention to practical consciousness. This may imply methodological considerations (owing to the limited accessibility of practical consciousness to techniques based on verbal or written expression) and an account of the boundedness of the knowledgeability of actors.

3   Human agency and social structure are intimately related to one another. The description and interpretation of agency needs to be rooted in an understanding of the interplay of individual capabilities and competences (available resources and individual constraints) and structural presuppositions (formal and informal rules and social constraints).

3.1   Environmentally harmful agency in the light of the theory of structuration

The study of environmentally sound agency implies the study of routine agency in confrontation with demands expressed in public debate for altering or abandoning specific individual practices. Day-to-day activities such as managing a household, shopping for everyday commodities, driving to work, or meeting with friends are typically of a highly routinized nature. At present many individuals are faced with the dilemma of becoming or being aware of the undesirable properties of routine activities which hitherto not only were taken for granted but also are part of flows of action which in many respects are constitutive to the normal course of life in society. This growing awareness is at the same time undermined by the experience of undue personal inconvenience and by contradictory information. Who is in fact responsible for specific environmental problems, what is effectively to be done and by whom, how are unintended or undesired effects of strategies for change to be assessed? These are questions commonly without an unambiguous answer, leaving many a concerned person in a muddle of guilt, anxiety and confusion.

According to Giddens' differentiation of consciousness the problem of adapting individual practices to the needs of ecological sustainability can be described as:

(a) A conflict between practical and discursive consciousness.
(b) A conflict within discursive consciousness.

Both cases will be discussed below in order to demonstrate the hermeneutical value of Giddens' concepts of routinized agency, discursive and practical consciousness and structures (as rules and resources recursively implicated in the reproduction of social systems) when applied to the investigation of environmentally harmful agency.

3.2 Environmentally harmful agency seen as a conflict between practical and discursive consciousness

In facing the environmental problems, not only individuals but also human civilization as a whole is confronted with a fatal misconception on which much of its development since industrialization seems to be based. The place of humankind within nature is in need of revision in the light of human-made phenomena such as the ozone hole, the greenhouse effect and large-scale technical disasters.

On the level of general causalities the ecologically (self-)destructive tendencies of industrial and post-industrial societies are widely understood. This understanding is given clear expression, for example, in the numerous opinion polls ranking the environmental crisis high on the list of most urgent contemporary problems. As has been noted time and again, the widespread public concern stands in sharp contrast to the very faint traces indicating a reorganization of individuals' day-to-day life. Notwithstanding the questionable fact that such statements frequently oppose problem assessments on a social level to the intentions or routines followed on an individual level, when starting from a routine-based conception of human agency, there is only limited value in seeking explanations from discursively reported knowledge and attitudes. Daily enacted routines may be in opposition to knowledge about unintended consequences and possible remedies, but nevertheless they are appropriate and successful with reference to the institutions to which they relate and the social contexts by which they are encompassed. If now, as a result of public information campaigns, certain practices are stigmatized as ecologically – and thus socially – irresponsible, the consequences for and the reactions of the individual are potentially negative in several respects.

Routinely followed practices are linked to conventions concerning the legitimacy of goals and the means used in pursuing these goals. Rules, mostly of the informal kind, regulate day-to-day conduct, entailing positive or negative sanctions. These rules are incorporated in the knowledgeability that individuals tacitly have of the social context in which they live. As Giddens notes: 'we can say that awareness of social rules, expressed first and foremost in practical consciousness, is the very core of that "knowledgeability" which specifically characterizes human agents' (p. 21f).

In terms of the evolution of human societies, many of the presently valid rules which are relevant to the ecological impact of human activity are linked to the achievements of the industrial era. They are firmly rooted in the dramatically increased opportunities that individuals have for the organization of their lives, forming constitutive elements of the 'modern way of life'.

Seen against this background, the current imperatives of ecologically sound agency appear in a different light. Conduct guided by the rules underlying the 'affluent society' is contrasted by demands more or less unrelated to the web of the recursively intertwined and stabilized relations, obligations and dependencies characterizing and enabling this modern lifestyle. The routinely followed or accepted rules of conduct may not be – and commonly are not – compatible with the requirements of ecological sustainability. Nevertheless they still have a strong position in the social organization of industrial and postindustrial societies. This allows for paradoxical situations such as persons who (as employees) are dependent on partaking in the production of goods which they (as consumers) are supposed to reject for reasons of environmental protection.

Consequently, ecological imperatives and proposals for action need to be analysed on the grounds of their compatibility with the routines constituting the normal course of everyday life. Comparative empirical research could show to what extent actual resistance to (ecologically motivated) demands to alter modes of conduct is an expression of the degree of interconnectedness that a practice has acquired with respect to the encompassing social system. A manifest example is the range of different experiences connected to programmes promoting recycling on the one hand and those attempting to induce a shift from private to mass transportation on the other hand.

The tight interrelatedness of routines and rules is but one aspect impeding the application of problem-relevant knowledge to everyday practices. In addition to the 'substantially "given" character of the physical milieux of day-to-day life' (p. xxv), the options which are open to an individual depend strongly on the resources that he or she disposes of.[6] To give an example: before routinized practices come into play, energy conservation in buildings obviously depends on the owner/tenant status as well as on various decisions taken at early stages of planning a building. Similarly, authority relations in husband/wife or employer/employee associations deserve particular attention, when the consistency of concern and action is at stake. Evidently, the far-reaching effects of (intended or unintended) control, which are understood as 'the capability that some actors, groups or types of actors have of influencing the circumstances of action of others' (p. 283), are of great significance to the understanding of many instances of environmentally relevant conduct.

Furthermore, important conditions of action are constraints concerning health, money and access to information together with limitations deriving from varying characteristics of the context of agency (availability of convenient facilities or suitable alternatives, for example, recycling centres or context-compatible, ecologically sound consumer goods). These need to be taken into account when intending a realistic assessment of potential alternatives for environmentally harmful agency.

Another example of potential collisions of practical and discursive consciousness can be found in the way in which routines relate to the mental constitution of human beings. 'Routinization is vital to the psychological mechanisms whereby a sense of trust or ontological security is sustained in the daily activities of social life. . . . the apparently minor conventions of daily social life are . . . essential . . . in curbing the sources of unconscious tension that would otherwise preoccupy most of our waking lives' (p. xxiii-xxiv).

Thus, routinization can be seen as a mechanism of psychological economy which separates the 'taken-for-granted' from instances of immediate concern or interest. This understanding sheds a different light on efforts attempting to lift habitual practices of subordinate relevance to the level of discursive consciousness. If countless routine activities have to be treated as problems demanding critical attention, the 'mental division of labour' indicated above is – at least temporarily – fundamentally questioned. On the practical level it, therefore, seems rational to insist on an open and transparent dialogue between scientists, entrepreneurs and politicians over general ecological parameters which are relevant to the whole life cycle of consumer goods, rather than expecting consumers to make the 'right' choice based on contradictory information in front of the ever-changing supermarket shelf. More evidence for this point of view will be furnished in the next section.

3.3 Environmentally harmful agency seen as a conflict within discursive consciousness

The concepts of routine and practical consciousness undoubtedly are of importance in accounting for the persistence of patterns of agency. However, it cannot be denied that mastering the environmental problems requires efforts involving discourse on a broad social basis. If the argument in the previous section can be taken as stated, rules and resources structuring much of everday life in society must themselves become the object of public debate and occupation.[7] For the present, this aspect clearly has been pushed aside by the emphasis placed on altering specific practices and habits of everyday life. What, therefore, can be said about the conditions under which individuals deliberately seek to convert their concern into action?

While in traditional (pre-industrial) societies the contexts of perception, experience and agency were by and large identical, in modern, highly differentiated and globally interlaced societies information and knowledge on the one hand and immediate perception and practice on the other diverge widely. Consequently, what a person knows and believes about environmental problems depends largely – more than in many other domains of political concern – on mediated information. While being aware of its fundamental significance to their well-being, most individuals lack the ability to evaluate this information independently.

Given these circumstances, ontological security deriving from the foreseeable nature of everday life is on the decline, in favour of anonymous professional expertise which is attempting to counterbalance the uncertainties that accompany fundamental activities such as eating, drinking, breathing or exposing one's skin to the sunlight. What is to be feared and what is to be hoped are defined by experts, negotiated in political controversy and selectively brought to the lay public's attention by the mass media. Concern is predominantly directed at sinister threats lurking unseen behind the façade of the ups and downs of modern life.

In this kind of situation, individuals who are aware of their share of the problems and who are looking for opportunities actively to divert the frightening prospects that are forecast, are confronted with a confusing array of 'truths'. A closer look at the topics related to environmental problems which are presented in the mass media reveals a crude mix of disastrous events to which individuals seem destined to fall victim, recommendations of often questionable practicability or credibility, general information on the functioning of ecosystems unrelated to individual day-to-day experience, and problem interpretations by experts which diverge at times on even the most basic assumptions. Most individuals characteristically have neither the experience nor the competence routinely to process and filter these masses of 'information' according to criteria of relevance and credibility. In a world where all truth is relative, the limitations of a policy-approach based on exhortations to insightful, reasonable agency must be duly recognized and taken into account.

4 PERSPECTIVES FOR FUTURE RESEARCH

From the foregoing discussion the conclusion can be drawn that many illusions, simplifying assumptions and – not rarely – hypocritical assertions seem to be involved in the debate on the potential and necessity of the individual to desist from environmentally harmful practices or habits. Applied studies with a down-to-earth approach could help in enhancing the adequacy and efficacy of related policy strategies, bringing to the fore another important feature of this debate: the passing on of responsibility.

If, as described above, most individuals – besides being totally involved in context-related and context-conforming routines – frequently are not in the position to reach rationally founded decisions independently, the fundamental question arises as to what extent they can be held responsible for the (ecological) effects of their actions. Can persons who are unwittingly part of economical, political and ecological processes of global dimensions be called to account for the far-reaching (unintended) consequences of their everyday conduct? Answering this question will presume consideration of:

1 The bearing of individuals' decisions for the living conditions, opportunities and actions of others (owing to a difference in access to allocative and authoritative resources, that is, the potential for control).
2 The personal variation in awareness and comprehension of the (economical, political, ecological) consequences of alternative actions owing to differential access to relevant information.
3 The structures of information production, processing and dispersion and the conditions under which decisions are taken and consequences must be accounted for.

Clearly, thus far the environmental concern approach has contributed little to issues such as these. Its focus is the individual exposed to the environmental problems as they are represented and constantly redefined in processes of public opinion formation. However, without proper consideration of the above issues, the project of building an ecologically sustainable society runs the risk of lacking reference to a less surfacial, more complex reality.

A clear-headed approach to environmental policy, envisaging the active participation of the general public, not only presumes a differentiated understanding of the production and handling of information and the distribution of resources but also needs to refine its concept of individual responsibility and collective, democratic control in the light of a global economical and technological self-dynamics unfolding at unprecedented orders of magnitude.

Beyond these structurally rooted variations of conditions of agency, human existence is characterized by a decline of control over some of the most basic realms of individual reproduction. The conduciveness to health of foods, activities and environments has more and more been subjected to the judgements of anonymous specialists who are to be trusted in a quasi-fatalistic manner (cf. Beck 1986). Thus, a politically relevant aspect for future developments might be the tapping of the potential to compensate for this loss on local and regional levels. The daily experienced surroundings are more easily comprehended and so autonomy – and thus responsibility – may be regainable to a certain degree.

In the endeavour to find a way out of the human-made environmental threat, social scientists, too, can assume responsibility by playing an active role in making more transparent the basis on which problems and their causes are defined and strategies developed and evaluated. They can contribute to an informed, broad-minded discussion of the options available and the implications involved. To meet this end there is a need to provide adequate, integrative descriptions of the states and processes characterizing social and individual change in respect to environmental quality.

To conclude, four suggestions will be put forward for taking environmental concern research closer to social and political reality:

1 The environmental concern approach has a clear tendency to analyse and

interpret the environmental problems as problems of individual cognitive processes in their interplay with the normative context. If it wants to render its contributions more pertinent to environmental policy, it might profit from a conceptual expansion in order to encompass socio-structural aspects in a broader sense. A possible way of achieving this is increased attention for agency as the mediating element between cognition, person-specific constraints and opportunities, and the presuppositions contained in the social system.

2 Applied research on environmental concern commonly is biased towards a rationalistic understanding of agency. Its interpretive and explanatory power could gain from making due allowance for the diffusely motivated, experience-guided nature of much of human agency.

3 The environmental concern approach presumes an individual who is – within the limits of social norms – free to act according to his or her knowledge about, and understanding of, the world. This paper has attempted to assemble some arguments which demonstrate that this standpoint needs differentiation. This seems advisable also with respect to some serious misinterpretations entailing an overly idealistic free-will conception of human agency. First, it diverts from the fact that in reality even democratic societies are not made of homogenous aggregates of individuals with equal opportunities, but instead consist of members with unevenly distributed access to information and power. Secondly – and more closely concerning the environmental concern debate – it fosters a disparaging notion of human beings, as people who do not seem to be willing or able to learn when faced with the awkward ecological consequences of their actions. If their behaviour is considered to be irrational (as is a popular belief in certain circles), this implies that they lack political majority. This line of reasoning conveniently supplies the justification for isolated decisions and actions to be taken by experts and officials in key positions. If scientific research does not want to back such simplifying (and dangerous) conclusions, it cannot exclude the ideological dimension of the environmental problems and reduce them to a matter of attitude and lifestyle change on an individual level.

4 An action- and experience-based approach, as proposed in Giddens' theory of structuration, may be inspiring as to the direction in which existing conceptual gaps could be filled and weaknesses overcome. It offers a sound basis to enhance the value of social scientific contributions to the topic of an ecologically sustainable social organization. In doing so, it might help in dealing with some of the difficulties and limitations of integrative scientific work in the domain of individual–society interaction.

5 NOTES

1 Cf. the reviews published in 1987 by Buttel and by Stern and Oskamp.
2 According to opinion polls, in Switzerland since the mid-1970s environmental protection has occupied first place, almost without exception, on the list of social problems (Walter-Busch 1989). A variety of other findings from current surveys on environmentally relevant attitudes and behaviour in Switzerland can be found in Dyllick 1990.

   For the USA of the 1980s, a constant rise in the support of radical ('regardless-of-cost') measures to protect the environment was revealed by polls which in 1989 were representative of 80 per cent of the population (Ruckelshaus 1989: 117).
3 The reverse direction of operation is also investigated in individual cases (cf. Weigel 1985: 72–4).
4 Ecological psychology defines 'environment' in a broader sense as the individual perceptual and behavioural context.
5 'Where what agents know about what they do is restricted to what they can say about it, in whatever discursive style, a very wide area of knowledgeability is simply occluded from view' (Giddens 1984: xxx).
6 Giddens (1984: 33) distinguishes between allocative resources as 'forms of transformative capacity generating command over objects, goods or material phenomena' and authoritative resources as 'types of transformative capacity generating command over persons or actors'; cf. Lawrence (this volume) for a more extensive discussion of the notions of rules and resources within the theory of structuration.
7 Cf. Evers and Nowotny (1987) on the process of individuals becoming aware of their role as potential co-creators of society.

6 REFERENCES

Balderjahn, I. (1986) 'Das umweltbewusste Konsumentenverhalten', PhD thesis, Berlin: Technical University Berlin.

Beck, U. (1986) *Risikogesellschaft – Auf dem Weg in eine andere Moderne*, Frankfurt a/M: Suhrkamp.

Buttel, F.H. (1987) 'New directions in environmental sociology', *Annual Review of Sociology* 13: 465–88.

Dierkes, M. and Fietkau, H.-J. (1988) *Umweltbewusstsein – Umweltverhalten*, Mainz: Kohlhammer.

Dunlap, R.E. (1985) 'Foreword', in D.B. Gray (ed.) *Ecological Beliefs and Behaviors. Assessment and Change*, Westport, CT: Greenwood Press.

Dunlap, R.E. and van Liere, K.D. (1978) 'Environmental concern. A bibliography of empirical studies and brief appraisal of the literature', *Public Administration Series*, Bibliography P-44.

Dyllick, T. (1990) 'Ökologisch bewusstes Management', *Die Orientierung* (Schweizer-ische Volksbank, Bern) 96.

Evers, A. and Nowotny, H. (1978) *Über den Umgang mit Unsicherheit – Die Entdeckung der Gestaltbarkeit von Gesellschaft*, Frankfurt a/M: Suhrkamp.

Fietkau, H.-J. (1984) *Bedingungen ökologischen Handelns. Gesellschaftliche Aufgaben der Umweltpsychologie*, Weinheim: Beltz.

Fietkau, H.-J. and Glaeser, B. (1981) 'Wie umweltbewusst sind Landwirte? Über-legungen und empirische Befunde', *Zeitschrift für Umweltpolitik* 4: 521–44.

Fietkau, H.-J. and Kessel, H. (1981) *Umweltlernen. Veränderungsmöglichkeiten des Umweltbewusstseins*, Königstein/Ts: Hain.

Fietkau, H.-J., Kessel, H. and Tischler, W. (1982) *Umwelt im Spiegel der öffentlichen Meinung*, Frankfurt a/M: Campus.

Fishbein, M. and Ajzen, I. (1975) *Belief, Attitude, Intention and Behavior*, Reading, MA: Addison-Wesley.

Giddens, A. (1984) *The Constitution of Society*, Cambridge: Polity Press.

Giddens, A. (1989) 'A reply to my critics', in D. Held and J.B. Thompson (eds) *Social Theory of Modern Societies: Anthony Giddens and His Critics*, Cambridge: Cambridge University Press.

Gray, D.B. (ed.) (1985) *Ecological Beliefs and Behaviors. Assessment and Change*. Westport, CT: Greenwood Press.

Grob, A. (1990) 'Meinungen im Umweltbereich und umweltgerechtes Verhalten. Ein psychologisches Ursachennetzmodell', PhD thesis, Bern: University of Bern.

Held, D. and Thompson, J. B. (eds) (1989) *Social Theory of Modern Societies: Anthony Giddens and his Critics*, Cambridge: Cambridge University Press.

Kessel, H. and Tischler, W. (1984) *Umweltbewusstsein. Ökologische Wertvorstellungen in westlichen Industrienationen*, Berlin: Edition Sigma.

Kruse, L., Graumann, C.-F. and Lantermann, E.-D. (1990) *Ökologische Psychologie: Ein Handbuch in Schlüsselbegriffen*, München: Psychologie Verlags Union.

Langeheine, R. and Lehmann, J. (1986a) *Die Bedeutung der Erziehung für das Umweltbewusstsein*, Kiel: IPN, University of Kiel.

Langeheine, R. and Lehmann, J. (1986b) 'Stand der empirischen Umweltbewusstseins-forschung', in R. Günther and G. Winter (eds) *Umweltbewusstsein und persönliches Handeln*, Weinheim: Beltz.

Milbrath, L.W. (1984) *Environmentalists: Vanguard for a New Society*, Albany, NY: State University of New York Press.

Nöldner, W. (1984) 'Psychologie und Umweltprobleme. Beiträge zur Entstehung umweltverantwortlichen Handelns aus psychologischer Sicht', PhD thesis, Regensburg: University of Regensburg.

Raaij, W.F. van (1979) 'Das Interesse für ökologische Probleme und Konsumentenver-halten', in H. Meffert *et al.* (eds) *Konsumentenverhalten und Information*, Wiesbaden: Gabler.

Ruckelshaus, W.D. (1989) 'Toward a sustainable world', *Scientific American* 261 (3): 114–20B.

Sommer, R. (1987) 'Dreams, reality, and the future of environmental psychology', in D. Stokols and I. Altman (eds) *Handbook of Environmental Psychology*, New York: Wiley.

Spada, H. (1990) 'Umweltbewusstsein: Einstellung und Verhalten', in L. Kruse *et al.* (eds) *Ökologische Psychologie*, München: Psychologie Verlags Union.

Stern, P.C. and Oskamp, S. (1987) 'Managing scarce environmental resources', in D. Stokols and I. Altman (eds) *Handbook of Environmental Psychology*, New York: Wiley.

Tampe-Oloff, M. (1985) 'Zur Komplexität als Hindernis problemorientierter Reaktion auf das Waldsterben', PhD thesis, Freiburg i.B.: University of Freiburg.

Tampe-Oloff, M. (1986) 'Hindernisse einer problemorientierten Reaktion auf das Waldsterben im individuellen und politischen Bereich', in R. Günther and G. Winter (eds) *Umweltbewusstsein und persönliches Handeln*, Weinheim: Beltz.

Urban, D. (1986) 'Was ist Umweltbewusstsein? Exploration eines mehrdimensionalen Einstellungskonstruktes', *Zeitschrift für Soziologie* 5: 363–77.

Urban, D. (1987) 'Die kognitive Struktur von Umweltbewusstsein. Ein kausalanalytischer Modelltest', unpublished manuscript, Duisburg: University of Duisburg.

Walter-Busch, E. (1989) 'Wertewandel bei Bevölkerung und Unternehmen', Report 2 of the National Research Programme *City and Traffic*, Zürich.

Weigel, R.H. (1985) 'Ecological attitudes and actions', in D.B. Gray (ed.) *Ecological Beliefs and Behaviors. Assessment and Change*, Westport, CT: Greenwood Press.

Winter, G. (1981) 'Umweltbewusstsein im Licht sozialpsychologischer Theorien', in H.-J. Fietkau and H. Kessel (eds) *Umweltlernen*, Königstein/Ts: Hain.

16

THE 'CONCRETE MIND' HEURISTIC

Human identity and social compound from things and buildings

Alfred Lang

1 ON THE RELATIONSHIP BETWEEN THE INNER AND THE OUTER WORLD AS SEEN BY PSYCHOLOGY

It appears that psychology has its origins in the private experience of people. When they comprehended that an outer world might exist independent of them and of their experience, that is, the so-called objective world, the task of understanding the inner or private 'world' arose. Thus the mind–body problem, for generations, has been one of the unresolved great questions. Reasonable as it seems to assume influences going from the outer to the inner world, the reverse, although an undeniable fact, somehow contradicts basic tenets of scientific wisdom of a materialistic era. But the problem is a pseudo-problem, because there are not two realms of reality to combine, but simply two 'languages' to translate between them.

On becoming an empirical science, psychology has naturally been caught by modern scientific doctrine and has restricted its endeavour to a cause–effect way of looking at the world of people. Consequently, it thinks of behavioural acts as functions of situations. The mind is thought of as a mediating structure which has been built up by the impact of situations and is thus capable of carrying the influence of past situations upon present and future acts. Its mental or experiential aspects were declared a mere epi-phenomenon by the behaviourists, notwithstanding the public prevalence of a subjectivist or unscientific psychology and periodic subjectivist waves also in the academic realm. Unfortunately, this materialistic and positivist conception of the mind–body problem has led to an image of man as an exclusively reactive system.

Yet, it is one of the most obvious characteristics of living entities, including simpler organic systems and humans as individuals and societies, that they are *to some extent resisting the influences* from their surroundings and that they even shape the world around them. In this paper, I try to reconsider the balance between humans and their surroundings. I propose a more 'symmetrical' relationship which is capable of furthering the value of both. Humans are neither the godlike rulers of religious traditions nor the perfect puppets of a world of scientific myths.

2 ON THE LINKS BETWEEN HUMANS AND THE WORLD

The reactive image of humans is, in my opinion, one of the reasons that scientific psychology has nothing relevant to say to questions such as: why is it the case that people design and construct things and build houses and cities? There is a large number of related questions, a few of which might be listed as follows:

(a) Why do people appropriate spaces and accumulate objects?
(b) Why do people prefer some places and things, hate or neglect others?
(c) Why do people build and design differently in particular regions of the world and periods of history?
(d) In general, why do people spend a considerable proportion of their time and effort in changing their surroundings?

Ordinarily, explanations of purpose or attainment are offered. Sociological answers suggest benefits in group life. Biological answers refer to survival value and might in particular point out the short-term profits of reducing rather than submitting to selection pressures. But both of these explanations are unfit to explain particular manifestations of the built or the designed, because they are of a functional type and thus of an arbitrary nature. If one is ultimately interested to give a cause–effect type explanation, that is, an explanation that specifies the antecedents rather than the possible (not the necessary!) consequences of some phenomena, then a psychological perspective of explanation is indispensable. It is always individual people, albeit within social groups and/or traditions, who do the actual designing and constructing.

The above questions ask for the necessary and sufficient conditions of the following:

(a) Why some part of the surrounding world (as given, independent of people) is 'turned' into the 'environment' (of some person or group) by human beings and other living systems.
(b) How the environment of some person or group 'operates' on that person or group.
(c) How the separateness of living systems from the surrounding world, as well as their connectedness with it, is accomplished.

These general questions encompass the basic ecological problem. Naturally, this paper will give no answers but restricts itself to outlining a direction using facts and speculation in a free manner.

The questions have two branches, in that they refer to both an input and an output process. First, the study of perception of living systems demonstrates that any such system has evolved its proper way of gaining such information from the world that allows it to maintain at the same time its adaptedness to the real surrounding conditions as well as its relative independence from them. Unfortunately, in line with the reactive image of humans, traditional and modern study of perception in essence is trying to explain perception as a function of a given world, the so-called stimuli, and this in spite of the fact that stimuli must ultimately be described on the basis of other perceptions. Contrary to the prevailing opinion, perceptual systems do not simply image the stimulus or the world but instead construct their proper simplified and, as a rule, quite economical 'representation' of the world or, rather, produce their proper internal construction of something that refers to the surrounding world.

Second, as a rule the perceptual systems selectively refer to only those qualities of the world that have proven important, at least in some stage of the evolution of the respective phyla. In some phyletic strains, it seems to have been advantageous for the living systems to evolve beyond just acquiring matter, energy and information out of the world, and to act on sectors of that world, for example, by removing or collecting and composing material for burrows or nests, by cultivating symbionts, etc. The result is the creation of particular living conditions for those individuals, who thus improve on both their connectedness to and their separateness from the surrounding world by reducing their vulnerability. *Homo sapiens faber* is obviously the genus going rather far in that respect by creating material and symbolic culture as an enlarged and now quite indispensable living condition which we call the human environment.

All of this makes clear that the data of an empirical psychology should also include observations in the form of situations as functions of behavioural acts. At present, situations in psychology are studied (a) as stimuli and (b) as responses or indicators of behavioural acts. However and unfortunately, we find no systematic investigation, not even a taxonomy, of situations. If one claims a systematic nature of the mind and assumes ingoing and outgoing links between the mind and the surrounding world, then it is imperative to have at one's disposal also a systematic knowledge of those parts or aspects of the surrounding world which are or can become relevant for the individual in question, because those are in part the preconditions and in part the consequences of that particular mind or of that individual's behaviour. Obviously, natural laws do not specify everything characterizing situations.

To clarify terminology, I call the surroundings of a living system, be it individual or group, its 'environment', in so far as they have been relevant for that living system. The superordinate systems constituted of a living being together with its environment is called an 'ecological system'. Of course, a number of scientific and other disciplines study cultural artefacts. Usually they are separated from the persons involved and are dealt with as if environments were objective givens; so there is a high risk of losing sight of their being part and parcel of human beings (see also Lang 1985, 1988).

3 ON THE NOTION OF STRUCTURE FORMATION OR DYNAMIC MEMORY

It is not possible here to give more than a sketchy account of an eco-psychological perspective still in development. The central assumption refers to a general notion of structure formation, which can be specified as dynamic storage or memory, that is, creating a trace of some event, which in turn is capable of generating or influencing some further event. It is the prime function of structure formation for any living system to constitute both its separateness from and its connectedness to the surrounding world.

As far as morphological structure is concerned, this does not warrant discussion, because the organism is exactly that which separates the system from the rest of the world. Yet it should be remembered that there are, in all organisms, ingestive and eliminative substructures and, in higher organisms, additional sensory and executive subsystems, for keeping in touch – materially and informationally – with the surroundings.

The case is more intricate with information structures, but there are cogent parallels. For any living system, the capability of relying on an internal, space- and time-independent representation of all pertinent characters of the sur-rounding world is a very economical way of dealing with this world. Of course, this is adequate only as long as the surrounding world is not changing too much; yet this is exactly the precondition given for most plants and animals (except modern humans) in relation to their generation cycle. And such internal representations, together with perceptual subsystems for dealing with actual states of the surrounding world, make for an efficient connected-ness with or even for a kind of integration into the environment, considering physical support and locomotion, metabolism, and relation to con-specifics, prey or enemies.

At the same time, since the medium or carrier of this internal representation is necessarily something other than the world referenced itself, and since any storage system, including its way of encoding and its way of making use of the information stored, follows its own rules and laws (it is probably most reasonable to think of the memory as a sign system), separateness from the world or peculiarity is inevitably also constituted. In fact, since living information storage systems are always to some extent active, generative systems, I tend to think of the said separateness as a kind of autonomy. We have then to differentiate between two 'environments' of any individual: the external or real surroundings inasmuch as they are pertinent (I call them the 'ecological environment') and the internal representation thereof, which is often called the 'psychological environment' of the person in question. The two need not have identical contents. Both are constructs, not observables; they must be gauged by means of a third.

Structure formation is a very general principle. Phyletic history is also indirectly condensed in some structure, namely, the genom, and this is a generative memory capable of producing new autonomous and adapted organisms, morphologically and behaviourally. In addition, cultural productions, specifically environments modified by people, are structures incorporating a manifold history of human acts in situations. And ecological environments are themselves capable of generating or modifying new such acts and situations. These are at issue in the 'concrete mind' heuristic.

4 ONTOGENETIC MEMORY: THE INDIVIDUAL COGNITIVE AND ACTION STRUCTURE

On observing concrete everyday behaviour of living beings, it is easy to see that actions are determined to some extent by the characteristics of the actual situation, but they are also, and as a rule to a much greater extent, determined by internal representations of or 'knowledge' about the characteristics of the situation as well as 'knowledge' about one's action repertory. The advantage of the latter over the former lies in its pertinence far beyond the present inflow of information, in that the internal representations of the world also includes information on the possible future of the environment, on the possible consequences of one's decisions and undertakings, and so on. In addition it seems often more economical to store a readily available image of some complex situation rather than to analyse it anew every time that it occurs.

All of these internal conditions of acting are collectively termed the cognitive and motivational or action structure, ontogenetic memory or the mind. However, the mind is a construct, not an observable of an empirical science; only its 'utterances' by mediation of behavioural acts and/or productions can become part of its scientific data. Physically, ontogenetic memory must be some kind of structure formation (or modification of a given structure), although, in fact, we know almost nothing at present of how memory contents are stored, in physiological terms, in the brain. Ontogenetic memory in the main is commonly held to be a rather direct trace of the encounters of the individual with facets of the world and of the consequences of these encounters itself; the inclusiveness of ontogenetic memory is not well known so far. On the other hand, the genom incorporates the history of those mutational and recombination changes that have survived selection.

Note that 'memory', in the present use of the term, is more than the trace of one's past; it also integrates that past to a dynamic whole and thus also in some way implies the possibilities of the future. This may be on the simple assumption that some characters of the past will also be present in the future, or on the assumption that the world is following rules or laws. Inasmuch as natural laws are available in the memory, extrapolation from present states into the future is possible, and the same is true in a probabilistic manner for rules.

5 EXTERNAL MEMORY: OR THE 'CONCRETIZATION' OF THE MIND

It has been important to recall these memory analogues on the biological and the individual level. However, in the present context of pursuing the ecological question in view of the design and construction of things and places, the third kind of structure formation or external memory is to be focused upon.

Unquestionably, internal structures are in the main built up or differentiated by virtue of information coming through perception. The role of perception in that process is not generally clear. A case in point is the problem of unit formation. Usually we assume the world to be constituted of separate objects which are separately represented in our experience. However, exactly this assumption is already a result of perception. It is easy to demonstrate with simple experiments that perceptual and cognitive unit formation is for the most part a combined function of external givens and internal principles. Consider, for example, what constitutes a unit in language: perceived segmentation in the stream of spoken utterances is mainly determined by (inner) standards, largely inborn for the phonemes, and based on experience in a particular language for the morphemes. In written language, more seems to be given over to the external side, in that the letters and the words appear as separate; yet the reader needs to know the letters and must group the configurations of ink into letters and words. Figure–ground organization is basic for all perception; although as a rule the entities perceived or figures can be said to correspond somehow to the order found in the world, this is not necessarily so. Figures or perceptual–cognitive units of the psychological environment are rather the proper products of the organization of the mind, and yet they refer always to entities of the ecological person–environment system. The illusion of units being out there is obtrusive because we have no information except by way of perception. Even the most private emotion is a feeling of something; and, on the other hand, every 'objective' fact, say a physical measurement, is something conceived by a mind.

I therefore contend that it is perfectly arbitrary for psychology to conceive of ontogenetic structure formation as something exclusively internal based on inputs from outside (Lang 1985). So much of the production of the ecological environment, that is, the real surroundings of the person, is in fact structure formation based in part on internal states of the system in question. Most people of today's educated and industrialized societies are painfully incapable of living without a special kind of external structure formation, namely, the traces of their own and their ancestors' and contemporaries' acts left in written notes and pictures or in designed objects or built settings. The notion of external memory is on the one hand a triviality, indeed; on the other hand, it has not been psychologically investigated in its origin and impact for individual and social life.

It seems reasonable to point to the parallel between biophysical metabolism and biopsychological information exchange; no living system exists without permanent although varying inflow and outflow of both matter and energy, as well as information. Information flow is crucial for metabolism in higher animals. To place a frontier between the inside and the outside is perfectly arbitrary for both, particularly in view of the possibility of storage of matter/ energy and information both inside and outside of organisms.

The one system of thought that, in my opinion, comes closest to a systematic treatment of (information) structurizations in general, and thus has a potential for promoting a common language to deal with both the traditional mind and external concretizations, is general semiotics, especially in the tradition of C. S. Peirce. In order not to complicate matters too much I have withstood the temptation of writing this paper in semioticalese (for a general overview see Nöth 1985). An ecological psychology in semiotic form is, however, a challenge to be taken up.

The present chapter advances the 'concrete mind' in the form of a general heuristic, that is, it is neither a statement of fact nor a hypothesis which can eventually be confirmed or refuted, but rather it is intended as a sort of probe: as is the case with science in general, it can either succeed or fail in opening new insights or in procreating new investigations and fertile ways of dealing with the world (see Feyerabend 1989). The heuristic, of course, is in patent contradiction to many assumptions widely accepted in psychology and beyond, although it is perfectly compatible with much of its empirical evidence.

To summarize, living systems like cells, organisms, and groups of organisms such as societies each carry their own kind of information structure or memory. For the cell it is the genom; for organisms it is individual or ontogenetic memory proper, called the mind or brain. For persons and groups, small groups and larger groups including society, it is their ecological environment or culture, that is, the totality of material and symbolic things and structures which are interpreted, modified, formed, designed, or constructed. The three levels of memory are mutually interdependent, in that the mind is an elaboration of the genom and culture builds on both. In return, the mind of an individual is a factor for selective survival of a given organism and thus of its genom; and culture operates back on minds, to a lesser degree and indirectly (if we omit genetic engineering) also on genoms. All three are generative storage systems, that is, they do not wait to be asked for the information stored in them but function basically in their own manner, elicited by suitable events, or 'spontaneously', provided that favourable conditions prevail.

In the following sections an indication of elaborating the heuristic of the concrete mind is given. Of course this can be done no more than in an illustrative manner. It should be noted that culture is a generative memory not only for groups but also for any individual who produces, perceives, and otherwise uses these external structures. I shall touch upon both individual and social concretizations and functionalities; they are not meaningfully separated. In the functional approach of Section 6 the leading idea is the concept of identity, personal and social, which is attained by virtue of concretizations. The structural approach of Section 7 touches upon some selected concretizations in the double sense that they are examples or fields of the concrete mind and they should help to make the heuristic more tangible.

6 SOCIAL MEMORY OR CULTURE: CARRIER OF PERSONAL AND SOCIAL IDENTITY AND DEVELOPMENT

Traditionally, objects and spaces are conceived of as opposites of a subject. Categories such as material substratum, extension, quality, utility, value, possession, and the like are used to deal with them, and their qualities are mostly conceived of as belonging to the objects themselves. Ecologically it should have become clear by now that in the present heuristic many of them are seen as extensions rather than opposites of persons. If we compare the traditional internal mind with its external, concrete complement, a number of similarities and differences can be pointed out, of which I mention a few.

Internal and external memory are both conceived as being structured, organized complexes, that is, they are supposed to be composed of inter-related parts. Commonly some kind of Gestalt character (the whole is more or other than the sum of its parts) is attributed to the internal structures. About external structures nothing in this respect is very clear so far, although some degree of order is conspicuous beyond that provided by natural laws. The internal mind has a spatially concentrated existence, that is, the degree of connectedness between its parts seems to be very high. External objects, on the other hand, can be multiplied and spread over space, they can exist inside or outside of other objects and spaces; once created, few parts of external structures seem to be strongly interrelated. In temporal respects the mind is, in the course of its limited life-time, capable of being in an indefinite number of states, although apparently only one at a time; objects and spaces realize an enormous variety of 'life-times' or durations, some being transitory, some being permanent for all practical purposes (see also Boesch 1983, 1991). In any case, discrepancies or dialectics in temporal existence between persons and their external concretizations are most intriguing and deserve special attention for their potential to originate development (see below, Section 6.3).

Whereas the internal mind is private or directly open only to the person in question, most concretizations are public, that is, they are, as a rule, available both to that person and to others. The consequence is that one can expect any public concretization to have potential effects on any person present, whereas it is quite an exhausting task to transfer some content of one internal mind to another. The process is called socialization and happens mostly in face to face situations. In more formal education it has been widened to a one–many spread; but it is only after scripts and pictures have been attained that further amplification in scope has become feasible. Scripts and pictures and the modern media are all external concretizations. The point of the present heuristic is that in the history of humankind there were scriptures, if one can say so, long before written language was invented, namely, the manufacture of tools, of ritual objects, their placement, the structuring of space by walls and buildings etc., all in the service of individual and social processes of structure formation: 'Wenn wir Häuser bauen, sprechen und schreiben wir' (Wittgenstein).

I would like here to concentrate on the double function of concretizations in the service of individuals and groups and particularly point out their meaning for the self and for the social compound. In so far as concretizations enable a person to discourse with him/herself, they are a prodigious vehicle of self-cultivation (see Csikszentmihalyi and Rochberg-Halton 1981, Boesch 1991). On the other hand, these extensions of the individuals are at the same time a prolific social go-between; they carry a large portion of communicative intercourse and thus serve as a powerful glue for the social compound.

6.1 The concrete as a carrier of the social net (communication)

It is perhaps the most important impact of scientific thinking on social life that we believe every transfer of influence between entities including people to be based on an explicit, principally uncoverable act of communication, understood in a wide sense. In contrast to magic beliefs or to assumptions about hidden potencies etc., we have acquired a habit of wanting to find out about the causes of everything. So we ask what it is that makes group life possible.

Prime candidates are the following. Assuming that individuals are endowed with social instincts like other social animals is obviously not wrong; however, it is an insufficient explanation in view of the important role of ontogenetic memory. So socialization processes during a lifetime must be assumed. Among them social learning from models is considered important. Informal and formal education emphasizes mostly language-based information transfer. Yet all this seems insufficient given the overall high and efficient functionality of social life. It is trivial to add that a large part of social communication is carried by material forms, that is, objects and spatial arrangements, which bear meaning for everybody, sometimes immediately, sometimes only for the (internally) informed. However, most of these

257

processes do not occur in a conscious, reflected manner and therefore, probably, are not well investigated in that crucial role (see Boesch 1980, 1991). Symbolic carriers such as written material and pictures are a special case, much nearer to consciousness. One of the few research programs seeking to understanding the role of the physical preconditions for group life is Roger Barker's 'Ecological Psychology' with its concept of behaviour setting (see Schoggen 1989).

6.2 The concrete and the self or identity (cultivation)

The notion of self or identity of a person is generally believed to have two facets, one originating from within, the subjective 'I', and one originating from outside, the social 'Me'. Traditionally, both facets are understood as mental processes. The essence of the notion lies in its promise to bridge the aporetic gap of something being permanently in change yet remaining the same. Subjective identity is a miracle; words appear to disguise rather than to illuminate it. Social identity on the other hand must depend on communication in a group. In higher animal species where individuals are recognized among each other, morphological and behavioural peculiarities are the basis of social identity, that is, of the ability not only to recognize this or that particular individual but also to have reasonable anticipations of his or her preferences and reactions in certain situations and to have realistic expectations of successes and failures when dealing with him or her. In humans, additional characters or informative features must be assumed. A prime candidate beyond refinements of interactive vocal, verbal, and paraverbal communication (for further references see Nöth 1985) is again the complete and continuously changing set of physical accessories which can be put in connection with a particular individual.

Here again, research literature beyond collections of anthropological material is rather scarce (see, for example, Duncan 1982). Boesch (1980, 1991) is an original cultural psychologist working mainly with a phenomenological method on the basis of everyday experience and anthropological data. In the centre of his conception is the idea of an action-mediated relation between the individual and the cultural world. The concept of action includes inner experience; core processes are described as a mutual subjectivization and objectivization of persons and the world. Csikszentmihalyi and Rochberg-Halton (1981) have presented a multidisciplinary approach centred around the psychological notion of the self and they have given some of the rare empirical material on the concrete basis of identity for Western culture. Their core concept is that of 'cultivation' which refers to 'the modes of meaning that mediate people with objects' (p.173) and is thus an analogue to socialization mediating between persons. In other cultural and social sciences and also in semiotics, the aesthetic character of things or their functional and economic meanings prevail almost exclusively.

6.3 The concrete as a source of stability and change (development)

Perhaps the most intriguing and consequential feature of concretizations, or better yet of the relationship between external and internal structurizations, is their different temporal character. I can only repeat here my earlier arguments that the different time qualities of living beings and their environment constitutes an indispensable factor of psychological development (see Lang 1981, 1988).

Any system ruled by a single all-embracing principle such as closed system will necessarily tend to a stable state in the degree that its supreme governor is in power. In other words, there will be no development but rather circularity in such a system, because whatever happens in the sense of random events within the system or in the sense of a disturbance coming from outside will sooner or later come under the rule of the supreme governor and thus will be shut out in its effects. It is the prime merit of Darwin's theory of bioevolution to have pointed out the necessity of two independent sources or principles acting on the same entity in order for that entity to develop. A psychological system enclosed in the mind or brain could do best by staying as it is and refraining from letting in information that leads beyond the simplest necessities of safeguarding material metabolism and the minimum of social bonds. As a consequence, behaviours such as exploration, fantasy, invention and creation would be patently superfluous. Those are, however, true facts of our existence.

The concrete mind heuristic proposes exactly that creative relation of connectedness and independence of internal and external structures. As a rule the outer concretizations have a different time horizon than inner-psychic events or brain states. Some of the spaces and objects produced, such as houses, cities, traffic systems, survive the lifetimes of one or even many generations of individuals, if not in the particular single objects then at least in their fundamental structures. Thus, many a medieval layout of a city survives modern architecture, and the Shinto Shrines in Ise (Japan) are said to have been built exactly in the same formation over the last fifteen centuries, perhaps precisely because they are rebuilt with new material every twenty years. Other objects and spatial arrangements are made anew every day or week or year, for example, flower arrangements, seating arrangements at meetings, ladies' clothing. They thus demonstrate the persisting, durable character of the people involved by the same dialectic mechanism that in the former examples establishes the transitory and fleeting nature of all human affairs. The dialectics between short-term and long-term existences and changes in the two subsystems of ecological units, the internal and the external, are suggested to be the motor of human development, both in individuals and in societies.

7 SOME EXEMPLARY 'CONCRETIZATIONS'

The aim of this section is to illustrate very briefly major types of external structurizations. I purposely start and finish with topics such as knowledge and nature, that seem at first sight outside the realm of cultural concretizations at issue in this contribution. However, the point being made is that external memory is not something entirely new and, above all, it should not be treated in separation from traditional psychological conceptualizations. The emphasis, however, is on space and objects which are, in addition to conventional symbolic systems, the principal and most neglected fields of external memory. It is not possible to give full accounts.

7.1 Information, knowledge

In psychology, knowledge is a term originally used for that content of ontogenetic memory that is consciously or linguistically available to the individual. Recently the notion has broadened in scope in that it includes the totality of the cognitive and action potential of the individual, conscious or not, actual or latent. Like its broader equivalent, information, it is a seductive term. While the former puts its user in need of overcoming the connotations of static, fixed, passive, put aside for later use (the same is true for my evocations of terms such as storage or memory), the latter carries the burden of some very particular, technical theory, namely, that of communication between finite state systems. In everyday language both terms also refer to concretizations by symbol systems such as in books, pictures, or computers, for example, an encyclopedia or the knowledge base of an expert system.

Things might become a bit clearer when we emphasize this double nature of information or knowledge. It is an immutable fact that there exists no knowledge or information except in 'combinations' of meaning with a physical carrier, and that, in order to be used adequately, it must be specified in reference to (potential or real) senders and/or receivers of the meaning. The carrier or code must be physical in any case, either physiological as in the brain or physical as in linguistic, pictorial or other symbol systems which need acoustic waves or ink on paper or electronic bits in computers or further concretizations. The meaning, on the other hand, is necessarily relative to the beholder.

From the point of view of an observer dealing with symbol systems of any kind it is advisable, as we have seen, to distinguish between entities that belong to the surrounding world and entities that have become part of the environment of somebody. In parallel to the terminology proposed in the following three sections we might conceive of the world in general as constituted of formations of energy and matter, some of them naturally given, some the result of human activities. When these formations, as a result of an encounter, influence a living system, the latter becomes an interpretant and the former in turn appear to him or her as information or knowledge.

For reasons of space I cannot elaborate here what I call the semiotic–ecological approach. Its gist is the following: information or knowledge is not out there in a brain or in a book or in a room or street network, but it is always the result or sign of an encounter between two entities, a resultant of the meeting of a sender or reference source and a receiver or interpretant, whereby the receiver never (never!) simply takes the information given by the sender, but always creates a third, some combination of its own prior form and the influence received (refer back to Section 6.3 for the source of development). Thus, information and knowledge are both relational concepts of a triadic nature.

7.2 Space and place

The notion of place is most readily explained by borrowing from geography. Any encounter between an individual and other entities can be described in spatial terms. Space is a character of the perceptual–cognitive–actional organization of a perceiver and, therefore, systematically generalized, a useful descriptor of all matter–energy formations. It is impossible to meet non-spatially. However, space is not 'space'. What has proven useful for terrestrial physical, geometric or other endeavours, namely to conceive of the spatiality of the surroundings as a homogeneous, isotropic character, is not true of the spatiality of individual beings or different groups of individuals. Their space is anisotropic (it has thin and dense areas), structured (it has boundaries and containers), ordered (it has an inside and outside), and valued (it is a system of more or less important locations) and should be distinguished by naming it a 'place' (in somebody's environment). It might be added that the space of individuals and groups, that is, their place, is also meaningful: it has locations for this or for that; but perhaps it is better to deal with this aspect together with things because most meaningful spaces are constituted by sets of things. The concept of place (German: *Ort*) incorporates exactly all of these special characters emerging from the encounter of somebody with a section of what is generally conceived as space.

Place, as defined by the geographers, is 'a system of meaning. . . . Meaning is grounded in social relations, but is constructed within place (not projected into). . . . This is why place is discourse. Subjectivity is best viewed as part of the place–construction process' (Bourdoulay, in Agnew and Duncan 1989: 136; see also Hillier and Hanson 1984).

So, if space is a term of physical science, place is a psychological or sociological term. Space refers to the surroundings, place to the environment of somebody. The meaning of space claims to be above and beyond the peculiar perceptual–cognitive structures of any given individual, although it is supposed that everybody has a potential of acquiring that 'objective' meaning (e.g. three-dimensional, isotropic, meaningless) as a part of his or her personal knowledge. Thus, space is the place of a virtual group composed of every potential physical geometer. Place, on the other hand, emerges always as a particular meaning in every individual or group from and with its peculiar encounters with specific surroundings.

By way of example, an urban square is a place with a meaning that is moderately or widely shared by inhabitants and tourists. It incorporates a history of the city together with its inhabitants and guests, in that it concretizes peculiar social relations of former times. Various ruling or aspiring groups of former and present times display their signposts in the form of façades, monuments, furniture, etc.; others see and use it as a marketplace for exchange of goods and ideas, while they are, voluntarily or not, affected to a lesser or higher degree by the concretizations of the former; still others claim it as their parking lot and thus display (in the ethological sense) and exercise their power over pedestrians. The social nature of these public concretizations is obvious (see Lang 1987).

Another field where the mix-up of space and place of our civilization has created immense amounts of triumph and tragedy is the private and semi-private places of residence. I have argued elsewhere that dwellings should not be considered as objects but rather as functional 'organisms' for groups (see Lang *et al.* 1987). A general relation exists between certain qualities of the structures and the furthering or inhibition of certain social processes and structures. Places describe social relations and in addition have the power to govern the people exposing themselves. Dwellings in particular can be understood as regulators of autonomy and integration for individuals and groups. A psychology of the dwelling activity on the basis of the present heuristic is in progress. Similar considerations can be made in view of all spaces turned into place by architectural means.

7.3 Objects and things

In analogy to the space–place differentiation I propose to use the term 'object' for referring to entities characterized by their descriptive material, energetic, geometric, and surface qualities, which are accessible to a perceptual–cognitive system in general, be it of an everyday–linguistic or a scientific–systematic nature. The term 'thing' (German: *Ding*) on the other hand is used to refer to an object-person relation emphasizing the meaning of the object in question for the particular person or group. Particularly expressive and appellative characters are comprised. Things tell something to persons who are able to understand; they even do something with them, they afford their meaning. Their meaning, however, cannot be given except for specified persons or groups; they might be quite different for different people, although a certain degree of similarity or equivalence can be expected to prevail within a given culture.

Things fill the places of all people every day; yet psychologists have barely touched the topic (see Boesch 1983, 1991; Csikszentmihalyi and Rochberg-

Halton 1981; Graumann 1974). It might be a cardinal feature of this civilization that the Cartesian reduction of things to *res extensa* has lead to the present materialization of our life. The ultimate object or good, seemingly capable of replacing all other objects, is money; yet 'money is no object' (Rochberg-Halton 1986). We are possessed by things rather than that we possess them. We have such a limited understanding of things that most people do not even realize to what extent they serve as instruments of power. Things are usually represented in consciousness as separate objects with their appearance and function, not in their self-related and social meaning. It is understandable, therefore, that a psychology starting from conscious experience did not care for things. Things conceived as concretizations have an equivalent and parallel role in the inclusive psychological organization, both internal and external, as have conscious imagery or talk. So there is no need to re-represent things in consciousness; they are of a psychological nature already by themselves, and they operate often rather more dependably than thoughts or spoken words. Both of the latter are in need of conscious actualization in order to become functional, whereas actions with things are perfectly possible without even the presence of their originator. Things are the means of choice to turn space into place.

A further field where places and things are combined into intricate complexes of cooperating internal and external structures are the settings that ordinarily go under the name of institution. I already mentioned a particular general kind, the behaviour setting (Schoggen 1989). Fuhrer (1990) gives an empirical treatment of processes in institutional settings into which people are simultaneously socialized and cultivated by and for persons, places and things, including symbolic information.

7.4 Nature and settings in general

The term 'setting' is used here in the more inclusive sense of place together with temporal proprieties. It is an abstraction when we speak of knowledge, place or things without also considering their constancy or change over time which is one of their inevitable constituents. Things and places change over time together with and against the persons and groups whose environment they are (see Section 6.3 above). In fact, it seems that not a small part of traditional human cultural activity aims at freezing or conserving a given state of the mind proper in the form of external concretizations; examples range from monuments to souvenirs to photography. Our civilization in addition has produced quite a number of gadgets, such as clocks, machines, vehicles, which are exactly designed to evade time constraints by some form of attempted but often failed mastery over time.

The field where the temporal characters of our surroundings become most potent might be nature, that is, that general construction of human thinking which refers to everything given before humans touch upon it. Nature includes, of course, space and objects. I say 'construction' (of the objective-minded), because a little thought makes clear that it rarely exists anywhere as a reality, except perhaps in a few last mountain areas, in weather and climate, and on a cosmic scale in outer space. These examples already make clear that as soon as humans relate to nature, nature changes its nature, in that it attires meaning for humans: the mountains afford looking or climbing, the forests resting or walking, etc., and nature becomes a landscape; the weather is a threat or a delight, and sky and outer space turn into a huge projection screen for myths of all kinds. As the twentieth century nears its end, it has become clear that humans have left their civilized concretizations (debris) even in the sea, in deserts, in their small cosmic corner and also touched upon weather and climate. Thus, for our civilization, nature has become a setting full of concretizations of a particular kind; evidently nature is again no simple 'object'.

8 PERTINENCE OF THE 'CONCRETE MIND'

However, it seems that in our civilization, things and places and settings and even information are still considered prototypes of 'objects', separate from and opponents of subjects. But that is not a matter of fact, rather a matter of view, namely of the view of (self-declared) subjects. Such opposition is incorporated in our way of thinking, speaking, manipulating, making (ours is the making civilization!) to such an extent that we try to do whatever we feel like doing not only with practically every part of nature and culture but even with our own con-species, especially when they are symbolically represented in data base and similar substitute forms. On the other hand it might be likely that somebody is apt to acquire a new balance of autonomy and integration with his or her environment, somebody who has internalized, be it consciously or even better in the form of a general habit, that our cultural heritage including its natural base is to a cultural group (including him/herself) in an external form what his or her own mind is to his or her person as a unique being.

I found it revealing to confront the heuristic of the concrete mind with our traditional understanding of the 'environment' (namely that of the objective-minded). For example, I might resume the famous thesis of Aristotle (*De Anima* 3.8), claiming the hand as the archetype of bodily procreations, the psyche as the archetype of mental procreations. Aristotle's thesis incorporates the fundamental germ of Western civilization. I suggest that this is dubious and fatal, because it has led to an opposition between the public work of the hand and the private confines of the mind instead of articulating the complementarity of brain and hand, of the subjective and the objective. In addition, the confusion of the environment, one's own or that of science, with the world or surroundings common to everybody, not only has led to a grandiose failure of communication but also is the great myth of modernity which has unleashed and still is a source of unbelievable amounts of power.

In the present era the works of technology embrace every individual, with tools and vehicles and media and money and everything that money can buy, and put him or her at a particular place in a fully functioning society. What happens under such conditions, when, at the same time, the great collective myths, religious or secular, cease their daily enforcement of human communion? Naturally, the mind reacts, or counteracts, and designs magnificent mental constructions such as the person, equality, dignity, and it invents techniques such as the arts (*l'art pour l'art*), self-realization, psychotherapy and the like. But it is weak, this mind, when compared to the powerful works of materialistic technology and all of the other external concretizations of our time. I propose to develop a culture of complementarity of, rather than of opposition between, the designs of the hand and the constructions of the mind.

9 REFERENCES

Agnew, J.A. and Duncan, J.S. (eds) (1989) *The Power of Place: Bringing Together Geographical and Sociological Imaginations*, Boston, MA: Unwin Hyman.

Boesch, E.E. (1980) Kultur und Handlung: *Einführung in die Kulturpsychologie*, Bern: Huber.

Boesch, E.E. (1983) *Das Magische und das Schöne: Zur Symbolik von Objekten und Handlungen*, Stuttgart/Bad-Cannstatt: Frommann–Holzboog.

Boesch, E.E. (1991) *Symbolic Action Theory and Cultural Psychology*, Berlin: Springer.

Csikszentmihalyi, M. and Rochberg-Halton, E. (1981) *The Meaning of Things: Domestic Symbols and the Self*, Cambridge: Cambridge University Press (German ed: *Der Sinn der Dinge*, trans. W. Häberle [ed.] with a foreword by A. Lang, München: Psychologie Verlags Union, 1989).

Duncan, J.S. (ed.) (1982) *Housing and Identity: Cross-Cultural Perspectives*, New York: Holmes & Meier.

Feyerabend, P. (1989) *Irrwege der Vernunft*, Frankfurt a/M: Suhrkamp.

Fuhrer, U. (1990) Handeln-Lernen im Alltag, Bern: Huber.

Graumann, C.F. (1974) 'Psychology and the world of things', Journal of Phenomenological Psychology, 4: 389–401.

Hillier, B. and Hanson, J. (1984) *The Social Logic of Space*, Cambridge: Cambridge University Press.

Lang, A. (1981) 'Vom Nachteil und Nutzen der Gestaltpsychologie für eine Theorie der psychischen Entwicklung', in K. Foppa and R. Groner (eds) *Kognitive Strukturen und ihre Entwicklung*, Bern: Huber.

Lang, A. (1985) 'Remarks and questions concerning ecological boundaries in mentality and language', in H. Seiler and G. Brettschneider (eds) *Language Invariants and Mental Operations*, Tübingen: Narr.

Lang, A. (1987) 'Wahrnehmung und Wandlungen des Zwischenraums: Psychologisches zum urbanen Platz', unpublished paper, presented at the Schweizerischer Werkbund, Bern, 25 February, Bern: Department of Psychology, University of Bern.

Lang, A. (1988) 'Die kopernikanische Wende steht in der Psychologie noch aus! – Hinweise auf eine ökologische Entwicklungspsychologie', *Schweizerische Zeitschrift für Psychologie*, 47 (2/3): 93–108.

Lang, A., Bühlmann, K. and Oberli, E. (1987) 'Gemeinschaft und Vereinsamung im strukturierten Raum: psychologische Architekturkritik am Beispiel Altersheim', *Schweizerische Zeitschrift für Psychologie*, 46 (3/4): 277–89.

Nöth, W. (ed.) (1985) *Handbuch der Semiotik*, Stuttgart: Metzler.

Rochberg-Halton, E. (1986) *Meaning and Modernity: Social Theory in the Pragmatic Attitude*, Chicago, Ill.: University of Chicago Press.

Schoggen, P. (1989) *Behavior Settings: A Revision and Extension of Roger G. Barker's Ecological Psychology*, Stanford, CA: Stanford University Press.

Part IV

THE REGIONAL DIMENSION

17

INTRODUCTION
TO PART IV

1 REGIONAL GEOGRAPHY: OLD AND NEW

Any concrete investigation of human ecological situations is connected to regional contexts. Similar to human ecology at large, there is a traditional background to the study of regions on the one hand and a *response-to-crisis* aspect on the other. To consider the tradition first, the concern with regions and also boundaries between regions has, of course, been the explicit homeground of geographers for a long time. Until the middle of this century, the philosophy of geography had been one of treating regions as complexes of interrelated phenomena having, in a way, the status of individuals. Then this orientation came under attack by the 'quantitative revolutionaries' who declared the notions of individuality and (total) interrelatedness to be unscientific. Consequently, they established what has become known as 'spatial analysis', which amounted to a search for laws (or at least regularities) analytically describable by the fundamentals of location and movement in space (and time).[1] The new geography had one thing in common with the old one, however: it restricted its view to phenomena of the biophysical (natural as well as human-made) environment. Consequently, the models of spatial analysis were largely devoid of any content in the psychological, social and also (human) ecological sense. Spatial analysis meanwhile has lost its status as the core of geography, but has managed to survive in joining forces with branches of economy, political science etc., to form what is known as 'regional science', an orientation which indulges in a kind of mathematical model Platonism.[2] The development of a new kind of regional geography can be seen as the result of at least three factors.

First, geographers, noting the ever-increasing rapidity of change of cultural landscapes, have found that they no longer can be content with simply registering the succession of land uses. They have concluded that they should be looking at human action and the societal forces responsible for this change (cf. Werlen, this volume). Consequently, a number of variants of 'psychological' and/or 'social geography' have developed since the seventies (see, for example, Johnston 1986). This carries over to the new regional geography

269

which tries 'to show how the specificity of place is preserved and modified within the generality of social change' (Gilbert 1988: 218, quoted in Taylor 1991: 186).

Second, the long-standing dogma that science should be looking for generalities and regularities or, stated conversely, that a concern with individual cases and events could not be a scientific concern, has waned considerably. The study of individual characteristics of regions has become not only admissible again but also compulsory. How else could one say anything of value about the regions in question? Support comes from a recently developed philosophy of science known as 'transcendental realism' (Bhaskar 1975, Harré 1986, Sayer 1984). One of its main theses is that the notion of causality should be separated from that of regularity. In this light a unique event can be an object of scientific investigation just as well as can repeatedly occurring events.

Third, similar to the ecological movement, to which scientific quarters reacted by developing a modern kind of human ecology, various kinds of regionalist movements have encouraged not only the comeback of a new regional geography but also the development of regional orientations in a number of other disciplines. It remains to be seen whether the decentring of geography away from the old exclusive fixation on the spatial aspects of regional phenomena together with the detection of regional aspects in other disciplines may lead to the development of an interdisciplinary regionally founded human ecology rather than to a competition between disciplines for a new territory.

As mentioned previously we experience today an upsurge of regionalist movements. In many instances the main motive may, as in the present Eastern European developments, be political, to be understood as an attempt of ethnic minorities at attaining self-determination in the face of a dominating larger nation or, conversely, of ethnic majorities at avoiding the possibility of becoming culturally corrupted by an inflow of foreigners (Lübbe 1990). However, in other cases, notably in Western Europe, a dominant driving force behind regionalism seems to be the prevention of a further erosion or, conversely, the re-establishment of an overseeable sense-giving lifeworld (Weichhart 1990). Bahrenberg (1987) sees the demand for regional autonomy as a programme for a 'post-modern' transformation of society. In either case, the issue is one of regional identity. The revival of regional cultures, therefore, can be seen as a process that is compensatory to the global process of modernization which has more and more levelled out regional differences. Thereby, as pointed out by Weichhart (1990), the compulsion present in an individualized modern society to redefine constantly one's personal identity stands in curious contrast and conflict to the fact that, owing to large-scale homogenization, differentiated phenomena with which one could identify have become scarce.

As will be argued by Ola Söderström in his contribution, the quality of social life conditions cannot be seen independently from the mental states of individuals living in that society, nor separately from the way in which they treat the environment. The 'three ecologies' (Guattari 1989), the social, the mental, and the environmental, go together. As a matter of fact, the concept of largely self-determined and self-sufficient regions, which in turn enable the development of social networks and individual identities, is becoming the focus of attention in debates about possible cures for the ecological crisis. Various aspects of the importance of the regional dimension will be taken up from different viewpoints in detail in the contributions contained in this part of the book. As the renewed interest in regions is very much at the heart of a new geography, it may not come as a surprise that all papers in Part IV are written by geographers. In the following we wish to bind together some of the arguments by highlighting three aspects in a more coherent fashion: the need for social theory, the question of identity, and the notion of regional decentralization as a building block, hopefully, for an ecologically sustainable society. The quest for personal identity and regional culture can be interpreted as manifestations of a certain degree of a backward orientation: this 'evolutionary reattachment', as we call it, will be the theme of a fourth section.

2 CHANGE AND CRISIS: THE NEED FOR SOCIAL THEORY

As pointed out earlier, in recognizing that the transformation of landscapes cannot be understood by looking at the biophysical environment alone, geographers have ventured away from their homeground of regions and boundaries towards the disciplines of psychology and sociology, which traditionally have had little interest in spatial and regional aspects. The development of various kinds of environmental psychology is largely of a more recent date,[3] and then the concern is mainly with the local surroundings of individuals. Sociology on the other hand has been rather reluctant to become involved in spatial or regional issues. At any rate there is the recognition that environmental change is an outcome of the dynamics of societal, in particular economic, development and that, therefore, one needs theoretical instruments with which to identify the mechanisms of change, to diagnose the present crisis, and hopefully to find ways to influence the further development. In short there is a need for social theory (or, perhaps better, socio-cultural theory).

In Part III of this volume we have pointed out that the theory of structuration of society by Giddens (1984) has been a central building block for our project of developing a concept of integrated human ecology. The reasons have been given there. Suffice it to repeat here that Giddens' theory is of interest because it can be interpreted as a substantial theory within the newly emerging evolutionary worldview (or the transactional worldview according to Weichhart, this volume). As such it affords an explanation of human societies as recursive systems. Such systems are an expression of the self-organizing capabilities underlying human social development. Consequently, the theory of structuration is at the same time a theory of social stability as well as of social change. Interestingly enough, however, Giddens himself does not seem to be particularly interested in the questions of change, at least not as it relates to the present ecological crisis. He does refer, however, to the possibility of contradictions within given social structures, a theme taken up in the contribution by Benno Werlen in his discussion of cultural identity.

It should be of interest, therefore, to try to complement the structuration theory by components from other social theories. Two of those are mentioned in the contributions of Part IV. Reference is made to the autopoietic system theory of society by Luhmann in the paper by Gerhard Bahrenberg and Marek Dutkowski as well as in the contribution by Dieter Steiner, Gregor Dürrenberger, and Huib Ernste, and to the theory of communicative action by Habermas in the article written by Ola Söderström. Luhmann (1982, 1984, 1989a, 1989b) understands his theory as a continuation of the system theoretic thinking of Parsons[4] on the one hand and as an application of the biological theory of autopoiesis by Maturana and Varela[5] to the social realm on the other. Luhmann's main point is that social systems are pure communication systems. They are autopoietic in the sense that communication events produce themselves, so to speak: a communication event happening now enables the occurrence of a consecutive communication event. Thus, Luhmann's stance is very extreme indeed in that he claims that social systems 'do not consist of psychic systems, let alone of embodied human beings' (Luhmann 1984: 346)[6]. Instead the elements of the system are communication events. It appears to be a theory drenched with a massive dose of cynicism and, to be frank, it is not a theory which one wishes to have at the core of a concept of human ecology. Nevertheless, Luhmann has one thing in his favour and that is, unfortunately, a definite degree of realism that applies to his notion of society as a series of functional subsystems, each of which operates according to a specific communication code such that there can be no meaningful discourse between systems. This would seem to be an excellent picture indeed of the characteristics of a modern Western society where there is a disorderly juxtaposition of subsystems. In other words, there is no hierarchical order and, therefore, there are no priorities set within a cultural framework. Indicating that there is no single subsystem in society that would represent society as a whole, Luhmann says: 'The society as a whole does not occur within society once again' (Luhmann 1989b: 23).[7] Or else one can point to the fact that, if there is a hierarchical order, it is, seen in an evolutionary perspective (cf. Steiner, this volume), an inverted one: a subsystem such as the economic one, which in evolutionary terms should be subsidiary, can take on a leading and dominant role.[8] To this extent, therefore Luhmann's theory has an explanatory value with respect to the situation of ecological crisis in which today's society finds itself.

Obviously the self-dynamic ascribed by Luhmann to the various sub-systems seems to leave humans behind as mere and perhaps reluctant observers. This is the point at which we can find a connection to the way in which society is portrayed by Habermas (1989)[9]. Communication in the Luhmannian sense has a purely technical meaning: information coded in a way appropriate to the subsystem in question is transferred over largely anonymous channels. It is what Habermas simply calls the 'system': the network of relationships within a society provided largely by the media of power and money and functioning according to an instrumental kind of rationality. In contrast, communication for Habermas is real human discourse in the sense that information is exchanged by interested and dedicated human actors, preferably in a situation of face-to-face contact. It is the ideal of communicative rationality which Habermas sees as constitutive for the human lifeworld: the day-to-day activities which bring humans in contact with each other such that relations of trust and confidence can be established. As pointed out by Söderström, the distinction between system and lifeworld by Habermas provides for a connection to Giddens' concept of system integration versus social integration. System integration makes the coordination of extended societies possible to start with. In particular, a global society is unimaginable without some means of system integration. On the other hand, Habermas diagnoses the fact that the system erodes or, as he calls it, colonizes the lifeworld as a pathology. Again we can talk here about a hierarchical inversion: instrumental rationality supersedes communicative rationality rather than the other way round. A call for a strengthening of the latter is timely indeed, and we will see below that there is hope for such a strengthening to be achieved through a process of regional decentralization. As Bahrenberg and Dutkowski remark in their contribution, the present-day functional differentiation of society (as described by Luhmann) should give way to a regional differentiation.

We can thus see the value of the ideas put forward by both Luhmann and Habermas for the project of an integrated human ecology. Open questions remain, however. With respect to Luhmann we have pointed out already that his theory is too heavily purely descriptive; any normative content is lacking. Conversely, a critical component is certainly not lacking in Habermas' thinking. A deficiency, on the other hand, can be pinpointed in the fact that the latter is willing to ascribe the possibility of communicative rationality to interactions between human beings only, whereas any dealings of humans with the biophysical environment acquire the character of instrumental rationality. This would seem to be rather crucial against the background of the ecological crisis, and it leads to the question of whether one should not think of the possibility, if not necessity, of communication with 'things' in the environment (Dews 1991). Surely this would involve other levels of human consciousness than just the discursive one, but then these other levels are also always involved in human verbal interaction.[10] And conversely, we all know that it is good to talk to animals and get some reaction on their terms from them; some even claim that plants feel well when they are talked to and, who knows, perhaps they talk back in their own way. This apparent deficiency in Habermas' concept of communicative action may have to do with another shortcoming (pointed out by Söderström), namely the complete absence of any notion of spatiality. To the extent that geographers on the one side still display problems in finding access to social theories, and sociologists on the other are still reluctant to take conceptual notice of space and environment, the attempted linking of geography and sociology is a task just begun. A contribution to such a link may be the topic discussed in the next section.

3 A FOUNDATION OF HUMAN EXISTENCE: IDENTITY AND IDENTIFICATION

To acquire an identity is a crucial precondition for any person to find his or her place in a human society and to feel secure. It can be understood as a process in which relations to other persons, which otherwise would remain external, become internalized. It refers to the way in which I am seen within society or, rather, the way in which I think others see me. As a result, I can enter into interactions with other persons and have some degree of confidence about the outcome of such interactions. Identity, therefore, is connected to the establishment of social bonds and, as such, it is clearly an aspect related to the lifeworld of individuals.[11] It is hardly possible to attain an identity within a system context.[12]

As stressed by Werlen in his contribution, to attain a wholesome identity presupposes the existence (if not in reality at least in the perception of the persons concerned) of non-contradictory socio-cultural structures within which relations can be placed and internalized. This, however, is precisely part of today's problem: the structures of our modern society are by no means free of contradictions. They can be seen as emanating from the disintegrated juxtaposition of largely independent subsystems as described by Luhmann. If, as he likes to see it, persons are mere bystanders and not participants in those subsystems, then this situation might not present much of a problem, but rather contribute to the entertainment of those bystanders. If, however, these same persons are engaged in several subsystems in a way that is linked to their survival, metaphorically and literally speaking, then to be exposed to the contradictions emerging from the structures of non-integrated or even competing subsystems may become deeply disturbing. Take as an example the situation still common for our society that hard-working men are also husbands and fathers and are hardly in a position to reconcile the work load put on them through their labour participation in the economic system with the expectations expressed by the members of their family.[13] There is a need in our society to have more than one social identity, one for each subsystem in which one is participating, and, as these subsystems do not stand in any meaningful relation to each other, these partial identities may contradict rather than complement each other. It is our suspicion that such a situation of contradictory multiple identity results in some kind of split personality, and that split persons cannot be expected to behave in a reasonable way with regard to their (natural as well as social) environment. Conversely, 'environmentally responsible action can be seen as one form of positive social behavior', because 'environmentally responsible action . . . serves to improve the common natural and sociocultural environment and thus contributes to the well-being of all human beings' (Kastenholz 1991: 230–1).

A remedy to this situation of split identities might be to re-establish the evolutionarily speaking correct hierarchy of subsystems in a way that, speaking with Luhmann, the informational codes used by a lower system are guided by those of a higher one. To shape ideas, simply consider the coarse differentiation made by Steiner (this volume) in the form of the hierarchical sequence abiotic – biotic – archaic (social in the narrow sense of the word) – political – economic with associated priorities. A further elaboration of this topic is not possible here, but we can recognize that it should be a very fundamental one for an integrated human ecology.[14] Clearly this is asking for a totally new kind of society, one which, for one thing, would require a rectification of the situation described in Steiner (this volume) as one of moral inversion. More realistic for the time being may be the changes proposed in the contribution by Steiner, Dürrenberger and Ernste. Recognizing the disunity of lives in our society as a problem, it is suggested that a process of flexibilization of structures of various subsystems might allow them to be meshed in such a way that individuals can shed their feeling of contradiction and instead acquire one of integrated identity. It is more realistic in that this is not simply a theoretical proposal, but a suggestion in line with certain observable trends. These are trends toward alternative ways of living, educating children and doing business. Particularly conspicuous is the development of 'flexible specialization' as a post-Fordist style of economy which seems to be based on a phenomenon of regional self-organization (see Ernste and Jaeger 1989, Ernste and Meier 1992), hence from here also emerges the connection to the topic of regional decentralization below.

Whether or not human individuals acquire a territorial or environmental identity similar to the way in which they acquire a socio-cultural identity, and whether at the collective level there can be such a thing as a regional identity, is a matter of debate. It would mean that humans establish (internalized) relations to 'things' (landscape features, artefacts, plants, animals) that are around them in their biophysical environment. Werlen for one holds the opinion that there is no such place attachment, at least not *per se*, that any such relations come about only in a social context. Consequently, what may appear as a place attachment is really a social attachment resulting from the fact that things in the environment acquire meanings of a socially representative character from the community occupying the place in question. Accordingly, things in the biophysical environment cannot have a meaning in themselves, only a socially given meaning. This seems to be a clear contradiction to the kind of relationships that exist between animals and their environment at a level of practical consciousness as described by Carello (this volume). Animals identify themselves with certain environmental features that constitute opportunities for them. They fit into spatial–environmental structures, so to speak. And Carello definitely thinks that this applies also to human beings. Opponents of such an idea would presumably point out that this might be true for archaic people still living in natural surroundings to which they are, for direct reasons of survival, intimately connected, but is no longer applicable to members of modern Western civilization who are exposed to a largely artificial, hence socio-culturally created environment.

Nevertheless, others, for example Weichhart (1990), argue in favour of the existence of direct, that is, not socially mediated relationships to the environment. Without denying the fact that many environmental features or objects are important in a social context because they manifest some symbolic value or represent locations where social interactions can take place[15], he insists on the existence of a separate, non-social dimension of human–environment relationship. He quotes psychological literature in support of this (for example, Boesch 1991), research that apparently has been disregarded by social geographers and sociologists alike. If we remember our childhood we will be inclined to think that such a separate dimension exists. True, this was at a time before we had become fully socialized. And yet it is revealing to make enquiries of any adult who, after moving from one place to another, had the feeling of being totally uprooted, and not only because the people in the neighbourhood were no longer the same.

It is true that probably in many situations it is hardly possible to separate the social from the non-social. And there may be a difference in the strength of such identities, depending on whether all one could do was try to fit passively into already existing spatial–environmental structures or whether one could actively contribute to the shaping of such structures (building one's own house, caring for a garden, etc.). All the more important is the question of how to further the possibility of identification (in the sense of Naess 1989) of individuals with their environment, in particular, because of the ecological crisis, with natural components of that environment. It seems to us that this question is very much related to another one (referred to again below), namely whether we can be open to the idea of being able to communicate with nature, more concretely with plants and animals, for example. These questions definitely do not refer to trivial problems. One handicap may simply be that an attempt at identifying oneself more closely with nature is in conflict with certain roles one is expected to play in society. In the end what we need is a congruence between individuals and social structures such that the environmental consciousness of the former is matched by an internalization of environmental features into the latter.

4 'RADICAL DECENTRALIZATION' AND 'PRACTICAL REASON'

At about the time when geography moved away from its traditional concern with regions the regional dimension was discovered in economics, a discipline which hitherto, at least in its neoclassical non-spatial thinking, had been assuming that occurring regional differences were temporary disequilibria which, given enough time, would level out. But then a number of heretics started to point out that the process of economic progress clearly was associated with the emergence of regional disparities. Theories of polarized development were developed (e.g. by Friedman 1972) in which it was shown that a flourishing of centres invariably was linked to a stagnation or even decline of peripheral regions. In the context of spatial analysis, geographers were eager to pick up the idea, to participate in the investigation of disparities in terms of locational advantages, labour supply, income levels, etc., and in the formulation of development strategies for the periphery. Meanwhile, with the advent of a consciousness for the ecological crisis, there has been a change of perspective: regions are no longer regarded only as areas to be developed, but also or predominantly as areas to be conserved and protected in terms of their nature and culture. There is an increasing conviction that strong regions (in terms of self-determination and self-sufficiency) are needed as a counter-weight to the trend toward a global society, in particular also that environ-mental problems emerging on a global scale cannot be solved unless a considerable degree of sustainability of a regional scope has been attained beforehand.

In other words, we have here the idea that an environmentally sustainable, post-economic society (cf. Steiner, this volume) can develop only on the basis of a regional foundation. The latter would have to be provided by the existence of regional subsocieties with a high degree of autonomy in all respects. In essence it is a continuation of an already older discussion about 'self-reliant development' (see, for example, Bassand et al. 1986). In the light of the concept of 'sustainable development' (World Commission on Environ-ment and Development 1987) this continuation, however, now carries with it strong environmental overtones. If we wish to depict the needed situation in system theoretic jargon we would say that a regional society, in order to have a high degree of autonomy, has to have a number of internal connections that far exceeds the number of external connections.

This should, first of all, have consequences for the social realm, evidenced by a considerable shift away from indirect anonymous interactions to direct face-to-face encounters. Social integration gains in importance at the expense of system integration, at least as far as internal connections are concerned.

This can foster a climate of trust and cooperation such that the often-heard interpretation of our predicament as a bad case of the prisoner's dilemma loses its conviction. Decisions can be arrived at jointly on the basis of a 'practical reason', as Dryzek (1987) calls his version of Habermas' concept of communicative rationality.[16] It is no longer the individual economic actors that are exposed to an all-side competition but only the region as a whole. In other words, the internal competition can be transformed into a competitive cooperation towards a common goal. As stressed by Jaeger (1991) an important prerequisite for such a scenario is, however, the regeneration of a culture of professionalism with high ethical standards, which supersedes the labour morale associated with the old Fordistic and Tayloristic style of economy. Through an all-side connectedness of the different sectors of society (not only within the economy) the setting becomes supportive for the emergence of 'innovative milieux'[17] with a high degree of tolerance for alternative styles of living. Again we might mention the flexibilization of structures discussed by Steiner, Dürrenberger, and Ernste in this context.

As a counterpart to the intensification of relations within regional societies we may envision more direct interactions with the environment, in particular, of course, with the natural components of the environment. This can happen if regions strive for a higher degree of self-sufficiency in terms of food supplies. To have an effect on the majority of people, however, food production should not be delegated exclusively to a particular segment of society, but as many as possible should be involved. This in turn would require a large-scale 'ecologization' of unproductive areas, in particular within urban agglomerations. In addition, regional self-sufficiency would also have to apply to energy consumption and waste disposal.[18] These and related topics are addressed in the presentation of an 'ecoregional strategy' in the contribution by Bahrenberg and Dutkowski. They themselves call the idea of ecoregions a utopia, but then we need utopias as guiding-stars, as regulative ideas.[19] What other options do we have? At any rate it is plausible that new environmental ethics can come about not by decree but only through the personal experience of individuals. It is also plausible that a more direct relationship between people and environment, including the more direct confrontation with possible problems, should further an improved ecological understanding and an increase in timely feedback information about the state of the environment (Dryzek 1987).

In summary, it is plausible that the delegation of large amounts of responsibility to lower (regional) levels will in fact make the people concerned more responsible actors. The call for 'radical decentralization' (Dryzek 1987), however, is not supposed to mean that any kind of regional specialization or division of labour and, hence, of interregional trade should be abolished. On the contrary, a certain degree of specialization will be needed exactly for ecological reasons in that this will enable a better degree of adaptation to regional conditions. However, such specializations should not degenerate into one-sided dependencies. For example, the Alps in Central Europe should not be turned into mere playgrounds for the people from urbanized areas seeking relaxation and entertainment (Bätzing 1991). Also, a boundless interregional transportation of goods cannot be tolerated, again, of course, for ecological reasons.[20] The interregional mobility of persons is another problem with an ambiguous nature. It hampers the growth and stabilization of social and environmental identities, which is negative in terms of the argument put forth earlier: that such identities may be a crucial prerequisite for the existence of environmental concern. On the positive side is the fact that such mobility may support the growth of interregional, intercultural understanding. In order for this aspect to be at all important, however, we have to have a situation in which regional cultures can survive and are not trampled upon by an all-invasive and faceless global culture. Cultural diversity (see, for example, Elsasser *et al.* 1988) may become essential as a pool of potential creativity for the future cultural evolution in a way similar to the importance of biodiversity for a continuation of biological evolution. On the other hand, we recognize, of course, the need for negotiations, agreements and decisions on a global scale. To strike a balance between the global and the regional dimension in such a way that ecological sustainability is guaranteed may very well become a problem having the squaring-of-the-circle quality.[21]

5 EVOLUTIONARY REATTACHMENT

In a closing note we wish to consider again the statement made earlier about the need to respect the evolutionary hierarchy from the abiotic through to the economic level of phenomena. In reference to Figure 4.4 in the contribution by Steiner (this volume) we can envisage these levels as recursive systems that can be depicted metaphorically by circles within circles. To attain ecological sustainability a system represented by an inner circle must respect any limitations set by an outer system. Expressed in a positive fashion the true creativity of an inner system will manifest itself in the way in which it makes use of the degrees of freedom accorded without violating the limits. Here we simply wish to draw the reader's attention to the fact that much of what has been said above about human identity and regional cultures is reminiscent of certain aspects of archaic human life: the direct social contacts, the attachment to a region, the immediate experience of nature, possibly the development of new values of an ecological nature. The need for a reattachment of this kind has been pointed out before by Steiner (this volume).

Of course, if we put forward notions of a regional life embedded in a social network and in contact with nature, it is not primarily because we like to remember something that we have lost, but because it is something that may help to find a path towards sustainable development in a time of increasing globalization of social problems and ecological risks. In this context Peter Gould's treatise of the diffusion of AIDS is a description of a concrete risk which humankind is facing today. We might ask whether the advent of AIDS is not an expression of a nature that has started to strike back. Lest the reader condemn this as animistic thinking, let us point to the fact that this issue is being discussed quite seriously by, for example, Renggli (1992) in the following fashion: a time of crisis for an individual results in stress for that individual and, if it persists, may eventually lead to a psycho-somatically induced breakdown of the immune system. This model may be extended to situations of crisis for whole societies: the result may be a mass psychosis which finds its 'socio-somatic' expression in epidemics such as the black death towards the end of the Middle Ages or AIDS now.[22]

The occurrence of AIDS may also serve as a metaphor for a wide range of dangers in the form of unintended consequences of human actions that we are facing in a society with existing or emerging interregional, global networks. The role of boundaries becomes crucial. It is a real irony that the problem of AIDS results from human actions that may be taken as having an archaic quality: sexual promiscuity. Of course, it is not promiscuity limited to a small group of people and having the character of necessary social relations, but it is worldwide and largely anonymous and, in this sense, exactly not of an archaic quality. It is a case of 'love', otherwise supposed to be a cornerstone of social integration, becoming a medium of system integration. Among other things we may be reminded here of the problem of international sex tourism, an important aspect of which is, of course, that it takes advantage of situations of economic dependence. If we wish to adhere to a distinction made by Sayer (1984), there may be argument about whether we have here a case of contingent or necessary relations.[23] Strictly speaking, only the latter have the property of being structural and thereby contributing to system integration. Of course, we can think of a modern global society free from the phenomenon of sex tourism, but it is obvious that it would have to be a society with a constitution far different from what we have now. Presently, the way it works is that inequalities and economic dependence are unavoidable and in this sense phenomena such as sex tourism are a necessary result.

Whether we have to do with contingent or necessary relations, the example of AIDS, taken as a problem by itself and as a metaphor for other problems, demonstrates the importance of the responsibility of persons engaged in such relations. If ever we wish to attain the state of a human society which is truly sustainable in ecological, social, political and economic terms, then the responsibility of individuals must be one of its distinguishing foundations. But being responsible cannot only mean to do the 'right' thing at a particular moment in a particular situation. Responsibility must refer to a total view of life on this planet, in the past, now, and in the future. A decisive question is this: do we see ourselves as the managers of the world, or rather as transient participants in a larger evolutionary process?

6 NOTES

1 The first editor (D.S.), having had the opportunity in 1963 and 1964 to attend the courses in spatial analysis offered by B.J.L. Berry at the Department of Geography, University of Chicago, became one of those 'revolutionaries' for a time.

2 Cf. the remarks made in Introduction to Part II, Section 2.

3 Cf. General Introduction and Nauser (this volume).

4 See Parsons (1966) and Parsons and Shils (1951).

5 See Varela (1979) and Maturana and Varela (1980, 1985). A general overview can be found in Steiner (1989).

6 Translated from the German original by D.S.

7 Translated from the German original by D.S.

8 It is interesting to note in this context that, as discussed in Dürrenberger (1989), an older version of Parson's theory contains a normative cultural system which is superimposed on action systems (Parsons and Shils 1951), while a later version abandons this hierarchy and sees the cultural system on the same level as other action systems (Parsons 1966).

9 A brief appraisal of the value of Habermas' work specifically in a human ecological context is provided by Brulle (1992).

10 Non-verbal components of communication would seem to be of greater importance in pre-industrial societies than in our present civilization, cf. Note 20 in Steiner (this volume).

11 We note that the social psychologist Mead (1934) makes a distinction between a social identity of the sort just discussed, constituting what he calls the 'Me', and a personal identity, referring to the way in which I see myself, constituting the 'I'. He is interested in the mutual relationship between the two and sees the total identity of a person as resulting from the combination of 'Me' and 'I' to the 'Self'. This may remind us of Guattari's different 'ecologies' mentioned earlier; Mead is talking about two of them. We can readily surmise that there should also be a concept of 'environmental identity', a question taken up below.

12 As an example, imagine being connected to the social world solely by two-way gadgets such as telephones and telefaxes or, worse, by one-way gadgets such as radios and television sets.

13 We note at this point that there is a relation between the roles 'normally' played by adult males and females in our society and the way the two genders become socialized and can acquire an identity during their youth. This is a topic investigated by Carol Gilligan (1982: 8): 'For boys and men, separation and individuation are critically tied to gender identity since separation from the mother is essential for the development of masculinity. For girls and women, issues of femininity or feminine identity do not depend on the achievement of separation from the mother or on the progress of individuation. Since masculinity is defined through separation while femininity is defined through attachment, male gender identity is threatened by intimacy while female gender identity is threatened by separation.'

14 Cf. Introduction to Part I, Section 5.

15 Consider the link here to the notion of 'locale' by Giddens (1984) or that of 'territory' in Steiner, *et al.* (this volume).

16 The terms 'practical reason' and 'radical decentralization', used as the heading to Section 4, refer to the argumentation put forward by Dryzek (1987) in his search for a form of social choice mechanism that might ensure ecological sustainability.

17 Clearly, the term 'innovative milieu' is a slogan that has to be regarded with caution. The following question arises unavoidably: how much innovation do we

actually need? And how much can we actually bear? Do we have to have a new model of personal computer every second year? Only one thing is sure: we definitely need innovative ideas with respect to how we go about getting out of the ecological crisis with planet Earth and humankind relatively intact. Let us note in this context that Millendorfer (1987) associates the capacity for innovative thinking with a kind of mentality which he calls 'rurality' (*Bäuerlichkeit* in German) as it still exists in rural areas. This is highly interesting in connection with the question raised in Introduction to Part II, Section 3 : the need for re-establishing a tradition. Could it mean that the 'conservative progressivity' contained in Millendorfer's rurality is a phenomenon capable of linking tradition with innovation? Indeed, as Furger (1992) points out, innovativeness is a characteristic of regions in which there are functioning stable personal and professional contact structures.

18 It may be appropriate to remind the reader of the question raised earlier: is it not so that the possibility of communication with the environment is neglected in Habermas' theory of communicative rationality? We feel that it is and propose to enlarge the notion of 'practical reason' to one that includes human–environment interactions.

19 Compare also with Bätzing (1991) who discusses this problem based on the concrete example of the Alpine region.

20 For example, a stop must be put to the utter nonsense associated with the following situation: potatoes grown in Germany are shipped on trucks to Italy where they are washed, because labour is cheaper there, and then shipped back to Germany for consumption.

21 'Sustainable Development: die Quadratur des Kreises?', lecture by Carlo Jaeger in the environmental sciences interdisciplinary seminar at the ETH in Zürich, 10 December, 1991.

22 Note, however, that population biological factors also are likely to be responsible for the occurrence of epidemics. Many pathogenic micro-organisms require a large contiguous population in order to exist. Consequently, it is plausible that infections were never a major problem for archaic societies in a primeval setting of small population size and regional boundedness (Boyden 1987).

23 It seems that there are more and more instances in a modern society in which this distinction is by no means clear. Take the example of car accidents: there is no accepted social structure which provides for cars bumping into each other as a form of interaction. Yet it happens regularly, car drivers know that they take a risk when they are on the road, insurance companies thrive on exactly this risk, and the repair work done on cars in workshops and on people in hospitals is added as value to a nation's GNP. Indeed, risks and accidents have become a normal feature of life in Western society (cf. Perrow 1988 and Beck 1986).

7 REFERENCES

Bahrenberg, G. (1987) 'Unsinn und Sinn des Regionalismus in der Geographie', *Geographische Zeitschrift* 75 (3): 149–60.

Bassand, M., Brugger, E.A., Bryden, J., Friedman, J. and Stuckey, B. (eds) (1986) *Self-reliant Development in Europe. Theory, Problems, Actions*, Aldershot: Gower.

Bätzing, W. (1991) *Die Alpen. Entstehung und Gefährdung einer europäischen Kulturlandschaft*, München: C.H. Beck.

Beck, U. (1986) *Risikogesellschaft. Auf dem Weg in eine andere Moderne*, Frankfurt a/M: Suhrkamp.

Bhaskar, R. (1975) *A Realist Theory of Science*, Leeds: Leeds Books.

Boesch, E.E. (1991) *Symbolic Action Theory and Cultural Psychology*, Berlin: Springer.

Boyden, S. (1987) *Western Civilization in Biological Perspective. Patterns in Biohistory*, Oxford: Clarendon Press.

Brulle, R.J. (1992) 'Jurgen Habermas: An exegesis for human ecologists', *Human Ecology Bulletin* 8: 29–40.

Dews, P. (1991) 'Die ökologische Relevanz des Unterschieds zwischen System und Umwelt bei Habermas', manuscript of a lecture given at the Department of Geography, ETH, Zürich.

Dryzek, J.S. (1987) *Rational Ecology. Environment and Political Economy*, Oxford: Blackwell.

Dürrenberger, G. (1989) 'Menschliche Territorien. Geographische Aspekte der biologischen und kulturellen Evolution', *Zürcher Geographische Schriften* 33, Zürich: Verlag der Fachvereine.

Elsasser, H., Reith, W.J. and Schmid, W.A. (eds) (1988) *Kulturelle Vielfalt, regionale und örtliche Identität – eine sozio-kulturelle Dimension der Raumplanung?*, Wien: Institut für Raumplanung und Agrarische Operationen, Universität für Bodenkultur.

Ernste, H. and Jaeger, C. (eds) (1989) *Information Society and Spatial Structure*, London and New York: Belhaven.

Ernste, H. and Meier, V. (eds) (1992) *Regional Development and Contemporary Industrial Response. Extending Flexible Specialisation*, London and New York: Belhaven.

Friedman, J. (1972) 'A general theory of polarised development', in N.M. Hansen (ed.) *Growth Centers in Regional Economic Development*, New York: Free Press.

Furger, F. (1992) 'Ökologische Krise und Marktmechanismen. Umweltökonomie in evolutionärer Perspektive', PhD thesis, Zürich: Department of Geography, ETH (publication forthcoming).

Giddens, A. (1984) *The Constitution of Society. Outline of the Theory of Structuration*, Berkeley and Los Angeles: University of California Press.

Gilbert, A. (1988) 'The new regional geography in English- and French-speaking countries', *Progress in Human Geography* 12: 208–28.

Gilligan, C. (1982) *In a Different Voice*, Cambridge, MA: Harvard University Press.

Guattari, F. (1989) *Les trois écologies*, Paris: Editions Galilée.

Habermas, J. (1989) *The Theory of Communicative Action, 1: Reason and the Rationalization of Society. 2: The Critique of Functional Reason*, Cambridge: Polity Press.

Harré, R. (1986) *Varieties of Realism. A Rationale for the Natural Sciences*, Oxford: Blackwell.

Jaeger, C. (1991) 'The puzzle of human ecology. An essay on environmental problems and cultural evolution', habilitation thesis, Zürich: Department of Environmental Sciences, ETH (publication forthcoming).

Johnston, R.J. (1986) *Philosophy and Human Geography. An Introduction to Contemporary Approaches*, London: Edward Arnold.

Kastenholz, H. (1991) 'From obligation to motivation: a human–ecological approach to the greenhouse effect', in M.S. Sontag, S.C. Wright and G.L. Young (eds) *Human Ecology. Strategies for the Future*, Fort Collins, CO: Society for Human Ecology.

Lübbe, H. (1990) 'Nationalismus und Regionalismus in der politischen Transformation Europas', *Neue Zürcher Zeitung* (230), 4.10.90.

Luhmann, N. (1982) 'Autopoiesis, Handlung und kommunikative Verständigung', *Zeitschrift für Soziologie* 11 (4): 366–79.

Luhmann, N. (1984) *Soziale Systeme. Grundriss einer allgemeinen Theorie*, Frankfurt a/M: Suhrkamp.

Luhmann, N. (1989a) *Ecological Communication*, Cambridge: Polity Press.

Luhmann, N. (1989b) 'Ökologie und Kommunikation', in L. Criblez and P. Gonon (eds) *Ist Ökologie lehrbar?*, Bern: Zytglogge.

Maturana, H.R. and Varela, F.J. (1980) *Autopoiesis and Cognition. The Realization of the Living*, Dordrecht: Reidel.

Maturana, H.R. and Varela, F.J. (1985) *The Tree of Knowledge*, Boston, MA: New Science Library.

Mead, G.H. (1934) *Mind, Self, and Society*, Chicago, Ill.: University of Chicago Press.

Millendorfer, J. (1987) 'Für eine bäuerliche Landwirtschaft', *Agrarische Rundschau* (3/4): 65–72.

Naess, A. (1989) *Ecology, Community and Lifestyle. Outline of an Ecosophy*, trans. and rev. D. Rothenberg, Cambridge: Cambridge University Press.

Parsons, T. (1966) *Societies. Evolutionary and Comparative Perspectives*, Englewood Cliffs, NJ: Prentice Hall.

Parsons, T. and Shils, E.A. (eds) (1951) *Toward a General Theory of Action*, New York: Harper & Row.

Perrow, C. (1988) *Normal Accidents. Living with High-Risk Technologies*, New York: Basic Books.

Renggli, F. (1992) *Selbstzerstörung aus Verlassenheit*, Hamburg: Rasch & Röhring.

Sayer, A. (1984) *Methods in Social Science. A Realist Approach*, London: Hutchinson.

Steiner, D. (1989) 'Zur autopoietischen Systemtheorie', in G. Braun and R. Schwarz (eds) *Theorie und Quantitative Methodik in der Geographie*, Berlin: Department of Geography, Free University of Berlin.

Taylor, P.J. (1991) 'A theory and practice of regions: the case of Europe', *Society and Space* 9 (2): 183–95.

Varela, F.J. (1979) *Principles of Biological Autonomy*, New York and Oxford: North Holland.

Weichhart, P. (1990) 'Raumbezogene Identität. Bausteine zu einer Theorie räumlich-sozialer Kognition und Identifikation', *Erdkundliches Wissen* 102, Stuttgart: Franz Steiner.

World Commission on Environment and Development (1989) *Our Common Future* (the Brundtland Report), Oxford and New York: Oxford University Press.

18

AN ECOREGIONAL STRATEGY TOWARDS A FAULT-TOLERANT HUMAN–ENVIRONMENT RELATIONSHIP

Gerhard Bahrenberg and Marek Dutkowski

1 INTRODUCTION

During the last two decades the relationship between humans and their natural environment has attracted much attention in modern societies. The *Waldsterben*, the poisoning of water, air and soils, the global warming of the atmosphere by the greenhouse effect with its possible accompanying climatic changes, the risks of atomic energy and genetic technology, the destruction of the tropical rainforests, and the extinction of a large number of species are topics widely discussed in the public. Not surprisingly there exists a widespread feeling that modern society faces an ecological crisis with which it may not be able to cope.

Although ecological crises are by no means a new phenomenon in the history of humankind,[1] the present environmental problems and their perception show some peculiarities that deserve attention. For the first time in history it seems that not only specific cultures and societies are endangered but the whole world population: the species *Homo sapiens*. In addition, it is assumed that the environmental changes may lead to political and economic destabilization on a global scale with unforeseeable effects.[2]

From this point of view there is obviously a need for strategies to overcome the ecological crisis. Before discussing them it may be useful to regard some basic characteristics of modern ecological risks in order to clarify the potential range and limitations of different strategies. These characteristics are the subject of the first part of the paper. The second part is devoted to the strategies. We shall argue in favour of an additional strategy which we call an 'ecoregional' strategy. Basically, this strategy aims at strengthening local/regional power in political decision-making processes and the ability of households to become self-sufficient in basic goods. Part three investigates possible steps towards the ecoregional strategy.

GERHARD BAHRENBERG AND MAREK DUTKOWSKI

2 CHARACTERISTICS OF MODERN ECOLOGICAL CRISES

What distinguishes modern ecological crises of developed industrial societies from earlier, pre-industrial ones is – among others – the way in which they are perceived. Formerly, nature was held to be external to humans. People believed that natural environmental disasters and catastrophes were not caused by humans but sent by God(s) or other external forces; that they could not be avoided, and their effects at most be restricted in range. In contrast to this, present environmental risks are attributed to human actions which in principle can be changed, that is, they can be avoided by proper actions and adequate measures of control, political strategies and decision-making. Thus, environmental policy can be seen as the most recent step in the process of enlightenment.

On the other hand, the environmental problems we are facing today have many characteristics that make any comprehensive, straightforward policy almost impossible. The human impact on the natural environment occurs mainly through interventions in the circulation of matter and the flow of energy which are inevitably connected with all human activities. We permanently release matter and energy into the environment and extract matter and energy from it. This holds true for all societies in history. But specific to the present industrial society is the quantity and quality of these interventions. The amount of matter being released by the world population has exceeded the buffering capacity of the natural environment whose material composition changes irreversibly, at least in terms of short- to medium-term time scales (cf. the increase of carbon dioxide in the atmosphere during the last 100 years). By 'quality' we mean the production and deposition of synthetic materials whose behaviour within the circulation of matter is not totally understood (e.g. the so-called CFCs). In addition, their behaviour cannot be understood in advance because of the large time and spatial scale of their potential effects. At best, we can gain *ex post* knowledge about the changes we induce in our natural environment but what we would need is an a priori knowledge.

It should be noted that our lack of knowledge becomes the more dangerous the more the processes of world population growth and of the global uniforming of manufacturing, of life and consumption styles continue, since these processes multiply possible negative effects. In view of these circumstances one can imagine how difficult it is to formulate a policy to solve our environmental problems or to avoid new ones.

This task becomes even more difficult if one takes into account the structure of society that has evolved during modernity. In order to explain this point we refer to Luhmann's *Ökologische Kommunikation* (1986). His reasoning is as follows: modern society, as opposed to segmentarily structured archaic societies and hierarchically stratified traditional societies, is characterized by an increasing functional differentiation into subsystems. Such subsystems are, for instance, economy, law, science, politics, religion, education. Each subsystem uses a specific code for communication which is in its simplest form a binary code. For example, the code of the economic subsystem is based on the medium of money. It has the two values 'having money' and 'not having money'. The subsystem science is based on truth as its medium in so far as it differentiates between 'true' and 'false'.

The codes of the respective subsystems express their specific abilities to perceive differences. This means that each subsystem can observe only those differences which are formulated in its specific code. These observations of a subsystem (for instance of its respective environment) are extremely reductionist; on the other hand this reduction of complexity is a necessary prerequisite of its observations. In other words, the scientific subsystem can tackle a problem only if this problem has been translated into statements to which the values true and false can be attributed. Accordingly, the economic subsystem can deal with problems only if they have been transformed into the 'language of prices'.

The respective codes allow the subsystems to operate quite efficiently and to increase their problem-solving capacity. They also facilitate the problem-solving of society if the problems at hand can be formulated in such a way that they fit into the realm of one subsystem. On the other hand, there exist many problems that extend across several subsystems. They can be solved only through rather complicated coordination and cooperation processes among the different subsystems involved. According to Luhmann, environmental problems belong to this second category. Since there is no specific subsystem which is able to recognize these problems they are difficult to detect, and even if they are detected that does not mean very much. For example, if the scientific subsystem believes in the harmfulness of the CFCs for the ozone layer in the stratosphere and the induced increase of UV-C- and UV-B-radiation with its toxic effects on human health (that is, if the scientific subsystem holds these effects to be true), this does not mean anything to the economic subsystem. Above all, it does not lead to a reduction in the production of CFCs. It also does not mean anything to the political subsystem which operates with the code 'having power' – 'not having power'. Only if the politicians believe that a reduction of the CFCs (which can be achieved by laws or other regulations) and its accompanying socio-economic effects would not diminish their chances of re-election (and their retention of power) can the different subsystems start to think of appropriate means to reduce the production and use of CFCs. In order to reach this a complicated coordination process of many subsystems is necessary. This is the reason for the long time-span between the recognition of a problem and its solution.

The situation becomes even more complicated for two reasons. First, the solution of a problem nearly always creates new unforeseeable problems. Quite typically ecological problems are often unintended consequences of actions which had quite different goals. The industrialization process was mainly based on the use of fossil energy. Nobody did and could foresee that this would change the CO_2-concentration in the atmosphere so that a global warming might occur. Second, ecological problems rank differently in different societies. In general they are long-term problems and cannot receive much attention from people who have to struggle for their daily survival.

3 AN ECOREGIONAL STRATEGY

If the brief analysis given above is accepted it should be clear that there are no simple, immediate and definite solutions to our ecological problems. Above all, slogans such as participative (instead of exploitative) behaviour towards nature, holistic thinking etc., which are popular in some circles of the 'green movement', are far too simple in order to be of any help. And it is difficult to see how they can be operationalized. To us, the term 'fault-tolerant' behaviour towards nature seems to be more appropriate as a guiding principle of human–nature relations. It was introduced by von Weizsäcker and von Weizsäcker (1984) and is characteristic of open systems. It emphasizes the need for human actions towards nature that are in principle reversible if they prove to be harmful and if they show negative, unintentional effects. The goal cannot be to avoid faults at all, but to keep the outcome of faults as small as possible. Of course, the question arises of how such a principle can be pursued by society.

In his latest book Sieferle (1989: 204 ff) discusses – in a chapter under the title 'The crisis of nature and the policy of the whole' (authors' translation) – principal orientations and limitations of environmental policy, that is, a policy which is directed towards the preservation of nature as a basis of human life. According to Sieferle, the limitation of any environmental policy is to be seen in the 'medium range' of modern technology. On the one hand our technology is powerful enough to exceed the natural buffering capacity of the biosphere, on the other hand it is not competent enough to manipulate the ecosystem so completely and effectively that it could maintain a healthy environment for human life. In other words, the degree of complexity that technological systems can handle is too low compared with the high complexity of nature.

In view of this situation there are two principal options. First, we can try to improve scientific knowledge and technological competence up to a level where we are eventually able to control natural systems to such a degree that ecological catastrophes can be avoided. In principle, this would mean to replace nature by technology (we are already planning and designing 'biotopes'). It is this strategy that we pursued hitherto and there is no doubt that it did not work too badly. However, there are some problems connected with it. First of all, it assumes that technical systems can be completely controlled. As recent catastrophes in the chemical and nuclear industries have shown this assumption is not very realistic. Second, it is doubtful whether we shall ever be able to accumulate enough knowledge and technological competence to control systems which are as large and complex as, for instance, the climate. Thirdly, our knowledge may come too late in order to correct processes in our natural environment that have already become irreversible.

To some degree one can evade these deficiencies by political and legal measures. For example, Beck (1988) pleads for admitting only the production of those goods which are proven to be harmless. Although such a procedure would improve our environmental policy it fails to observe the hypothetical character of our knowledge. Following this procedure the production of CFCs would have been allowed at least until 1974 when Molina and Rowland discussed their possible threat to the ozone layer. Before 1974 CFCs were regarded as completely harmless because of their chemical stability.[3] In addition it is questionable whether such a political and legal regulation could work. We already have many environmental laws which are probably permanently violated without prosecution.

Finally, strengthening political and legal control of production processes may well lead to an over-bureaucratization and in the end to a totalitarian state which would strongly oppose our ideas of a democratic society.

The second option open to us is a strategy which tries to keep human intervention in the natural environment within the limits of its buffering capacity. Since the relevant thresholds are unknown it is only possible to minimize human interventions. We believe that this could be achieved by an 'ecoregional strategy'. It consists mainly in decentralizing political decisions by strengthening individual and local/regional autonomy by reducing the size of the political decision-units and by shifting responsibilities from higher to lower levels of government. We discuss the arguments at the aggregate and at the individual, household level.

3.1 The aggregate level

1   Some of the most severe risks in terms of environmental damage are caused by big technology and large-scale industries. These industries need a large and fairly homogeneous market in order to operate successfully. The necessary homogeneity weakens if a large society is split into many small ones with specific demands.

2   Technologies for providing goods and services are different according to whether they are developed for small or large markets. An energy policy aimed at supplying a population of fifty million with cheap and secure energy may favour the use of atomic energy. If the population to be served is only one million, the decision would not be the same.

3   The innovative potential of many small societies is probably greater than that of one large society, in terms not only of technology but also of establishing a new social order, of creating new value systems, etc.

4 The trend towards supraregional and supranational organizations and political decision-units impedes the application of already existing and established environmental techniques. The discussion in the European Community about the introduction of means to reduce the emission of motor vehicles is a good example. A united group of formerly autonomous states can move forward only as fast as the slowest members are prepared and willing to do. In West Germany, cities cannot prescribe an intra-urban, area-wide speed limit of 30 km/h as a rule, because only the federal government has the competence to enact general regulations of speed limits.

We have sacrificed on the altar of unity and uniformity and bigness our opportunities and abilities to act quickly, to invent and implement new solutions to perceived problems, to adjust our rules and regulations to new needs and necessities. Of course, it is not possible to find an optimal size of a society. Kohr, who was one of the first to discuss this topic (see Kohr 1957), argues for an optimal size of 2–300,000 people, taking into account social, economic and political factors. This optimum can be stretched to a critical size of fifteen million by technical, organizational and educational means (see Kohr 1983: chap. 2). But these numbers should not be regarded as absolute limits. Depending on the degree of economic and administrative integration, that is, on the degree of autonomy and autarky, they can be altered.

If one takes into account that the citizens of such a society should have a fair chance of participating in public affairs the cited number may even be too big.

Regardless of the exact size, a small society needs enough natural resources to become self-sufficient, which is a prerequisite for autonomy. This means that it should occupy an area large enough to provide it with essential goods, especially water and food. The size of that area cannot be determined in general since it depends on the specific natural condition and on the available knowledge and technology. We call such an area, including the population that it sustains, an 'ecoregion'.

3.2 The individual level

In order to keep a balance with their natural environment, ecoregions should consist of households that are autonomous and self-sufficient at the individual level as well.

To make this clear, we refer to a critique of the modern urban and housing policy:

> Our city planning and house-building is oriented towards a model of life which is based on a strict separation of work, dwelling and leisure time. The design of settlements and the layout of dwellings are characterized by the effort to eliminate the obligation to work. Dwelling is

pure consumption. . . .. If work, i.e. the preparation of meals, the making and cleaning of clothes, the fabrication of furniture, the heating of the flat, is eliminated from the household it will be organized by industries which are more productive and which follow their own profit-seeking logic. Tinning and other food factories, textile industries, dye-works and dry cleaners, furniture and household appliance industries, power stations will develop. Packing and chemical industries will grow. The household becomes a transit station where consumer goods flow in and rubbish and garbage flow out.

(Häussermann 1987: 16; authors' translation)

This situation is problematic not only because of the garbage being released by the households but also, more importantly, because the households have lost all of their independence and autonomy. Since they have become totally dependent on the wages that they receive from their jobs they have a vital interest in the development of the economy. This gives the economic subsystem an overriding power over the other subsystems of society. It makes people engage in the maintenance of industries with unhealthy jobs and with products that are proven to be harmful to the environment.[4]

According to Häussermann two things must change: the integration of the single household into a network of big technology and industries for the supply of essential goods and our consumption and working ethic.

The question is how these changes can be achieved when living in a flat of 50–100 m² living space on the fourth floor of a residential building or in a small terraced house with a backyard of 50 m². To reduce the integration of the household and to develop a new consumption and working ethic, a physical setting is needed with large enough dwelling units and utilizable land for annexes, small workshops and vegetable and fruit gardens. These conditions can be fulfilled only in rural areas or in small cities with a low population density.

In summary, in order to increase the autonomy of an ecoregion it is necessary to improve the autonomy of its members, the households. They have to become 'prosumers' – to use a phrase by Toffler (1980: 272ff) – instead of being just consumers.

4 PRACTICAL STEPS TOWARDS AN ECOREGIONAL STRATEGY

Many people will characterize the ecoregional concept as utopian. But there are chances to realize it and first steps can be observed all over the world.

First of all it should be noted that apart from functional differentiation there exists a 'territorial principle' in political decision-making which may be utilized. This principle can be found in almost all constitutions of the world as well as in international law.

291

It says that only those people can participate in a decision-making process who live in the territory where the problem to be solved is located. This principle is applied independently of the specific form of participation, be it direct or by representatives. To give just two arbitrary examples: whether or not the central business district in the city of Bremen should be easily accessible for private automobiles must be decided by the citizens of Bremen; it cannot, should not and need not to be decided by the citizens of München or by the federal government of West Germany. The energy policy of Japan has to be formulated by the people of Japan, not by the people of the USA.

The territorial principle is effective at all possible aggregation levels, from the local, through the regional and national, to the international level where it is negatively defined as the non-intervention principle. Obviously it is a basic element of democracy and of rationality in political decision-making.

An important and implicit assumption underlying the territorial principle (cf. Hard 1982: 155) is: the objects of political decisions can be located in space and time, they are always concrete and specific in historical and spatial terms. Apart from both a general competence of all people and the sectoral competence of experts (which is 'produced' in the functional subsystems of society) there exists a specific regional/local competence of the inhabitants of the territory where the object of a political decision is located. It is therefore acknowledged that competence is founded – among other things – on the territorial membership of people.

In a way, the territorial principle constitutes only a formal right. During recent social evolution it has been eroded from two sides. First, we can observe a growing dominance of general expert knowledge over territory-specific knowledge. This has led to a situation where people have been declared incompetent by experts (cf. Illich et al. 1979). Second, administrations at the lower levels often have 'deliberately' given away competences and responsibilities to units at higher scales. For example, the fringe conditions that are relevant to a farmer in northern Germany are defined by neither a regional nor national authority but by the commission of the EC. These shifts seemed necessary for the sake of economic efficiency and growth and of equalizing living conditions.

However, both processes seem to be increasingly criticized and successfully opposed. People have the feeling that large decision-units are too ponderous, too bureaucratic and that the solutions they offer are simply not adequate to local and regional situations since they cannot take into account their specific potentials and constraints. It is also obvious that the larger political decision-units are, the more general their regulations have to be and thus the scope for action at the lower levels of the political and administrative system also becomes broader.

We are witnessing today social movements that strive for the strengthening of regional and local competence and power. People fight successfully against noxious facilities by implicitly or explicitly basing their arguments on the territorial principle. The strongest resistance against nuclear power plants and industries with toxic products and/or wastes comes from the local initiatives of people who feel that their life might be negatively affected by these facilities.

Nowadays there exist many movements for improving regional consciousness, self-reliance and autonomy; people start to defend their right to decide locally on land use and zoning. This 'regionalist struggle', as Eisel (1987) calls it, is accompanied by many practical steps towards more regional and local independence. Programs have been started to improve regional autonomy in economic terms by raising the regional potentials and creating intraregional business cycles (cf., for example, Bassand *et al.* 1986, Bundesforschungsanstalt für Landeskunde und Raumordnung 1984, Hahne *et al.* 1986, Morris 1982, and Mose 1989). At the local level people develop new lifestyles and new economic systems in order to find new adjustments to labile natural environments (for the Alps, see the comprehensive overview of Haid 1989). Producer–consumer cooperatives have been established to facilitate the marketing and supply of healthy food (see, for instance, Brink 1986). Of course, these are only first steps, but they show that the erosion of local and regional autonomy is by no means an irreversible process.

At the individual level, autonomy is much more difficult to realize. At least the self-supply of food does not seem to be an unrealistic aim. This is all the more important since nutrition is one of the basic human needs and since large-scale agriculture contributes significantly to environmental damage.

According to reports by Hitschfeld (1984) on several experiments with food self-supply, a household in West Germany of 4–6 members needs about 0.25–1.0 ha garden land for complete provision of all necessary foods, even without any fertilizers and pesticides and without any import of energy (except sun radiation). The range results from variations of soil and climatic conditions and from the kind of food being produced (exclusively vegetarian food needs less space than does keeping animals). In the case of a pure vegetarian diet, the time necessary for the gardening totals to 1.25 hours per day for the whole household. This is a rather low figure even if one takes into account that it is an average across the year and is somewhat higher in the summer. Since the time of the experiment, working-time has decreased in West Germany by about 25 per cent which makes 1.25 hours per day almost a *quantité négligéable*.

Of course, an ecoregional strategy cannot replace other means and instruments of environmental policy. To be sure, we need scientific knowledge and technological competence to deal with the ecological crisis. It does, however, provide a framework which allows for a new ethic towards nature, for a new working and consumption ethic and for some kind of self-restriction which modern industrialized societies may have to adopt in any case if they are to deal successfully with the environmental problems they have brought about.

There is no guarantee that an ecoregional strategy can avoid ecological catastrophes but it can help to restrict them in size and scale and it may prevent them from becoming global disasters.

5 NOTES

1 Examples of early historical ecological crises are discussed by Martin (1973), Harner (1977), and Jacobsen and Adams (1958); cf. also the summary by Hoffman (1980). It may well be that even the so-called 'neolithic revolution' (Childe 1951) resulted from an ecological crisis of the hunter–gathering systems when these were unable to produce enough food for growing populations (cf. the interpretation by Cohen 1977). Herrmann (1986) covers ecological problems in the Middle Ages and examples of environmental devastations, reaching from the Middle Ages into this century, are dealt with in Sieferle 1988.
2 See for example Bach and Lesch (1989: 516) with regard to the assumed 'climate catastrophe'.
3 Ironically it is exactly their chemical stability which lets them reach the stratosphere where they presumably weaken the ozone layer.
4 In addition, this explains why in West Germany the workers' unions had so many difficulties with adopting a policy whose main goal is the protection of the environment. Such a policy was first formulated and claimed for by the Green Party.

6 REFERENCES

Bach, W. and Lesch, K.-H. (1989) 'Auswege aus der Klimakatastrophe', *Ökologische Konzepte* 15 (30): 3–9.
Bassand, M., Brugger, E.A., Bryden, J., Friedman, J., and Stuckey, B. (eds) (1986) *Self-Reliant Development in Europe*, Aldershot: Gower.
Beck, U. (1988) *Gegengifte: Die organisierte Unverantwortlichkeit*, Frankfurt a/M: Suhrkamp.
Brink, A. (1986) *Die Wendland-Kooperative*, Beiträge zur räumlichen Planung 15, Hannover: Fachbereich Landespflege der Universität Hannover.
Bundesforschungsanstalt für Landeskunde und Raumordnung (ed.) (1984) *Endogene Entwicklungsstrategien?*, Informationen zur Raumentwicklung 1(2), Bonn-Bad Godesberg Selbstverlag.
Childe, V.G. (1951) *Man Makes Himself*, New York: Mentor.
Cohen, M.N. (1977) *The Food Crisis in Prehistory: Overpopulation and the Origins of Agriculture*, New Haven, CT: Yale University Press.
Eisel, U. (1987) 'Landschaftskunde als "materialistische Theologie": Ein Versuch aktualistischer Geschichtsschreibung der Geographie', in G. Bahrenberg *et al.* (eds) *Geographie des Menschen – Dietrich Bartels zum Gedenken*, Bremer Beiträge zur Geographie und Raumplanung 11, Bremen: University of Bremen.
Hahne, U., Ossenbrügge J., Hawel, B.W., Radtke, G.P., Baumhöfer, A., Schulz, W. and Stadt & Land E.V. (1986) *Ökologische Regionalentwicklung – Theoretische und pragmatische Beiträge*, Raumpolitische Argumente 4, Kiel: Stadt & Land, Gesellschaft für raumpolitische Forschung, Planung und Beratung.
Haid, H. (1989) *Vom neuen Leben: Alternative Wirtschafts- und Lebensformen in den Alpen*, Innsbruck: Haymon.
Hard, G. (1982) 'Länderkunde', in L. Jander, W. Schramke and H.-J. Wenzel (eds) *Metzler Handbuch für den Geographieunterricht*, Stuttgart: Metzlersche Verlagsbuchhandlung.
Harner, M. (1977) 'The enigma of Aztec sacrifice', *Natural History* 86: 46–51.

Häussermann, H. (1987) 'Ökologische Stadtentwicklung', *Impulse aus der Forschung* 3: 14–17 (Universität Bremen).

Herrmann, B. (ed.) (1986) *Mensch und Umwelt im Mittelalter*, Stuttgart: DVA.

Hitschfeld, O. (1984) 'Die Schaffung landwirtschaftlich–gärtnerischer Nebenerwerbs-stellen: Ein sicherer Weg aus der Krise', unpublished manuscript, Berghaupten.

Hoffman, M.P. (1980) 'Prehistoric ecological crises', in L.J. Bilsky (ed.) *Historical Ecology: Essays on Environment and Social Change*, Port Washington, NY: Kennikat Press.

Illich, I., McKnight, H.J., Zola, I.K., Borremans, V., Caplan, J., Shaiken, H. and Huber, J. (1979) *Entmündigung durch Experten. Zur Kritik der Dienstleistungsberufe*, Reinbek b. Hamburg: Rowohlt.

Jacobsen, T. and Adams, R.M. (1958) 'Salt and silk in ancient Mesopotamian agriculture', *Science* 128: 1251–8.

Kohr, L. (1957) *The Breakdown of Nations*, London: Routledge & Kegan Paul.

Kohr, L. (1983) *Die überentwickelten Nationen. Rückbesinnung auf die Region*, Salzburg: Winter.

Luhmann, N. (1986) *Ökologische Kommunikation. Kann die moderne Gesellschaft sich auf ökologische Gefährdungen einstellen?*, Opladen: Westdeutscher Verlag.

Martin, P.S. (1973) 'The discovery of America', *Science* 179: 969–74.

Molina, M.J. and Rowland, F.S. (1974) 'Stratospheric sink for chlorofluoromethanes: Chlorine catalyzed destruction of ozone', *Nature* 249: 810–12.

Morris, D. (1982) *Self-Reliant Cities: Energy and the Transformation of Urban America*, San Francisco, CA: Sierra Club Books.

Mose, I. (1989) 'Eigenständige Regionalentwicklung – Chance für den peripheren ländlichen Raum?', *Geographische Zeitschrift* 77 (3): 154–67.

Sieferle, R.P. (1989) *Die Krise der menschlichen Natur: Zur Geschichte eines Konzepts*, Frankfurt a/M: Suhrkamp.

Sieferle, R.P. (ed.) (1988) *Fortschritte der Naturzerstörung*, Frankfurt a/M: Suhrkamp.

Toffler, A. (1980) *Die dritte Welle: Zukunftschance – Perspektiven für die Gesellschaft des 21. Jahrhunderts*, München: Bertelsmann.

Weizsäcker, C. von and Weizsäcker, E.U. von (1984) 'Fehlerfreundlichkeit', in K. Kornwachs (ed.) *Offenheit – Zeitlichkeit – Komplexität: Zur Theorie der Offenen Systeme*, Frankfurt a/M: Campus.

19

ON REGIONAL AND CULTURAL IDENTITY: OUTLINE OF A REGIONAL CULTURE ANALYSIS

Benno Werlen

1 INTRODUCTION

Cultural and social geographers are traditionally interested in the spatial differentiation of societies and cultures. Their related research interests are in defining socio-cultural regions (cf. Bartels 1968), in order to indicate clearly defined cultural spaces. However, this research encounters a major problem: only material things can be unequivocally localized within a physical space; 'immaterial' social norms and cultural values cannot. Any social or cultural research which embarks upon the use of territorial categories (area, region, etc.) without beforehand refuting the above argumentation remains unconvincing (cf. Werlen 1989a). By the same token, any attempts at reconstructing so-called 'regional identities' are questionable. If only material things can be spatially localized, then the argument for the existence of regional identities would mean that the inhabitants of a certain area could have an identity with material entities. The empirical demonstration of this proposition would pose a real problem for any regional analyst.

Despite these problems, it is obvious that cultures are spatially differentiated. Thus, it is not surprising that in the context of cultural identity, territorial categories continue to play an important role. The difficulties experienced by immigrants in retaining their cultural identity within their new society is a sufficient indication of this.

In what follows I shall argue that the clarification of regional conditions of cultural identity should be substituted for a spatially centred search for regional identity. As Giddens, together with Schütz, is one of the few social scientists to include this problem seriously in his *oeuvre* (cf. Giddens 1988), his works of social theory are of particular relevance to regional geography. This paper should be seen as an outline of the sense in which the findings and discoveries of regional culture analysis can be related to the problem of cultural identity. This also involves a clarification of what is to be understood by 'culture'.

First of all, I would like to summarize briefly some of the main points of

Giddens' theory. Linked to this and as a background to the question of when 'someone' can be identical with 'something', I will define the cultural understanding of an action-centred theory of society, before reconstructing the regional aspects of cultural identity.

2 SOCIAL AND CULTURAL RESEARCH IN THE THEORY OF STRUCTURATION

In his outline of the theory of structuration, Giddens asks how one can best investigate the relationship between agency, structure, social system and social reproduction within the framework of social scientific research (Giddens 1984a: 207). In this, the meaning assigned to 'culture' takes on a less important role. His main aim is to draw attention to the 'duality' of social structures and to incorporate the corresponding consequences into social theory. Before I take issue with some aspects of regional culture analysis from the standpoint of action theory, I would like to review very briefly the general framework of Giddens' conception. Following on from this, I hope to link some elements of action theory in social geography with Giddens' theory of structuration.

Giddens' work aims to illuminate the mediation between socio-cultural reality and the subject; that is, between socio-cultural structures and action, in order to avoid the inadequacies of objectivism as well as subjectivism. Thus, he wants to solve the problem of mediation between the macro and the micro levels in order to overcome the gap between structural and interpretative analysis. To solve this problem, 'we must go beyond positivism as well as interpretative sociology' (Giddens 1984b: 8), that is, we must overcome the gulf between objectivist–structuralist and subjectivist–hermeneutic concepts of cultural and social research (Giddens 1984a: 1ff). Epistemologically, this aim implies the overcoming of the dualism of the structural–causal and interpretative–teleological method by developing an additional integrative methodological approach. One of Giddens' main points is the thesis of the dualistic view of socio-cultural structures, and it is to this which I shall now turn.

The concept of 'duality of structure' is, according to Bernstein (1986: 242), the 'central vision' and, according to Held (1982: 99), the 'key concept' in Giddens' theory of structuration. With this concept he wants, by means of an internal critique, 'to overcome categorically the one-sidedness of . . . objectivist theories on the one hand and of . . . subjectivist theories on the other, in order to be able to "formulate" his own theory' (Kiessling 1988: 173).

Action and structure should be understood in terms of a dialectical process of mediation, as moments of one and the same socio-cultural reality. Giddens provides the following definition of 'structure' (1976: 121): 'A structure is not a "group", "collectivity" or "organization": these have structures.' Structures also do not have a subject: they are to be understood as systems of semantic rules (structures of *Weltanschauungen*), as systems of resources (structures of domination) and as systems of moral rules (structures of legitimation). These structures only become real through actions and are only reproduced through actions. 'The proper locus for the study of social reproduction is in the *immediate process of the constitution of interaction*' (Giddens 1976: 122) and 'to enquire into the process of reproduction is to specify the links between "structuration" and "structure"' (ibid., p. 120). Structuration refers to 'the dynamic process whereby structures come into being' (ibid., p. 121). These structures only become manifest in social interaction. In accordance with the threefold notion of structure, three different kinds of interaction are to be distinguished: 'communication', 'power' and 'morality'. For Giddens, 'duality of structure' therefore means 'that social structures are both constituted by human agency and yet at the same time [they] are the very *medium* of this constitution' (ibid., p. 121).

Thus, it should be clear that according to Giddens the social world is constituted through actions taking place in concrete situations of interaction. In these processes of constitution, agents, in the framework of certain (available) resources, refer to specific semantic and moral rules, which achieve an integrated application in the act of constitution. If the modalities of production and reproduction are related to an 'integrated *system* of semantic and moral rules, we can speak of the existence of a common culture' (ibid., p. 124).

Accordingly, 'cultural identity' expresses itself in the manner of execution of structuration processes (i.e. in the manner of application of the different rules) without giving rise to an 'opposition between structural principles'. If we then hypothetically pursue Giddens' argument, 'cultural identity' is achieved when the agents in the structuration processes can reconcile in practice, without contradiction, both the intersubjectively shared semantic and moral rules, and their own subjective knowledge.

However, we must stress firstly that 'to know a rule is not to be able to provide an abstract formulation of it, but to know how to apply it to novel circumstances, which includes knowing about the contexts of application' (Giddens 1976: 124). Second, a distinction between 'conflict' and 'contradiction' now becomes necessary, a distinction which is also of particular importance for regional research and regional politics (ibid., p. 125f). In order for 'conflict' to arise, the interactive partner must be referring to the same rules. 'Contradiction', on the other hand, involves an opposition between different semantic and/or moral rules. Thus, according to Giddens' terminology, 'conflicts' can arise within the framework of the application of the same semantic and moral rules. This should particularly be the case when differences arise in the realm of resources. By comparison, 'contradictions' can only emerge when structural oppositions occur on the level of rules (even when 'no' differences are apparent in the realm of resources). Thus, it is in this sense that we must understand Giddens' assumption that cultural identity implies a contradiction-free acceptance of the entire range of rules. This terminological framework therefore allows for a clear distinction between inter-cultural contradictions and intra-cultural conflicts.

With this discussion we can localize the 'cultural realm' in the social world by means of the categories of structuration theory. However, the conceptual differentiation is not yet altogether sufficient with regard to the area that we are addressing. I shall now attempt to delimit the problem of cultural identity in a more specific context by introducing some of the basic categories of action-oriented social geography.

3 THE UNDERSTANDING OF CULTURE IN SOCIAL GEOGRAPHY

From any action-centred perspective, 'culture' is to be understood as an analytical category (Saunders 1987: 64) and not as an object of investigation *sui generis*. To begin with, in comparison to structuration theory, 'culture' can be regarded more generally as an important aspect of the consequences of past actions and of the conditions of present and future actions of individuals. This specification relates to the evaluated and evaluating dimensions of individuals' ways of acting as members of a society. This shows that what is designated as 'culture' in general refers neither to the sum of material artefacts, nor only (as, for example, within the framework of structuration theory) to immaterial systems of values or rules. We can only understand and adequately interpret any artefact if we recognize the meaning that the human act of production affords it; in short, if we have experienced its purpose and its symbolic content. The explanation and understanding of 'culture' thus presupposes a knowledge of both aspects. However, 'earlier' systems of values and rules can persist in material artefacts and in their patterns of arrangement in the physical world – particularly when one ignores their social content – to the extent that these systems can offer considerable resistance to socio-cultural change.

Culturally specific ways of acting are perceptibly expressed in certain customs, usages and social conventions (Weber 1968: 29). However, like other forms of action, they refer to a certain orienting frame of reference. The aspect of this frame of reference which Giddens designates as systems of semantic and moral rules encompasses above all typical values and the patterns of interpretation and the norms which are founded in these values. Therefore, the members of a society express their adherence to a certain cultural realm when as a group they consciously, less consciously or routinely refer to these values, thus effecting the corresponding structuration processes and consequently producing and reproducing the cultural content.

The manner of value orientation for a certain action is also a central criterion by which certain actions can or cannot be associated with a certain culture. By 'value' I mean in this sense a principle of allocation, a rule for the interpretation of circumstances and events. Correspondingly, we should understand the process of evaluation as an allocation of meanings to events, physical objects and social circumstances of a situation, meanings which for their part are linked to allocative and authoritative resources in Giddens' sense (cf. Giddens 1984a: 33ff, 258ff). Thus contradictions, in the sense explained earlier, appear if an agreement over values and evaluations cannot be reached and the unquestioned ground of ontological security, on which the cultural identity is based, is removed.

Giddens explains the idea of ontological security as the 'confidence or trust that the natural and social/cultural worlds are as they appear to be, including the basic existential parameters of self and social [or cultural] identity' (Giddens 1984a: 375). This indicates that 'cultural identity', in the sense of an action-centred perspective, refers to the reciprocal nature of constitutions of meaning of several subjects in interactional situations (i.e. to patterns of interpretation as they are shared intersubjectively).

The application and maintenance of standards of value, or rather of the system of semantic and moral rules must not be seen in a 'power-free' context.[1] It must be seen as closely linked with the social stratification or class structure (cf. Giddens 1981a), which according to Giddens is itself constituted through allocative and authoritative resources. Giddens explains what he means by these terms: 'Allocative resources refer to capabilities – or, more accurately, to forms of transformative capacity – generating command over objects, goods or material phenomena. Authoritative resources refer to types of transformative capacity generating command over persons or actors' (Giddens 1984a: 33). From this we can go on to say that those principles of allocation of contents of meaning which are secured by domination and legitimation have a greater chance of enjoying pre-eminence and survival.

This sheds a different light on Müller's thesis (1987: 78) which says that chiefs (personalities carrying out the function of leadership) 'embody their group's capacity for living and are the central expression of its identity'. Now it seems that it is less the symbolic meaning which secures the identity of an individual with the group but rather the fact that, on the basis of authoritative resources, leaders are in a position to implement principles of allocation of contents of meaning for the actions of a majority of individuals. In this way, they secure their position on the one hand, and on the other they stabilize the reciprocity of the constitutions of meaning. This example leads us to conclude that 'perfect' power is obtained when 'subordinates' constitute the meanings of 'the world' by means of communicative acts in such a way that the existing structures (of domination) are reproduced unaltered.

We can thus speak about one 'culture' in particular when, through actions – in the sense of structuration processes, values and the norms and schemes of interpretation grounded in them refer to each other in (a more or less) concordant mutual harmony. In spite of all of the 'standardization' of patterns and expectations of action in traditions, institutions and mass communication, every 'culture' remains subject to the interpretation of actors. In every action, 'culture' just as much as 'power' is contained within structuration processes and is not external to them. For this reason, culture must not be understood as something static, as macro-analytical theories of structure often lead us falsely to believe; on the contrary, culture is subject to constant processes of interpretation and re-interpretation. Culture is embedded in the flow of action and is expressed in its 'becoming' and 'passing away' (Schütz 1982: 47). Put less cryptically, what is meant by 'culture' is expressed both in the generation of acts of structuration of social reality and in their execution and (intentional or unintentional) consequences.

So-called 'material culture' (mobile and immobile material artefacts and the values attributed symbolically to them) can never, as we have already indicated, be analysed separately from so-called 'immaterial culture'. The former must always be understood as a consequence (intentional or unintentional) of meaningful action, which for its part is to be conceived of as an interpretation of 'immaterial culture'. 'Material culture', the main object of traditional regional research, is thus always to be understood as a consequence of action which cannot be researched separately from the meaning of actions (which have produced it) and especially not separately from the significances which these material artefacts receive in the current execution of actions. Therefore, the so-called 'cultural landscape', for example, cannot be examined to any meaningful extent 'in itself' as an object of cultural geographical research. Rather 'cultural landscape' should be – in a regional culture analysis – interpreted in categories of action theory, or more precisely in the categories of structuration processes of the socio-cultural reality. Any materiality transformed by actors is only understandable and explainable in terms of the meaning and purpose of the transformation: the users of these material artefacts must always – if the use of these artefacts is to be successful – decipher the meaning of the transformation before it can become relevant to action. This also means that in the situation where actions in the sense of structuration processes take place, this symbolically meaningful component of the artefacts can turn into an aspect of action orientation and cultural identity.

From this action-centred perspective, artefacts, whose totality are thematized in traditional cultural and social geography in the sense of 'cultural landscape' or rather as one 'spatial artefact', are to be conceived as the (intended/ unintended) result, as the expression of culturally specific, meaningful action. In the situation of action, the artefacts' content of meaning can be recognized more or less appropriately by the actor with reference to certain intentions. In that sense, the use of the artefact is to be conceived as an anonymous interaction.

4 IS REGIONAL IDENTITY POSSIBLE?

The fact that cultures, particularly in pre-modern times, were clearly spatially definable has led numerous researchers to use regional analysis or rather territorial categories to analyse 'cultural identity'. They are intent on defining homogeneous cultural spaces, and on establishing cultural landscapes as an object for regional analysis (cf. Bartels 1968, 1970). The first question which arises on the horizon of a regional culture analysis is the following: can cultures be defined by means of territorial categories and is it appropriate to regard the cultural landscape as an object for regional analysis? If the answer to this question is no, then we must explain what regional culture analysis can contribute to contemporary cultural research.

If we accept the action-centred perspective of cultural and social research, then, as already indicated, neither 'space' nor 'cultural landscape' in itself is a suitable object of analysis in cultural–geographical research, but we must look at actions in the context of specific socio-cultural and material conditions. In this perspective, 'space' can only mean 'concept' of space. Any concept of space can only provide a pattern of reference in which entities, which are problematic and/or relevant, can be structured and localized but not immediately explained (Werlen 1988a: 161–8).

If we designate 'space' or 'cultural landscape' as an object for cultural geographical research, then we also take on those problems which are typical of every kind of holism and which are thus opposed to an action-centred perspective.[2] Any attempt to grasp the immaterial socio-cultural world of values, norms etc. by means of territorial categories leads on the one hand to an inappropriate homogenization of the socio-cultural world and on the other hand to an inappropriate 'collectivization'.[3] By contrast to a holism of the social world there is – on closer examination – still the danger of a further, untenable operation of thought, leading to a naïve essentialism which allocates to 'space' or 'cultural landscape' its own constitutive force. As the history of the discipline shows, numerous researchers fell victim to this danger. In such works of research they maintain not only that collectives can act but also that 'landscapes' and 'spaces' must be assigned a capability to act and produce.

Otremba (1961) documents the consequences of this manner of thinking most clearly. If these consequences, which I shall discuss, are still acceptable even for only a very few researchers these days, they nevertheless continue to be relevant. They also remain a latent danger of any 'spatially centred' or 'cultural landscape centred' regional research.[4]

Otremba (1961: 33) proceeds from the assertion that geographical research is confronted by three levels of productive forces in the 'world of appearances'. After the physical forces of the world and those of the 'human spirit', 'on a third level the spaces themselves begin to operate on the chess-board of the earth. Only on this field of play is it possible to recognize the changing force and the value of spaces. Watching the competition, the mutual completion, the temporally limited predominance and the "distant effect" of the spaces, we have the key to the understanding of their complete personality' (ibid., p. 133), 'and in the consideration of the value of spaces as personalities in the society of spaces there lies an unending, continually self-renewing task' (ibid., p. 135). 'All spaces . . . affect one another' (ibid., p. 134). However, this is not all. These spaces, according to Otremba's construction, take effect also in the determination of human activities. He regards Jessen's work, *Die Fernwirkungen der Alpen* ('The Distant Effects of the Alps', 1949/50) as proof of this hypothesis. Although this spatially deterministic holism certainly becomes weaker in later scientific concepts of space in cultural and social geography (cf. Bartels 1968: 160ff, 1970: 34ff), it has never fully been abandoned.[5]

The acceptance of an action-centred perspective, in contrast to the conception I have just described, means that 'space' can only be conceived of as shorthand for problems and possibilities which occur in the performance of actions with regard to the physical world in connection with the corporeality of actors and their orientations within the physical world. Thus, it cannot be meaningful to proceed from the assertion that 'space' or materiality already have a meaning 'in themselves', a meaning that is constitutive of social facts. They only become meaningful in the performance of actions under certain social conditions. This means that the symbolic or rather 'social charging' (Klüter 1986: 2f) of spatially regionalizable and localizable circumstances (as is done, for example, in connection with so-called 'regional identity' research) and the social meanings of material artefacts cannot be grasped appropriately from a spatially centred point of view, but rather in categories of action theory in specific socio-cultural and physical contexts.

For example, when Müller (1987: 66) maintains that territories 'represent the *spatial aspect* in the ensemble of means of expression which describe the whole, the *identity* of a group', he falls prey to exactly that misunderstanding which characterizes the space-centred perspective. One cannot have identity *per se* with territory. Even if certain material entities in a certain spatial arrangement have the same meaning for the majority of individuals and thus become the vehicle of cultural identity, this cannot be adequately argued and justified by the investigation of territory alone. As Simmel shows, one will find a more appropriate means of access if one examines the evaluation processes at work in and through the actions of individuals.

Simmel (1903) analyses the basic structure of those processes which provide certain places, artefacts in a particular position in space, and designations of place with a symbolic content of meaning. According to Simmel's account, immobile artefacts structure the physical world's realm of action in such a way that they become spatially fixed pivots of social relationships. Thus, through the practice of action, this place attains 'a particular foothold in the consciousness. In memory the place usually develops a stronger associative force than time, because it is sensually more vivid' (Simmel 1903: 43). In other words, the context of action and the meaning of actions are, in the memory of the actor, transferred to the place, the artefact or the designation of place, in which or through which the action has happened, 'with the result that, in memory, the place becomes inseparably linked with this action' (ibid.). If the meanings of the actions, which become interactions in particular physical spaces in the world – for example, in a church – persist as much as the meeting-place, all participating actors would imbue this place with the same symbolic content. This transferred contents of meaning would then awake in the actors who interact in this place 'the consciousness of belonging' (ibid., p. 41). The place becomes symbol and occasion for the remembrance of those actions which were performed here by several individuals with the same meanings. Since the place 'in itself' does not already have this meaning, one cannot, via an analysis of place, disclose the culturally specific meanings of actions.

5 REGIONAL ASPECTS OF CULTURAL IDENTITY

These results also show that it cannot be meaningful to look for the 'regional identity' of someone. Instead of searching for regional identity, it is more appropriate to research the regional aspects of cultural identity. By this I mean that in the framework of cultural geography one should increasingly be concerned with the sense in which the corporeality of the actors is important in the reproduction of cultural values and the constitution of intersubjectively shared patterns of interpretation, which constitute the basis of cultural identity.

There are two complementary possibilities here. The first lies in the application of Schütz' 'theory of life-forms' (1981) to the problem of cultural identity. If we put the constitution and application of intersubjectively shared patterns of interpretation in the centre of our deliberations, then first, with regard to a regional culture analysis, we must address the question of what conditions are necessary for these patterns to arise at all. If a subject wants to learn the intersubjectively valid rules of interpretation that exist within a socio-cultural world, then it is necessary for him or her to verify his or her interpretations and evaluations. This means that the constitution and application of intersubjective meaning-contexts depend on the possibilities of testing the validity of certain allocations of meaning. Consequently, the first condition of intersubjective constitutions of meaning can be seen in the immediate possibility of verifying subjective interpretations in terms of subjective validity.

Thus, the basis of any social communication is to be seen in the integration of subjective meaning into intersubjective meaning-contexts. According to Schütz (1982: 127), this means that every constitution of meaning is based on a subjective stock of knowledge. An intersubjective constitution of meaning of circumstances, or rather a reciprocity of constitutions of meaning, pre-supposes at least in part a uniformity of the stocks of knowledge. We can conclude from this, as the second condition of the constitution of inter-subjective constitutions of meaning, that commonly shared experiences can be regarded as an important element in the development or maintenance of cultural identity.

If we accept both of these conditions, it becomes clear that the subjective experience of things cannot be regarded as sufficiently certain, or rather, cannot attain the status of intersubjectivity, as long as it is not verified by an alter ego. The intersubjectivity of meaning in the co-existent socio-cultural and physical world can thus only be constituted on the basis of social inter-actions. The attainment of certainty about intersubjectively valid constitutions of meaning is possible above all in the immediate face-to-face situation. Here the bodies of the actors face each other directly as fields of expression of the consciousness of ego and alter ego. This makes it possible to support communication through subtle symbolic bodily gestures, thus limiting the number of misinterpretations. Moreover, it becomes possible immediately to query any remaining obscurities, and to refer to the commonly shared context, which allows one to immediately verify and, if necessary, correct the mutual symbolizations and interpretations. Accordingly, co-presence be-comes the central situation in which the immediate verification of the content of communication becomes possible. Co-presence is the prerequisite con-dition of ontological and interpretative security on which both the more abstract and the more anonymous allocations of meaning are based. All forms of mediated experience of the social world, which extend to anonymous institutional realities (e.g. houses of parliament, economic system) must, according to Schütz (Schütz and Luckmann: 1974: 61ff, 73), be seen as derivations of immediate experience.

Given the acceptance of the particular importance of face-to-face inter-actions, the basic spatial–regional structure of everyday courses of action can, according to Schütz, also be significant for the biographies of different subjects. He asserts that, owing to the specific forms of the 'spatial' and social conditions of everyday life, the respective actors' stocks of knowledge always experience a specific biographical articulation. Formally, Schütz distinguishes, within this framework of general conditions, between the following aspects in the analysis of biographies within these general conditions: every actor is born into an historical situation, each with a specifically formed stock of knowledge of his or her interaction partners in actual reach, who for their part are imbued with similar encounters with their ancestors. These con-ditions are imposed on every individual subject. The respective biographical articulation of the stock of knowledge limits the possibilities of life-fulfilment within a situation. Thus 'the social structure is open to [every individual] in the form of typical biographies' (Schütz and Luckmann 1974: 95).

The socialization of every individual in immediate encounters establishes his or her typical biography within certain degrees of probability. The formation of the different stocks of knowledge and the biographies based on them are differentiated in many ways through the body-boundedness of the actors and their actual and potential reaches in physical and regional as well as social respects. To the earliest face-to-face situations, new ones are added continually, leading to further differentiation of biographies, depending on the actual and potential reaches of ego and alter ego within the physical and social world.

Besides giving access to the basic conditions of cultural identity, Schütz also opens up, at least hypothetically, the possibility of a deep analysis of the regional disparities in individual life opportunities. The meaning of the spatial dimension for different biographies appears again in the body-boundedness of the actors, or rather in the localization of the body within the physical world and in the primary meaning of the actual encounter in acquiring an adequate knowledge of physical and social aspects of the existing world. The actual we-relationship is always linked to a spatial common ground and it is from this form of relationship that 'the intersubjective character of the life-world is originally developed' (Schütz and Luckmann 1974: 97). Thus, the body has to be understood as a 'mediating link' [*Funktionalzusammenhang*] (Schütz 1982: 41) between the physical world, the subjective knowledge/meaning and the intersubjective socio-cultural world. Through this we can hypothetically explain the formation of 'spatially differentiated' intersubjective socio-cultural worlds of meaning, which were clearly definable particularly in pre-modern times. Since the very first sedimentations of experience in the stock of knowledge usually remain important throughout one's whole life, the physical and social place of primary socialization also remains of continuing significance in modern societies.[6]

The second promising basis for the development of a regional culture analysis from the perspective of action theory is provided by Giddens' concept of 'social integration' (1984a). He defines this term as 'reciprocity of practices between actors in circumstances of co-presence, understood as continuities in and disjunctions of encounters' (Giddens 1984a: 376). Giddens regards the processes of social integration as the basis for ontological security, which one can also interpret as the basis of cultural identity. 'Social integration' is distinguished from 'system integration' and Giddens defines the latter as 'reciprocity between actors ... across extended time–space, outside conditions of co-presence' (ibid., p. 377), also saying that 'system integration refers to connections with those who are physically absent in time or space. The mechanisms of system integration certainly presuppose those of social integration, but such mechanisms are also distinct in some key respects from those involved in relations of co-presence' (ibid., p. 28). I cannot go into these distinctions in greater detail here, although they represent an important deviation from and, under certain conditions, an extension of Schütz' conception.

It is much more significant that Giddens, like Schütz, considers situations of co-presence as the basis for ontological security, or rather as the precondition for the possibility of intersubjective constitutions of meaning, which form the essence of cultural identity according to the arguments we have developed here. Since co-presence is linked to the position of the body in the context of the physical world, we must not underestimate the significance which accrues to the consideration of actors' corporeality in the reproduction of cultural values and the constitution of intersubjectively shared patterns of interpretation, at least within the framework of the two theories I have discussed. If these considerations are accepted, they certainly provide an interesting foundation for the production of hypotheses in a regional culture analysis. As I cannot discuss in detail a concept of action-centred regional cultural research here, as a conclusion I should like to indicate several consequences of the above-mentioned approach in comparison with traditional cultural geography.

Besides the different kinds of values and patterns of interpretation reproduced in co-presence (Saunders 1979: 206–31) which, mediated through the corporeality of the actors, lead to regional differentiations of socio-cultural worlds, we should also investigate the consequences of these contexts for intercultural communication. In addition, we should analyse which regular or occasional opportunities of co-presence exist for which actors in an intra-regional context,[7] or rather, which preconditions for the maintenance or recovery of cultural identity (or identities) exist or are absent in modern lifeworlds. Finally, we should explain the role played by meanings preserved in artefacts within the framework of socio-cultural reproduction processes.

6 NOTES

1 Cf. Claessens and Claessens (1979: 21f). Surprisingly, Giddens does not take this into account when he speaks about 'common culture'.
2 On the incompatibility of holistic positions with the basic tenets of structuration theory, see Giddens (1984a: 207ff). On the relationship between structuration theory and methodological individualism, see Werlen (1989b).
3 The most prevalent form of this approach is the nationalist or regionalist argument which talks about the will or the opinion of, for example, the Swiss or the French, the defeat of Switzerland by Austria, or the particular characteristics of the inhabitants of Cambridge or of Oxford. In these cases the unsuitability of the homogenization usually expresses itself in the form of generalizing prejudices, and the holistic component is expressed in the implicit assertion that a collective can act 'in itself'.
4 The further danger of this is, for example, expressed in connection with the so-called 'cultural geographical doctrine of forces' in Wirth (1979: 229ff) and in the (scientific) spatial approach in general (cf. Werlen 1988a: 233–51).
5 See the current debate in Germany about 'regional consciousness' or 'regional identity', in Bahrenberg (1987), Blotevogel et al. (1986, 1987, 1989), Hard (1987a, b, c), Heinritz (1989).
6 This fact could be a reason why in the everyday praxis of communication the spatial provenance of the interaction partner is seen as a more important aspect for mutual standardization. Seen in general terms, this is probably a more important reason for the obstinate insistence on territorial categories in the analysis of social circumstances.
7 In the sense of a political and/or systematically administrative unity.

7 REFERENCES

Bahrenberg, G. (1987) 'Unsinn und Sinn des Regionalismus in der Geographie', *Geographische Zeitschrift* 75: 150–60.

Bartels, D. (1968) *Zur wissenschaftstheoretischen Grundlegung einer Geographie des Menschen*, Wiesbaden: Franz Steiner.

Bartels, D. (1970) 'Einleitung', in D. Bartels (ed.) *Wirtschaft- und Sozialgeographie*, Cologne and Berlin: Kiepenheuer & Witsch.

Bernstein, R. (1986) 'Structuration as critical theory', *Praxis International* 5: 235–49.

Blotevogel, H.H., Heinrich, G. and Popp, H. (1986) 'Regionalbewusstsein: Bemerkungen zum Leitbegriff einer Tagung', *Berichte zur deutschen Landeskunde* 60: 104–14.

Blotevogel, H.H., Heinrich, G. and Popp, H. (1987) 'Regionalbewusstsein: Überlegungen zu einer geographisch-landeskundlichen Forschungsinitiative', *Information zur Raumentwicklung*, 7/8: 409–18.

Blotevogel, H.H., Heinrich, G. and Popp, H. (1989) '"Regionalbewusstsein": Zum Stand der Diskussion um einen Stein des Anstosses', *Geographische Zeitschrift* 77: 65–88.

Claessens, D. and Claessens, K. (1979) *Kapitalismus als Kultur*, Frankfurt a/M: Suhrkamp.

Giddens, A. (1976) *New Rules of Sociological Method*, London: Hutchinson.

Giddens, A. (1981a) *The Class Structure of the Advanced Societies* (revised edn), London: Hutchinson.

Giddens, A. (1981b) *A Contemporary Critique of Historical Materialism: Power, Property and the State* 1, London: Macmillan.

Giddens, A. (1984a) *The Constitution of Society: Outline of the Theory of Structuration*, Cambridge: Polity Press.

Giddens, A. (1984b) *Interpretative Soziologie: Eine kritische Einführung*, Frankfurt a/M: Campus.

Giddens, A. (1988) 'The role of space in the constitution of society', in D. Steiner, C. Jaeger and P. Walther (eds) 'Jenseits der mechanistischen Kosmologie – Neue Horizonte fur die Geographie?', *Berichte und Skripten* 36, Zürich: Department of Geography, ETH.

Hard, G. (1987a) '"Bewusstseinsräume": Interpretationen zu geographischen Versuchen, regionales Bewusstsein zu erforschen', *Geographische Zeitschrift* 75: 127–48.

Hard, G. (1987b) 'Das Regionalbewusstsein im Spiegel der regionalistischen Utopie', *Informationen zur Raumentwicklung* 718: 419–40.

Hard, G. (1987c) 'Auf der Suche nach dem verlorenen Raum', in M.M. Fischer and M. Sauberer (eds) Gesellschaft, Wirtschaft und Raum, *Mitteilungen des Arbeitskreises für Neue Methoden in der Regionalforschung* 17, Wien.

Heinritz, G. (1989) 'Weissenburg in Bayern oder Weissenburg in Franken? Wie relevant sind Untersuchungen zum Thema Regionalbewusstsein?', *Frankenland* 41: 31–9.

Held, D. (1982) Review of Giddens, A. (1981b) in *Theory, Culture and Society* 1: 98–102.

Jessen, O. (1949/50) 'Die Fernwirkungen der Alpen', *Mitteilungen der Geographischen Gesellschaft München* 35.

Kiessling, B. (1988) *Kritik der Giddenschen Sozialtheorie: Ein Beitrag zur theoretisch-methodischen Grundlegung der Sozialwissenschaften*, Frankfurt a/M: Lang.

Klüter, H. (1986) 'Raum als Element sozialer Kommunikation', *Giessener Geographische Schriften* 60.

Müller, K.E. (1987) *Das magische Universum der Identität: Elementarformen sozialen Verhaltens – Ein ethnologischer Aufriss*, Frankfurt a/M: Campus.

Otremba, E. (1961) 'Das Spiel der Räume', *Geographische Rundschau* 13 (4): 130–5.

Saunders, G. (1979) 'Social change and psycho-cultural continuity in Alpine Italian family life', *Ethos – Journal of the Society for Psychological Anthropology* 8 (3): 206–31.

Saunders, G. (1987) *Soziologie der Stadt*, Frankfurt a/M: Campus.

Schütz, A. (1981) *Theorie der Lebensformen*, Frankfurt a/M: Suhrkamp.

Schütz, A. (1982) *Life Forms and Meaning Structure*, London: Routledge & Kegan Paul.

Schütz, A. and Luckmann, T. (1974) *Structures of the Life-World*, London: Heinemann.

Simmel, G. (1903) 'Soziologie des Raumes', *Jahrbuch für Gesetzgebung, Verwaltung und Volkswirtschaft im Deutschen Reich* 1 (1): 27–71.

Weber, M. (1968) *Economy and Society*, New York: Bedminster Press.

Werlen, B. (1988a) 'Gesellschaft, Handlung und Raum: Grundlagen handlungstheoretischer Sozialgeographie', *Erdkundliches Wissen* 89, Stuttgart: Franz Steiner.

Werlen, B. (1989a) 'Sozialforschung in territorialen Kategorien?', *Soziographie* 2: 13–25.

Werlen, B. (1989b) 'Kulturelle Identität zwischen Individualismus und Holismus', in L.K. Sosoe (ed.) *Identität: Evolution oder Differenz*, Fribourg: Universitätsverlag.

Wirth, E. (1979) *Theoretische Geographie*, Stuttgart: Teubner.

8   FURTHER READING

Blau, P.M. (1977) 'A macrosociological theory of social structure', *American Journal of Sociology* 83: 25–38.

Giddens, A. (1979) *Central Problems in Social Theory: Action, Structure and Contradiction in Social Analysis*, London: Macmillan.

Werlen, B. (1987) 'Zwischen Metatheorie, Fachtheorie und Alltagswelt', in G. Bahrenberg, J. Deiters, M.M. Fischer, W. Gaebe, G. Hard and G. Löffler (eds) *Geographie des Menschen: Dietrich Bartels zum Gedenken*, Bremer Beiträge zur Geographie und Raumplanung 11, Bremen: University of Bremen.

Werlen, B. (1988b) 'Von der Raum- zur Situationswissenschaft', *Geographische Zeitschrift* 76 (4): 193–208.

20

FLEXIBLE STRUCTURES FOR A REINTEGRATED HUMAN AGENCY

Dieter Steiner, Gregor Dürrenberger and Huib Ernste

1 INTRODUCTION

Today, environmentally motivated geographers have started to look at regions in an integrated way which, in the sense of the human ecological triangle (cf. Steiner, this volume), attempts to relate persons, society and environment to each other. A decentring of geography away from the old exclusive fixation on the spatial aspects of regional phenomena may be the beginning of a tradition of human ecology without a very clearly defined disciplinary orientation. Such a human ecology, if alimented by contributions from other traditional disciplines such as psychology, sociology and economics, can become a model for a flexibilization of structures (the topic of this present contribution, albeit in the realm of social practice) in the domain of science, allowing for a linking of disciplinary paradigms, theories and methodologies and for an interdisciplinary conversation. However, a geographical viewpoint in the traditional sense should survive for a good reason: regions organized in a way that provides for a sufficient degree of self-determination, which in turn enables the developments of social networks and individual identities, are becoming the focus of interest in a world which otherwise tends to globalization and anonymity.

The present contribution discusses some of these aspects in an evolutionary perspective. Thereby, the disunity of human lives in today's society, which manifests itself in a spatial as well as in a temporal sense, is diagnosed as a (problematic) outcome of social development that may help to explain the (largely unintended) destructiveness of humans towards the environment. A flexibilization of structures that permits an improved meshing of domains associated with different social subsystems is seen as a possible remedy that might help to provide for a better integrity and continuity of human life. Examples are taken primarily from a Central European context.

2 FACETS OF TODAY'S HUMAN EXISTENCE

2.1 Differentiation and integration in the cultural evolution

For an attempt at pinpointing the problem, it is helpful to examine the cultural evolution as a process of increasing differentiation on the one hand and of increasing integration on the other. In terms of differentiation we note the development from archaic through political to economic societies, each level being associated with newly emerging structures (for further details see Steiner, this volume). As a result, the life of people in today's society takes part in three different basic domains: the private (the archaic sphere, so to speak), the political (our political heredity) and the vocational (the cultural base of economic activities). It is true that the degree of differentiation growing in the course of the cultural evolution is being counterbalanced by an increasing integration. In fact, it must be, otherwise there would be no society at all. In dealing with the aspect of integration it is useful to refer to the distinction between social integration and system integration as made by Giddens (1984) in his theory of structuration.[1] For the present argumentation we are concerned with social integration, which means direct human interactions in face-to-face situations. An individual is integrated into social systems by means of interactions. Such interactions are shaped by traditions, that is, rules that enable the recurrent reproduction of social systems by human actions.

Seen this way we can say that there is not only more than one (sub)system in today's society but also more than one life in the life of one person. It is plausible, therefore, that we should be concerned about the kind of lives that human agents have in modern society, and also that we should suspect that a tendency towards disintegration of these lives may have a crucial bearing on the relationship between persons and environment. If the individuals are torn between, say, private life and vocational life, and these two are ecologically speaking not compatible, they can hardly provide for integration between society and environment. We think that only morally 'whole' persons can develop a responsible behaviour towards the environment.[2] MacIntyre (1985), in his study of moral theory, describes the present language of morality as being in a state of disorder and, accordingly, the practice of morality in a unified way as being hampered by social obstacles, the disunity of human lives:

> The social obstacles derive from the way in which modernity partitions each human life into a variety of segments, each with its own norms and modes of behaviour. So work is divided from leisure, private life from public, the corporate from the personal. So both childhood and old age have been wrenched away from the rest of human life and made over into distinct realms. And all these separations have been achieved so that it is the distinctiveness of each and not the units of the life of the

311

individual who passes through those parts in terms of which we are taught to think and to feel.

(MacIntyre 1985: 204)

It is true that these different life segments do not stand isolated, but are bound to each other by external purposes. This, however, is simply another aspect of the problem. It means that most activities of most people today do not have an end in themselves, but are oriented towards something outside of them. Examples are: the journey to work (the expression speaks for itself), work being done to earn money, regenerating holidays to become fit again for more work, going to school 'to learn for (the later) life'.

The differentiation of personal lives finds its expression in spatial as well as temporal patterns. The former has been the concern of geography, whereas the latter has been the subject of research mainly in sociology and ethnology (see, e.g. Kohli 1978a). Eventually we should be interested in how spatial and temporal patterns combine.[3] A spatio-temporal framework can be developed by using the notions of lifespace and lifetime. The space in which and the time through which an individual lives is socially produced. In other words, the physical structures occurring within his or her lifespace are a concrete expression of existing (and sometimes past) social structures (see, e.g. Knox 1987),[4] and the social status that an individual has during his or her lifetime is equally well governed by these structures. For a human being living in an archaic society the lifespace is undivided and the lifetime is continuous without any major breaks. Both are provided by the situation of overall social integration: there is an all-side interaction between the members of such a society regardless of sex, age and possibly acquired status. The lifespace of individuals in modern society discriminates between a number of mutually exclusive territories requiring repeated sequential moves in space by these individuals; their lifetime is discontinuous in the sense that it is broken down into a number of life phases with sharp breaks between them.

2.2 The spatial dimension: territories

The three types of society emerging in the course of the cultural evolution (archaic, political, economic) are, organizationally speaking, dominated by a kinship system, a political system, and a vocational system, respectively. In today's modern economic society all three systems appear as major sub-systems in which individuals acquire social positions and identities and entertain social relations.[5] Accordingly, a person has three major part-identities: a first within a kinship subsystem as a mother, an uncle, a niece, etc., a second within a political subsystem as a citizen, a party member, a government minister, etc., and a third within an economic subsystem as a shoemaker, a nurse, a pilot, a programmer, etc. Clearly however, in accord-ance with the dominant status of the economic subsystem, the latter type of

312

identity, the vocational one, is for most people (at least the male portion of the population!) the most important. Nevertheless, the problem, it would seem, is that an individual in today's society cannot attain one overall identity, and conflicts or confusions may also arise between identities attached to the different domains.[6]

According to the notion of social production of space, the three different types of social domains all find their own spatial expression. Here we may use the concept of territory (Dürrenberger 1989a,b). This has first of all a biological connotation: animals (singles, couples or larger groups) usually claim a certain space, that is, a piece of the environment in which they live, as their own and they may defend it against intruders. Such a territory serves as a lifespace that offers food and opportunities for mating and raising offspring (Malmberg 1980). With the emergence of humankind out of the biological evolution, the question arises of how far territorial principles still play a role in human societies. One has to assume that a residual of the biological (genetic) foundation is still at work, albeit hidden under superimposed social structures, such that in the cultural evolution territorial behaviour has become territorial agency. Surely, however, territory for humans is still a part of environmental space that is more or less reserved for a group of people: it is used collectively for particular types of social interactions and it is delimited against other spaces.[7]

In modern society, therefore, the lifespace of a person is, in general, split into three types of territories: private, public and vocational. This is in contrast to the lifestyles of the earlier phases of cultural evolution. Within archaic societies everything that matters for life happens within a certain space that is shared collectively by all members of a local group. Essentially everything occurs within the private (in the sense of the extended family) domain, and accordingly there is only a private territory.[8] We may say that this situation still reflects a genuine unity of person, society and environment. In a political society the domain of public affairs gets separated from the private domain, and correspondingly it gets its own spaces assigned. Within a city or a nation state there are territories which are exclusively public, such as the places where the dealings of the government take place. There are also territories constituting an interface between the private and the public domain, the traditional marketplace being an example. Here an economic exchange takes place, but it is still directly from private producers to private consumers, and there is not yet an economic subsystem as such. This changes in the economic society. The production of material goods becomes more and more concentrated in special enterprises. What once was integrated under one roof, namely the private and the vocational domain, as was the case with a traditional handicraft, becomes spatially separated. Special shops, factories and, later, offices become vocational territories.

A characteristic of modern society, especially in an urban setting, is, of course, the often very wide separation of private and vocational territories and the ensuing commuter traffic. This really develops into the tragedy of the modern city in two ways: first, as already indicated, the commuters spend a substantial portion of their lives on an activity that is not a goal in itself, but has an external purpose. Second, the people living or working in the city find themselves in sharply defined private or vocational territories engulfed by traffic which has destroyed public space.[9] The absence of natural components within the private, and the loss of usable public space around both the private and vocational territories, has to be compensated at the weekend by tours through the countryside, producing more traffic.

2.3 The temporal dimension: life phases

The repeated movements between the different territories find, of course, their temporal expression also in that people's lives become heavily structured by daily and weekly rhythms. These short-term temporal aspects are certainly of interest within a discussion about the possibilities of reintegration by a flexibilization of structures. However, as these can be handled directly, together with the question of providing continuity between territories, we will in this section be concerned with the long-term temporal aspects as they apply to the life phases of individuals.

These phases constitute discontinuities (see, e.g. Kohli 1978b). In a modern society the social structures are such that normally a child stays at home during the first few years of his or her life. There the usual situation is still that the child sees the mother working in the household, but does not see much of the father, and does not see him working. The first break comes when the child has to go to school. It is the child's first institutionalized acting in public settings. However, the young person is given final access to the political system only much later: when he or she has the privilege of voting and the duty of paying taxes. Consequently, the second break comes when the young person leaves school and starts participating in the economic system by joining the labour force. More precisely, some persons will not do this at all or not for long (usually the women: many of them still will go back to the private domain to raise children and do housework under circumstances that may not be entirely of their own choosing).

As people today have a higher life expectancy than a hundred years ago, there is a break at the time of retirement, that is, all of a sudden one is not supposed to take part in the economic system any more, except as a consumer (compare, e.g. with Attias-Donfut 1988). From the point of view of the family, the following kind of rupture seems to be of increasing relevance: the young tend to leave the home early, long before they possibly start their own family. As the parents have a higher life expectancy, a couple may still live together for a considerable period after the last child has left the home and before one of the partners dies (Matthes 1978).

There is a clear contrast between such temporal patterns and those in an archaic society in which the structures provide for only one kind of life so to speak (compare with Benedict 1938/78). The baby accompanies the mother to all sorts of chores by being carried on her back. When a young child can walk unaided, he or she will go with the parents, is instructed by them, and is soon given small responsibilities. Therefore, the child learns gradually what the adult persons are doing and one day is an adult too. This is not to deny that there are breaks between childhood, adolescence and adulthood. On the contrary, those breaks may be very clear cut. Reaching sexual maturity does indeed change life. But this break is usually mitigated by some kind of ritual which helps to master the psychological stress that may be associated with this transition (Benedict 1938/78). Thus, life remains embedded in the 'old' context of small group and kinship affairs. It has not become divided into separate domains and phases. In such a sense there is an overall continuity.

2.4 From differentiation to separation

That there are pitfalls associated with the evolution of society, and especially of modern society, cannot be ignored. A possible understanding of those pitfalls can be derived from Giddens' notion of system integration.[10] In contrast to social integration, system integration refers to indirect (and predominantly anonymous) interactions by means of some medium. In modern economic society the medium of 'money', as a system integrative counterpart to the social integrative rules of vocational traditions, has become highly dominant. By means of money the economic subsystem spans the whole globe: the economic society at large is virtually run by one of its subsystems. We have the curious (and dangerous) situation that our modern society is working at the global scale because one of its parts provides for an overall integration, while the other parts still work on smaller and mostly regional scales. In a sense, the economic subsystem has decoupled from the rest of society, differentiation has become separation that lacks in communication between the parts.

Luhmann (1986) has given a very drastic description of this situation. He sees modern society as consisting of a number of functional subsystems (such as politics, economy, science, families, religion), whereby each type is organized around its own specific sense, using its own specific code of communication. As a result, although the subsystems influence each other, a common conversation and understanding across the society as a whole about any topic (Luhmann himself uses the example of 'ecological communication') becomes very difficult indeed, if not impossible. This picture may be somewhat of an overstatement, but it is certainly not quite unrealistic.[11] It would seem to be urgent, therefore, to re-establish a (non-economic!) language that is common to all subsystems. It is our contention that this can be achieved only via a reintegration of human agency within a framework of flexibilized structures, as will be described below.

An obstacle on the way to such a reintegration is the fundamental female–male split in our society. So far we have been discussing persons mostly in general terms. As a matter of fact, it should be obvious that such a generalization, which does not explicitly consider the relationship between genders, is not tenable. It indeed seems to be the case that there is, in the course of cultural evolution, an increasing disconnection not only between social subsystems but also between genders. In the (original) archaic societies the women as bearers of new life naturally tended to have a more central position than the men (see, e.g. Fester *et al.* 1982). Only during the subsequent cultural evolution, which led to the emergence of political societies, did the men somehow manage to exclude the women largely from the now important public positions, thereby establishing a suppressive patriarchal regime. The male dominance was later carried over into the economic society in which again women were largely kept outside the sphere of professional activities. In other words, the life of women in the past has been basically restricted to the archaic component in our society, the private domain. Because of this confinement to one domain women should theoretically be able to lead a more wholesome life than men. In a sense this may be true, but it also means that women are excluded from all other societal domains and as such cannot be full members of the society.

It is true that today the boundaries between the two genders are becoming increasingly perforated. For example, female employment in the industrialized countries has drastically risen during the last decades. Nevertheless, we are far from a situation with equal chances for both sexes: in particular there is a labour market segregation in that women are largely confined to work in offices, shops, hotels, restaurants and hospitals (Charles 1987). To improve the chances of women the flexibilization of structures gains quite a specific importance. Such a transformation can happen effectively only if the men themselves will acquire more 'female properties' and do their share in this transformation.

3 POSSIBLE REMEDIES

3.1 Top down or bottom up?

In terms of differentiation and integration we can diagnose our predicament as follows. First, we have an uneven social differentiation into subsystems with the economic subsystem dominating the private and public realm. Second, there is a lack of communication between the subsystems. Differentiation increasingly develops into separation. Third, integration that counterbalances differentiation is largely provided by means of system integration, that is, sex, power, and money. To correct for this development, one has to foster social integration on the one hand, and to provide for integrating links between the different subsystems on the other hand. If this were possible, we could hope to solve the problem of false integration at the same time, in that this would bring economy 'back into society', that is, put it back into its proper place as a subordinated subsystem. But how can this be achieved? Given that we believe that what happens on the person–society side of the human ecological triangle is decisive to what happens to the environment, we could think of starting either at the level of the society or at the level of individual persons. Let us first look at the former possibility.

If with society we mean the reality of social structures in the sense of Giddens, then we are dealing with the problem of changing those structures. In examining the present situation we find, of course, that the environment (in particular its natural component) is largely missing as an entity in the social structures of the modern economic society or, if it is there, that it most likely appears in the form of an economic good such as a parcel of land. This is in contrast to an archaic society in which aspects of the environment are firmly embedded in the cultural superstructure and may be even of overriding importance.[12] To go for a solution, therefore, at the structural level in today's society typically means to 'internalize' the environment into the political or economic structures. This is, of course, exactly the object of such new fields as environmental law and environmental economics (Baumol and Oates 1975, Smith and Kromarek 1989). Having just mentioned the archaic situation, we can see a problem with this at once: given that the society consists of several subsystems, any solution of this type can only provide for a partial internalization. There are as many different possible 'addresses' for environmental concerns as there are subsystems. A complete solution capable of bridging over the 'esoteric' structures specific to the various subsystems depends on the existence of a superstructure with an 'archaic quality'. Such a structure is, as we know, missing from our modern society; we are living in a civilization without an overall culture.

The reverse strategy would be to try to induce changes at the level of individuals in such a way that their actions become more environmentally compatible. To the extent that particular actions are believed to be consequences of particular types of motives and attitudes, the task is assumed to be one of altering the consciousness of individuals. There are a number of studies which are attempting to do just this; they look for suspected determinants of human actions in order to be able to offer recommendations with regard to environmental education (see, e.g. Fietkau 1984).

Both of these approaches, the one from the top down and the one from the bottom up, contain some justifiable elements, namely to the extent that they represent partial truths. It is true that in order to move towards a more harmonious relationship with the environment both social structures and persons have to change. The two approaches fail, however, in that they are one-sided and incomplete. In the first case it is assumed that an internalization of the environment in political or economic terms will appropriately change the behaviour of individuals acting in those structures. While those individuals will indeed be constrained by new legalities in their actions, the structural changes are, on the other hand, completely devoid of any psychological content. Conversely, the educational approach is doomed to failure as long as persons are supposed to act within unaltered structures, albeit with a changed consciousness. It may help in a small way, namely to the extent that existing structural rules can be interpreted and used in somewhat different ways, but nothing dramatic can develop out of this. Both proposed solutions are thus reductionistic, the first in a political or an economic sense, the second in a psychological sense.

3.2 Fundamental learning in recursive social systems

We need a non-reductionistic approach which admits to the circularity or recursiveness of processes occurring between different levels of reality, here between those of persons and those of social structures (compare with Steiner, this volume). Changes of human agency presuppose changes in structures that permit such changes and vice versa. The principle of going in circles, that is, the recursive interplay between individual human agents and social structures, is the topic of the theory of structuration of society by Giddens (1984) (see Lawrence, Nauser, Werlen and Söderström, all this volume). This interplay also suggests the possibility of self-organizing mechanisms of social change. Whereas Luhmann's portrayal of society is utterly pessimistic, Giddens' theory may be a cause for optimism: if it is a fair picture of reality, it shows a way out.

Indeed, as indicated already, social structures have, of course, been changing throughout the cultural evolution. But it is clear that a structural change does not happen by itself, it requires actions by human beings, regenerative actions that depend on processes of 'fundamental learning' (Siegenthaler 1987) or 'social learning' (Milbrath 1989). While 'normal' learning takes place within established rules and enables a person to become more routinized in the fulfilment of its tasks, fundamental learning questions these very rules and leads to an alteration of them. In dealing with environmental problems so far, we have been trying to solve them by largely using the same mechanisms, that is, acting within the framework of the given types of rules, which are responsible for the problems in the first place. Luhmann (1989) also is concerned about this and asks himself how long we can possibly do this.

How do we effect a quantum jump to a level of fundamental learning? Here Siegenthaler (1987) places great importance on social movements, groups of people who start thinking alternatively and innovatively and eventually act accordingly. A new idea can take shape and become effective only within a social network in which the bearer of the idea finds him/herself supported and trusted by other individuals, such that, as Kastenholz (1991) puts it, obligation turns into motivation. This is a process of conscious self-organization and to further it Ladeur (1987) calls for an 'outside organization of self-organization', a seeming paradox. What he means is stimulation from the outside: institutions such as governments should take care to encourage alternative thinking by 'rebels' in science and elsewhere. New social movements should be taken seriously. A 'deblocking' of present structures is urgent. They have become too heavily solidified so that the self-dynamic of the various functional subsystems continues to follow well trodden paths.

4 REINTEGRATION THROUGH FLEXIBILIZATION

It is our contention that the present openness of the societal situation provides a chance to further a process of structural flexibilization (which, as a matter of fact, is underway already) in a more conscious fashion and that this process can indeed provide for both a greater number of and more varied face-to-face interactions on a regional level. Opportunities with regard to time that different people can share and experience together are called 'time windows' by Henckel (1989). Flexibilization should provide for either the expansion of present or the creation of new time windows. Consider as an example a man who is an employee and also the father of small children and who can now spend part of the normal day-time hours with them because he is working part-time or under a flexible working time scheme. Correspondingly, one could talk about 'space windows', places at which people can meet for particular types of interactions. An example here would be the regaining of public space in cities now submerged in traffic.

The talk about flexibilization refers to structures, and this raises the question of whether this is not again a top-down approach. We can readily see that this is not the case once we know that what we are going to address in the following are 'low-level' structures. By this we mean structures of small systems or institutions, close to the base of individuals, with small circles such that structures are changed because the people concerned want them to be. The institutions that we have in mind are households, firms and schools, thus one each from the three basic domains, the private, the economic, and the public. In an exemplary fashion, we will be interested in looking at possibilities which create conditions that let the lifeworlds of persons evolve within a more real lifespace and a more real lifetime, 'more real' referring to an experience of an existence that is more unified and more continuous. An important component thereby would seem to be an improved capability of persons for exercising a degree of self-control over their living conditions. We agree with Schubert (1989) that such a capability is a prerequisite for the emergence of a new environmental ethic.

4.1 Private households

A flexibilization of structures at the level of households can have three aspects: it may concern (1) the internal social organization, (2) external social

319

relations to other institutions such as other households, firms or schools,[13] and (3) external relations to the (bio)physical environment. These aspects cannot necessarily be separated neatly; often there is a subtle transition from the internal to the external as, for example, in interactions between certain types of households or community living and certain degrees of self-determination in housing construction and neighbourhood design.

Looking at the internal organization first, we note that in the historical past people lived in the context of extended families and larger households which usually comprised several generations and sometimes a wide range of relatives. This family system was responsible for most basic aspects of life: economic production, education of the children, care for the old and the sick. As we know, households have, in the course of the process of modernization and especially under conditions of urban living, subsequently been reduced to what is still today regarded as a standard: the small family consisting of father, mother and children. Marriage and family life have become totally 'privatized' (Perrez 1979) in that former traditional social functions have been transferred to external institutions.

A structural flexibilization means the emergence of new types of households that deviate from the small family model (for a general discussion see Sommer and Höpflinger 1989, Spiegel 1986). As a consequence of young people leaving home at an earlier age, of a reduced rate and a later age of marriage, and of an increased rate of divorce, we have experienced a sharp increase in the number of one-person households ('singles') in the last two decades or so. At the same time there has been a growth of households with unmarried couples and of couples without children. As a contrast to these developments, which have resulted in a further reduction of the small family, we have also had types of community living come into existence. The commune[14] is an alternative type of household that has existed for some time now. In it half a dozen or so people (often students), who usually are of young age and not relatives of each other, live together. However, for most participants in such cooperative households this is only a temporary style of life. At the root of all of these developments may be a desire to get rid of the last remaining shackles in the private sphere. In the case of the commune, freedom is gained by moving out of the parents' home, but a new form of social integration is looked for at the same time. In all cases more freedom may amount to improved chances for self-realization in the sense of mutually rewarding new forms of external social integration. Internal social integration, however, may loosen and end in a 'living apart together'.

More permanent, and therefore more relevant, may be larger schemes of cooperative living that explicitly seek to establish external social relations, that is, schemes involving a number of individual households which form a small community. The intention is to revitalize the ideas of the neighbourhood and of small networks. In a spatial sense one can understand such settlements as a mix or rather a mosaic of private and (semi-) public territories.

320

This is not without problems, however: Wanner (1987), in a recent survey of such settlements in Switzerland, shows that, not surprisingly, conflicts often arise as old and new styles of living collide. A process of learning is necessary to find out where the border between private and public should be. An interesting aspect of community living of this sort is that it may also form the basis for a partial reintegration of residential living and working, at least in the immediate sense that part of the work (for example, handicrafts, medical care) may be done by members for other members of the community. Moreover, schemes of this nature usually are not restricted to just the question of social organization but may also involve a control over design and construction of living quarters and immediate surroundings. This leads to the topic of external flexibilization with regard to the self-creation of the (bio)physical environment.

Let us turn, therefore, to look at some forms of self-determination regarding housing design, construction and renovation (see, e.g. Michel-Alder and Schilling 1984). Häussermann and Siebel (1987) point out that the total privatization of households, in the social terms mentioned above, is paralleled by a modern style of living that is oriented towards a complete eradication of work. Indeed, residential living has become largely consumptive, albeit not necessarily always by the tenants' own choice. As Lawrence (1989) shows with the example of England the notion of a 'model house' has been entertained since the middle of the last century. Such a house is supposed to adhere to absolute standards, that is, to provide quality housing. In reality it imposes a normative behaviour on tenants through implicit prescriptions (the space is partitioned such that the possible uses of the rooms are predetermined and informal work is largely impossible) and explicit regulations. However, 'trends indicate that there is a plurality (rather than a homogenization) of household structures, household activities and human values attributed to house and home. ... the housing "needs" of a resident, or of a household, do not correspond to a constant norm but are defined by a complex matrix of interrelated factors that change during the course of time' (Lawrence 1989: 129). Here, therefore, flexibilization means getting away from totally predesigned housing and setting up schemes whereby present inhabitants can do their own thing in renovating and changing residences and future inhabitants in designing and even constructing residences (Bassand and Henz 1989).[15] This may mean a reinternalization of productive functions into the private domain in the form of informal work or may require new kinds of relations at a partnership level with architectural firms and construction businesses to provide for a diversification in such a way that flexible uses of residences become possible (Wüest 1987).[16] Such settlement schemes usually also include the consideration of ecological factors with regard to not only the design of buildings and choice of house technique but also the rebuilding and caring for the immediate surroundings. For example, traffic-free residential streets[17] or gardens for partial subsistence may be established.

4.2 Public schools

Among the three kinds of base institutions considered here schools are the least flexible in that they are under more or less direct government supervision. The following remarks about possible flexibilizing reforms are thus concerned more with a discussion than with actual measures already taken. At least this is true for public schools. Private schools on the other hand traditionally have been playing pioneer roles in alternative forms of education.

An example of such a discussion is a proposal by the green party of the Canton of Zürich, Switzerland, for a radical organizational reform of the entire educational system. Among the measures proposed are the following: an offering of electives already for the lower school grades, an abolition of rigid timetables in favour of more project-oriented teaching, a fostering of common social experiences during the school hours, a flexibilization of age requirements regarding the entrance to the different levels (as a result a school class would generally comprise a mix of differently aged pupils), and the replacement of entrance exams by test periods and of graduation exams by project work. The plan also calls for the establishment of a school collective for each neighbourhood consisting of all teachers and all interested residents above sixteen years of age which would determine and revise content and form of teaching in the local schools.

In terms of external relations there are, particularly in connection with the question of environmental education, proposals to interface school teaching with exposure to problems in the outside world. For example, Dieckhoff (1989) points out that a school could take on self-chosen tasks in areas where the larger social system cannot or does not want to act. Such a task would have a value of its own as a practical, problem-oriented contribution, and it would go beyond the mere pedagogical function which school activities normally are supposed to have. An example of such a project actually carried out by 160 school classes in Switzerland was a mapping of tree lichens as indicators of air pollution. It was organized by the World Wide Fund for Nature (WWF) and its results are of practical use.[18]

Another idea is to provide for contacts with the realities of adult life. A paradox of our educational system is that the child is supposed to learn for life, but the kind of life that the child has is rather far removed from the 'real' life of adults. As Robert (1989) says: 'It seems that our schools submit to a constraint that becomes stronger the higher the schooling level is: distancing and alienation from life.' Being at school means learning for the later life and, therefore, there is no sense for the present in what one does here and now, it is always for the future. To cure this, a program has been started in the city of New York in which personalities from politics, science, administration, business and arts can 'adopt' a class for a year (Wyss 1989). They have meetings with the class at regular intervals and share with it knowledge and professional experience in various forms.[19] In many matters the knowledge and the experience of the usual school teacher is second hand only; orientations such as those mentioned connect to the old notion that life, rather than being taught, should itself be the teacher. Along the same line is the idea that every school teacher should have a leave of absence at regular intervals, during which time he or she would be expected to become exposed to other kinds of social activities.

In connection with flexibilization in the economic realm, to be discussed below, we can point to a tradition which provides for a relatively smooth transition between school and professional life. It concerns the institution of apprenticeship, a traditional cultural resource in Germany, Switzerland and Austria with roots going back to the Renaissance and the Reformation. It combines on-the-spot training for a particular trade with continued schooling. It can, in fact, be regarded as a prerequisite for an economic style that promotes flexible work arrangements.

Finally let us add a note on restructuring possibilities at the university level. A radical example in this respect is the medical school program at McMaster University in Hamilton, Canada, which has been in existence since 1969 (see, for example, Neufeld et al. 1989). Its main feature is that it relies not so much on lectures but rather on self-directed learning within the setting of small tutorial groups which encourages cooperation between group members. Right away particular kinds of health problems are at the focus of attention, first in the form of paper patients, later of real patients. In this way medical knowledge is acquired for an immediately obvious purpose, and it can grow in an interrelated fashion connected to case studies rather than in unrelated theoretical bits and pieces. The style of learning, therefore, is close to real life situations: 'information (or facts per se) is no substitute for true knowledge, nor is knowledge a substitute for wisdom. The tutorial method that inquires into, probes and scrutinises materials and evidence, emphasises the point that doubt is always possible and is, indeed, the way to more real knowledge' (Pallie and Carr 1987).

4.3 Economic enterprises

Similarly to households and schools we can think of enterprises as undergoing structural changes in an internal or external sense. Let us start with considering some internal, that is, organizational aspects. An old paradigm of organization in enterprises is associated with Taylorism and maintains that one should plan the production process on the basis of technological and economic considerations. From this the suitable type of organization would then follow almost automatically ('the technological imperative'). Notable properties of such an organization are usually an extreme division of labour and capital specialization, hierarchical control mechanisms, and reliance on labour morale. More recently, the concept of Taylorism has been challenged with the idea of socio-technical systems (see Trist 1981, discussed in Ernste 1987). As the name suggests, organization and technology are to be looked at and optimized simultaneously. Technology opens options in the designing of organizational structures and, with the new information and communication technologies, options have multiplied in recent years. By stressing the socio-technical nature of production processes, the human being moves back to the firm. Humanpower is not regarded as purely physical power that supplements technology. On the contrary, technology has to be organized in a way which allows for humane work; that is, work that contributes to the development of a worker's or employee's vocational identity and vocational ethics (Ernste 1989). The increased emphasis on vocational questions, implicit in the concept of socio-technical systems, has led to work improvements such as job enrichment and job enlargement, and to new work arrangements generally labelled as semi-autonomous groups (Ulich 1989). In fact, as shown by Grotz (1990) for the Federal Republic of Germany, the trend away from hierarchies and heavy division of labour leads to a recombination of previously separated work processes, thereby creating a rising demand for high quality vocational training.

Immediately connecting to the foregoing is a (partial) transformation of the internal organization of a large firm into a decentralized network of externally linked smaller firms (vertical disintegration). The advantage is that such networks can be variably composed: it is easier to change suppliers and customers. It is also possible, of course, that already existing or newly established small firms link up to each other in flexible ways of this kind. An important aspect of this development is the re-emergence of regional economies (Sabel 1989). A gain of flexibility is also associated with the technological trend to increased computer-aided automation: it is no longer necessary to amortize heavy capital investments in the form of machines by a mass production of the same kind of product over many years. Instead, as production becomes programmable, it can become custom-tailored and the output may consist of small numbers. These are essential features of what is called 'flexible specialization' (Piore and Sabel 1984), an economic style which may show a path towards a socially and ecologically sustainable regional development (Jaeger and Ernste 1989).

Another feature of the trend to flexible specialization is the more integrative interfacing of the professional domain of the firm with the private domains of the employees, for example by means of spatially or temporally flexible arrangements which may enable new alternative forms of work. For a long time urban architects and planners have been advocating the spatial separation of homes and work places, originally for good reasons. Today, as most industries have become clean or have given way to services, such reasons are no longer valid, and we are left with the problem of coming to grips with huge volumes of commuter traffic. With the advent of microelectronics and telecommunication the possibility of electronic telework has arisen and makes a certain decentralization of work places a feasible proposition.

Desirable, of course, are any developments that bring private and professional territories spatially nearer to each other. An extreme solution would be to bring the work to the homes of those working, which however, except for special circumstances, is not to be recommended (Jaeger *et al.* 1987). There are psychological and social reasons that speak against it: surely it would not make sense to replace the social integration in an enterprise by complete isolation. A better solution is provided by satellite offices, small branches set up by a firm at the location where the employees concerned are living.

It is interesting to note here that some of these developments are not really new. This is the case especially for some central European areas that have become industrialized in a rather diffuse manner in the past. A notable example here, particularly for its integration of private and economic domains, is the so-called *Terza Italia*, the northeast and central regions of Italy (see Fuà 1983). Here traditional small and medium-sized enterprises often have close connections to agriculture in that an extended peasant family is the nucleus around which various industrial activities develop. Also, informal work plays an important role. From the point of view of modernization such structures have long been regarded as being backward. Today, centralization is no longer considered to be desirable, necessary or unavoidable. Regions such as in Italy (similar structures in Europe exist in southern Germany, Switzerland, and parts of France) are now in a position to combine the advantages of both small-scale economic structures and advanced technology in a way that gives such regions a pioneering character for the development of alternatives to the presently dominant and ecologically devastating way of running the world economy.

In temporal terms various models of flexible working time, including possibilities of part-time work and job sharing are under discussion (Walter and Seiffert 1989). As Jaeger (1987) points out, a heavily schematized life lets creativity wither away. Ensuing conflicts between family and profession may be severely disruptive. A flexible working time, which means a better individual control over time, may help to smooth out discontinuities and to support a better personal self-development and self-fulfilment (Ley 1987).[20] Another aspect of flexibilization concerns the life working time, that is, the age of retirement. We have mentioned this in Section 2.3 as an event which may be experienced as an incisive discontinuity if not as a shock. To become retired may be acceptable to persons much more readily if such a fundamental step is not forced upon them, but is the result of their own decision. Consequently, the age of retirement should be flexible (Lehr 1988, Tschudi 1987).

5 CONCLUSION: THE REGION AS A LABORATORY

In starting to sketch a possible route to a reintegrated human agency, we return to the geographer's homeground: the region. An archaic society depends on social integration, that is, all interactions require the co-presence of the actors concerned, it is egalitarian with, however, a possible and usually temporary orientation of its members towards persons of natural authority, it exhibits a minimum degree of functional differentiation (i.e. there is only the division of 'labour' between females and males and no other distinction of subsystems), it is based on principles of reciprocity and cooperation, its members can organize their lives with a large degree of self-determination (there are activities, but no work in our sense), and it finds its moral orientation within a (religiously motivated) superstructure.[21] The character of archaic life, however, is certainly not fully described as consisting merely of the familiarity of the camp, the closeness of the fellow beings, and the comfort taken in rituals. There is the adverse dimension of moving about, sometimes into the unknown, towards unforeseen adventures and perhaps new discoveries. The daily or weekly rounds of searching for foodstuffs and the seasonal migrations from, for example, one water hole to the next, are at the root of this dimension. The life of the hunters and gatherers is as monotonous as it is varied. Human beings have lived for hundreds of thousands of years in this fashion, and from this lifestyle we seem to have inherited a psychological ambiguity which Tuan (1977) describes with the metaphors of 'place' and 'space'. The longing for stability, familiarity, closeness and security, in short the attachment to a place, is complemented by, and sometimes conflicts with, a lust for freedom, adventure, surprise and uncertainty, a desire for space.

As we have seen, a basic aspect of the past cultural evolution is the gradual retreat of social integration, absolutely and relatively speaking, relative with regard to system integration which has become the overriding principle. Situations of 'place' have, more and more, given way to situations of 'space'. Regional cooperation is successively replaced by international competition, particularly that of the world market, a state of affairs which for the economist Hayek (1988) is quite natural and unavoidable. In our cursory study of the present situation, however, we have stressed that the region may provide an increasingly important platform for social integration through flexibilizations which do not change the functional differentiation of society fundamentally, but provide for an overlap and interplay between subsystems in a way whereby a human actor will experience a greater degree of existential continuity. The physical result would have to be that associated territories – the places of social integration – get closer to each other and life phases merge into one another. Also, with a change in the way in which subsystems relate to each other, the now overpowering dominance of the economic subsystem may be relaxed. Eventually the terrain for a flexibilization of structures cannot be an isolated firm, a single household, or a particular administration, but must be a region, the smallest area that integrates territories, institutions and biographies to a truly cultural milieu.[22] Paths of sustainable development could be advanced and tested by some pioneering regions and then be emulated successively by other regions (Jaeger 1990). The region becomes the laboratory of human ecologists.

6 NOTES

1 For details see Part III and also the contributions by Werlen and Söderström in Part IV of this volume.

2 The notion of wholeness of a person is to be taken relatively; as many philosophers would point out, the human being is fundamentally a contradictory living being.

3 The concept of 'time geography' as developed by Hägerstrand and worked out further by others (e.g. Carlstein 1982) is an attempt at such a combination. However, its shortcoming is that it sees the physical environment as a mere backdrop against which individuals try to realize their 'projects' and, by doing this, are facing time constraints. More in line with an extended notion of such a time geography (as presented by Giddens 1984) we hold that spatial and temporal patterns are physical expressions of social structures and as such are as much governing aspects of individual lives as they are being produced and reproduced by the agents making use of them.

4 Cf. Introduction to Part III and in particular the contribution by Lang.

5 Perhaps, according to the list of different functional subsystems of Luhmann (1986), one could distinguish more types of domains such as, for example, education, science and law. It would seem to be possible, however, to collapse these additional subsystems into the vocational one. A teacher, a scientist or a lawyer do primarily act in vocational contexts.

6 For example, recurring political scandals attest to the fact that politicians are often not capable of keeping the political domain separate from either the private or the economic domain in a proper fashion.

7 This shows that our concept of territory is akin to the notion of 'locale' by Giddens (1984). To provide for continuity within an evolutionary perspective we favour the term 'territory'.

8 Another way of describing this situation is that there is a total fusion of the private, the public and the economic domain (if one wishes to make this distinction at all at this stage), whereby clearly economic aspects of life refer to subsistence activities and public aspects to incidents of, for example, territorial disputes with neighbouring tribes.

9 The loss of public space is just one, but perhaps the most obvious, evidence for the general decline of public life as described by Sennett (1977).

10 Söderström (this volume) develops a similar argumentation. He prefers, however, Habermas' theory of communicative action over Giddens' concept of structuration. In our view the picture provided by Habermas of a lifeworld being increasingly 'colonized' by the 'system' is mistaken: it seems to suggest that instrumental action (or system integration) coincides totally with the economic and the political domains of society, whereas communicative action (or social integration) is identical with the private domain.

11 For a fuller account of Luhmann's theory the reader is referred to the contribution by Bahrenberg and Dutkowski in this volume.

12 In a society with a totemistic religion, for example, certain animals are considered as relatives of human beings.

13 Here we will restrict our attention to relations to other households, as some remarks about the interfacing of households to firms and schools will be made in the following sections.

14 The type of household which in German is known as *Wohngemeinschaft*.

15 This is an example which demonstrates that flexibilization may be ambivalent in character: on the negative side is the fact that meanwhile a total commercialization of the 'do-it-yourself' movement has taken place (Häussermann and Siebel 1987).

16 In Switzerland, for example, the federal housing agency has expressed an interest

in such programs and has published a number of manuals to stimulate residents' participation in housing schemes (Meyrat-Schlee and Willimann 1984).

17 In urban areas, this may also be pursued independently of community type of housing projects (see Schilling 1987).

18 This reminds one of the early Land Use Survey of Britain in which school children were trained to do much of the work.

19 Reichert (this volume) examines the role of implicit knowledge, obviously mobilized in this kind of learning, as a way of knowing that transcends the boundaries of an individual and that contributes fundamentally to the identity of a person.

20 This, however, is another example of a flexibilization which may develop in a negative way. This is true if it will lead to an increase in night-time working, which obviously is incompatible with human biological rhythms.

21 Also, an archaic society in its original state does not know any epidemic diseases. This relates back metaphorically to the contrasting situation of a global society as described by Gould (this volume).

22 For a discussion of the role of regional cultures in attaining ecological sustainability and in forming cultural identities of individuals see Bahrenberg and Dutkowski (this volume), and Werlen (this volume).

7 REFERENCES

Attias-Donfut, C. (1988) 'Die neuen Freizeitgenerationen – Empirische Grundlagen und theoretische Überlegungen zu einer neu entstehenden Freizeitkultur', in L. Rosenmayr, F. Kolland (eds) *Arbeit – Freizeit – Lebenszeit. Grundlagenforschungen zu Übergängen im Lebenszyklus*, Opladen: Westdeutscher Verlag.

Bassand, M. and Henz, A. (eds) (1989) *Zur Zukunft des Wohnens, Empfehlungen. Schlussfolgerungen des ETH-Forschungsprojektes 'Wohnen 2000'*, Zürich: Lehrstuhl Architektur und Planung, ETH.

Baumol, W.J. and Oates, W.E. (1975) *The Theory of Environmental Policy and the Quality of Life*, Englewood Cliffs, NJ: Prentice Hall.

Benedict, R. (1938) 'Continuities and discontinuities in cultural conditioning', *Psychiatry* 1: 161–7, German translation: 'Kontinuität und Diskontinuität im Sozialisationsprozess', in M. Kohli (ed.) (1978) *Soziologie des Lebenslaufs*, Darmstadt/Neuwied: Luchterhand.

Carlstein, T. (1982) *Time, Resources, Society and Ecology*, Lund: CWK Gleerup.

Charles, M. (1987) 'Geschlechterspezifische Arbeitsmarkt-Segregation in der Schweiz', *Schweizerische Zeitschrift für Soziologie* 13 (1): 1–27.

Dieckhoff, K.-H. (1989) 'Zwischen Freizeitgesellschaft und Bürgerbewegung: Die Schule im Zugzwang der Umweltkrise', in L. Criblez and P. Gonon (eds) *Ist Ökologie lehrbar?*, Bern: Zytglogge.

Dürrenberger, G. (1989a) 'Vocational territories', in H. Ernste and C. Jaeger (eds) *Information Society and Spatial Structure*, London: Belhaven Press.

Dürrenberger, G. (1989b) 'Menschliche Territorien. Geographische Aspekte der biologischen und kulturellen Evolution', *Zürcher Geographische Schriften* 30, Zürich: Verlag der Fachvereine.

Ernste, H. (1987) 'Büro-Standorte und Informationstechnik', *Zürcher Geographische Schriften* 28, Zürich: Department of Geography, ETH.

Ernste, H. (1989) 'The corporate culture and its linkages with the corporate environment', in H. Ernste and C. Jaeger (eds) *Information, Society and Spatial Structure*, London: Belhaven Press.

Fester, R., König, M.E.P., Jonas, D.F. and Jonas, A.D. (1982) *Weib und Macht. Fünf*

Millionen Jahre Urgeschichte der Frau, Frankfurt a/M: Fischer.

Fietkau, H.-J. (1984) *Bedingungen ökologischen Handelns. Gesellschaftliche Aufgaben der Umweltpsychologie*, Weinheim: Beltz.

Fuà, G. (1983) 'Rural industrialization in later developed countries: the case of northeast and central Italy', *Banca Nazionale del Lavoro Quarterly Review* 147: 351–77.

Giddens, A. (1984) *The Constitution of Society. Outline of the Theory of Structuration*, Berkeley, CA: University of California Press.

Grotz, R.E. (1990) 'The demands of technological change on vocational training and retraining: The West German experience', *Tijdschrift voor Economische en Sociale Geografie* 81 (3): 170–81.

Häussermann, H. and Siebel, W. (1987) *Neue Urbanität*, Frankfurt a/M: Suhrkamp.

Hayek, F.A. (1988) *The Fatal Conceit. The Errors of Socialism*, London: Routledge.

Henckel, D. (1989) 'Zeitveränderungen und Auswirkungen auf die Stadt – Ergebnisse einer Untersuchung in der Bundesrepublik Deutschland', in A. Ducret, W. Dietrich, D. Henckel, M. Schulte-Haller and A. Viaro (eds) *Planification du territoire et temps de vie*, Lausanne: Institut de recherche sur l'environnement construit, EPF.

Jaeger, C. (1987) 'Telearbeit: Altes Recht für neue Arbeit?', in M. Rehbinder (ed.) *Flexibilisierung der Arbeitszeit*, Bern: Stämpfli.

Jaeger, C. (1990) 'Debt, conservation, and innovating sustainable regional development', *International Environmental Affairs* 2 (2): 166–73.

Jaeger, C. and Ernste, H. (1989) 'Ways beyond fordism?', in H. Ernste and C. Jaeger (eds) *Information Society and Spatial Structure*, London: Belhaven Press.

Jaeger, C., Bieri, L. and Dürrenberger, G. (1987) *Telearbeit – von der Fiktion zur Innovation*, Zürich: Verlag der Fachvereine.

Kastenholz, H. (1991) 'From obligation to motivation: a human–ecological approach to the greenhouse effect', in M.S. Sontag, S.D. Wright and G.L.Young (eds) *Human Ecology: Strategies for the Future*, Fort Collins, CO: Society for Human Ecology.

Knox, P.L. (1987) 'The social production of the built environment. Architects, architecture and the post-Modern city', *Progress in Human Geography* 11 (3): 355–77.

Kohli, M. (ed.) (1978a) *Soziologie des Lebenslaufs*, Darmstadt/Neuwied: Luchterhand.

Kohli, M. (1978b) 'Erwartungen an eine Soziologie des Lebenslaufs', in M. Kohli (ed.) *Soziologie des Lebenslaufs*, Darmstadt/Neuwied: Luchterhand.

Ladeur, K.-H. (1987) 'Jenseits von Regulierung und Ökonomisierung der Umwelt: Bearbeitung von Ungewissheit durch (selbst-)organisierte Lernfähigkeit – eine Skizze', *Zeitschrift für Umweltpolitik und Umweltrecht* 10 (1): 1–22.

Lawrence, R.J. (1989) 'Public and private responsibilities in the definition of housing quality', *Proceedings of Quality in the Built Environment Conference*, London Open House International.

Lehr, U. (1988) 'Arbeit als Lebenssinn auch im Alter. Positionen einer differentiellen Gerontologie', in L. Rosenmayr and F. Kolland (eds) *Arbeit – Freizeit – Lebenszeit. Grundlagenforschungen zu Übergängen im Lebenszyklus*, Opladen: Westdeutscher Verlag.

Ley, K. (1987) 'Flexible Arbeitszeiten', in M. Rehbinder (ed.) *Flexibilisierung der Arbeitszeit*, Bern: Stämpfli.

Luhmann, N. (1986) *Ökologische Kommunikation. Kann die moderne Gesellschaft sich auf ökologische Gefährdung einstellen?*, Opladen: Westdeutscher Verlag.

Luhmann, N. (1989) 'Ökologie und Kommunikation', in L. Criblez and P. Gonon (eds) *Ist Ökologie lehrbar?*, Bern: Zytglogge.

MacIntyre, A. (1985) *After Virtue. A Study in Moral Theory*, London: Duckworth.

Malmberg, T. (1980) *Human Territoriality. Survey of Behavioural Territories in Man with Preliminary Analysis and Discussion of Meaning*, The Hague: Mouton.

Matthes, J. (1978) 'Wohnverhalten, Familienzyklus und Lebenslauf', in M. Kohli (ed.) *Soziologie des Lebenslaufs*, Darmstadt/Neuwied: Luchterhand.

Meyrat-Schlee, E. and Willimann, P. (1984) *Gemeinsam Planen und Bauen. Handbuch für Bewohnermitwirkung bei Gruppenüberbauungen*, Bern: Bundesamt für Wohnungswesen.

Michel-Alder, E. and Schilling, R. (1984) *Wohnen im Jahr 2000*, Basel: Lenos.

Milbrath, L. (1989) *Envisioning a Sustainable Society. Learning Our Way Out*, Albany, NY: State University of New York.

Neufeld, V.R., Woodward, C.A. and MacLeod, S.M. (1989) 'The McMaster M.D. program: A case study of renewal in medical education', *Academic Medicine* August: 423–32.

Pallie, W. and Carr, D.H. (1987) 'The McMaster medical education philosophy in theory, practice and historical perspective', *Medical Teacher* 9 (1): 59–71.

Perrez, M. (1979) *Die Krise der Kleinfamilie*, Frauenfeld: Hans Huber.

Piore, M. and Sabel, C. (1984) *The Second Industrial Divide*, New York: Basic Books.

Robert, L. (1989) 'Gedanken zum Thema Schule und Ökologie', in L. Criblez and P. Gonon (eds) *Ist Ökologie lehrbar?*, Bern: Zytglogge.

Sabel, C. (1989) 'Flexible specialisation and the re-emergence of regional economies', in P. Hirst and J. Zeitlin (eds) *Reversing Industrial Decline?*, Oxford: Berg.

Schilling, R. (1987) 'Soziale Bewegungen verändern Stadt und Stadtgestalt', in M. Dahinden (ed.) *Neue soziale Bewegungen – und ihre gesellschaftlichen Wirkungen*, Zürich: Verlag der Fachvereine.

Schubert, H.J. (1989) 'Zum Zusammenhang von Ethik und Macht am Beispiel Eigenarbeit', in B. Glaeser (ed.) *Humanökologie. Grundlagen präventiver Umweltpolitik*, Opladen: Westdeutscher Verlag.

Sennett, R. (1977) *The Fall of Public Man*, New York: Knopf.

Siegenthaler, H. (1987) 'Soziale Bewegungen und gesellschaftliches Lernen im Industriezeitalter', in M. Dahinden (ed.) *Neue soziale Bewegungen – und ihre gesellschaftlichen Wirkungen*, Zürich: Verlag der Fachvereine.

Smith, T.T. and Kromarek, P. (eds) (1989) *Understanding U.S. and European Environmental Law. A Practioner's Guide*, Dordrecht: Kluwer.

Sommer, J.H. and Höpflinger, F. (1989) *Wandel der Lebensformen und soziale Sicherheit in der Schweiz. Forschungsstand und Wissenslücken*, Chur: Rüegger.

Spiegel, E. (1986) *Neue Haushaltstypen. Entstehungsbedingungen, Lebenssituation, Wohn-und Standortsverhältnisse*, Frankfurt a/M: Campus.

Trist, E. (1981) *The Evolution of Socio-Technical Systems*, Toronto: Ontario Quality of Working Life Center.

Tschudi, H.P. (1987) 'Flexibilität der Lebensarbeitszeit', in M. Rehbinder (ed.) *Flexibilisierung der Arbeitszeit*, Bern: Stämpfli.

Tuan, Y.-F. (1977) *Space and Place*, Minneapolis, MN: University of Minnesota Press.

Ulich, E. (1989) 'New orientations in management', in H. Ernste and C. Jaeger (eds) *Information Society and Spatial Structure*, London: Belhaven Press.

Walter, N. and Seiffert, E. (1989) 'Zeit für Arbeit. Ein Plädoyer für mehr Zeitsouveränität', *Neue Zürcher Zeitung* 251, 28/29.10.89.

Wanner, H. (1987) 'Die Entstehung neuer Wohn- und Siedlungsformen als sozialer Wandel', *Anthropogeographie* 7, Zürich: Department of Geography, University of Zürich.

Wüest, H. (ed.) (1987) *Siedlungsentwicklung nach innen. Ausmass, Inhalt und Bedeutung von Erneuerung und Recycling im Siedlungs- und Städtebau*, Rapperswil: Interkantonales Technikum.

Wyss, H. (1989) 'Für das Leben lernen – "Adopt a class". Ein Schulprogramm in New York mit Modellcharakter', *Neue Zürcher Zeitung* 136, 15.6.89.

21

SPECTACLE ECOLOGY, HUMAN ECOLOGY AND CRISIS OF URBANITY

Ola Söderström

1 INTRODUCTION

The exploration of human ecology, which is seen as an integrative theory 'reuniting' geography – a discipline torn apart between natural and human sciences – with the practical aim of facing environmental problems, is a recurrent theme of this volume.

This theme raises a series of questions, since it begins with a number of more or less explicit prior assumptions. This paper intends to discuss two of them, as they are, in our opinion, closely linked: first, the priority given to the environmental crisis as a practical aim of geography, and second, the relevance of human ecology as an integrative theory. These two postulates will first be questioned in general and then more concretely by viewing human ecology in the context of one of its fields of application: the urban field. With respect to the specific problems of the city, and in particular that which will be defined as the 'crisis of urbanity', human ecology will be evaluated as a 'globalizing' theory. The idea is not to make a case against human ecology, a theory sufficiently general, even vague enough, to be 'immune' to any such undertaking but to show, by analysing the different approaches to the urban phenomenon by authors who consider themselves human ecologists, some deficiencies which question the claim of globalization. An attempt will thus be made to show that the main deficiency in the approach to the urban field through human ecology consists in its lack of articulation to a coherent social theory. In order not to content ourselves with a critical position but rather to initiate a discussion on the form that such a conceptualization of the urban field should take, the reference to two social theories recently formulated – Giddens' theory of structuration and Habermas' theory of communicative action – will end this presentation, in an attempt to see to what extent they allow an understanding of the question of the crisis of urbanity.

This paper is to be considered mainly as a point of view on human ecology constituted by the approach of social urban geography. The attempt to evaluate the globalizing claims of human ecology will be restricted to this specific bias.

OLA SÖDERSTRÖM

2 APPARENT DIGRESSION: HUMAN ECOLOGY AND ECOLOGY AS SPECTACLE

In these post-Chernobyl years when we have learned to doubt even the quality of the atmosphere surrounding us, ecology has simultaneously become a political goal, a new market, an opening for the media and a spectacle guaranteeing the thrill of an impending apocalypse. In recent months we have thus seen the French broadcasting companies competing for the viewing public and exploiting this 'leading-theme' by organizing the kind of big-disaster-evening-shows in which the stars are experts who prophesy our end as a species if radical changes in our relation to the environment do not intervene.[1] All of this was without the slightest fear of paradox since one of these shows was sponsored by an important French firm – Rhône Poulenc – which found in it a way of recreating for itself a new ecological virginity that it badly needed. Such important events in the media end up creating a façade of consensus where all concerned – from the head of State to the new-born African – are presented as fighters for the ecological cause: the only decisive battle of the end of our century. The environmental crisis becomes a spectacle and, like any spectacle, it does not tolerate contradictions. Consequently, all and any fundamental interrogation as to the causes of the crisis evaporates, and this occurs to the benefit of a magical incantation for 'change'.

A more prosaic example will allow us to measure the nature of the ecological question today as it presents itself in its more restricted and everyday dimension. This concerns the application, in Switzerland, of the federal decree on impact studies of October 1988, which imposes the realization of this kind of study on numerous planned layouts (from high-ways to public dumps), and tends to open a whole market for existing or specially created firms in this domain. This regulation is there to prove that in certain countries ecological preoccupations now have a legal foundation that can oblige both the State and private enterprises to measure the fall-outs of their actions on the environment. This Swiss example effectively illustrates the mainstream acceptance of the ecological question, which now becomes an issue definitely grown beyond the confines of marginal political protest. A multitude of other examples could in fact be quoted as evidence of the institutionalization of this issue, still marginal twenty years ago, around a complex of economic and political interests. This movement is indeed reinforced by the electoral breakthrough of the 'green' parties which tend, as shown by the last European elections of June 1989, to establish themselves as an important and stable political force. The few allusions now made to the fate of ecology and its contradictions are of course well known material, but worth mentioning because other, perhaps even more serious problems lurk behind the commotion made over ecology today.

Of course, if there is reason to rejoice about the popularization and institutionalization of a question which is, as the consecrated formula states, 'vital for the future of humanity', eternal sceptics ask themselves whether, in effect, this crystallization of the public debate on the crisis of environment does not blind us to other, at least equally important issues. It is not fortuitous that Félix Guattari, because he is a psychiatrist and a philosopher, is one of those sceptics when he somewhat polemically writes: 'Our attention is focused on future disasters when the true disasters are actually there in front of our eyes in the form of degenerating social practices, stupefying mass-media, and a blind collective faith in 'market' ideology, which finally means yielding to the law of number, to entropy, to the loss of singularity, to general infantilization' (*Libération*: 'La grande peur ecologique', 30 June 1989, translation by the author).

To avoid the blinkers set by an ecology centred on the question of environment and confined to a technocratic perspective, Guattari (1989: 12) proposes 'an ethico-political articulation – [that he names] *ecosophia* – between the three ecological registers, the one of environment, the one of social relations and the one of human subjectivity'. It is certainly necessary to modify our relation to the environment in the light of the succession of 'accidents' caused by uncontrolled exploitation of natural resources, but it is equally necessary to rethink our social ecology, that is, to 'reconstruct the whole of the modalities of our being in a group', as well as our mental ecology by re-inventing 'the relation of the subject to the body, to fantasies, to time passing, to "the mysteries of life and death"' (Guattari 1989: 22). According to Guattari, this necessity of an articulation of the three ecologies finds its foundation not solely in the theoretical ambition to build up a holistic framework of analysis but essentially in the mere awareness that the three domains are not closed territories but are rather created jointly by means of our practices into 'existential territories'. The fact that these three domains are closely interdependent, because they are linked and created as such by our individual and collective practices, implies that the environmental crisis can only be solved through a modification of these practices, and therefore through a simultaneous modification of the relations within the three ecologies. The visibility and the concreteness of the decay of the physical environment as well as the interests which sustain 'spectacle ecology' should therefore not constitute a screen preventing us from discerning perhaps even more serious degradations of the mental and social environments.

In other words, a geography seeking an integrative theory, articulating individual, society and environment, should not limit its concerns to the environmental crisis. A reply to this could be that it is no real danger in the domain of research but only a mediating effect. Furthermore, that the practical goal of human ecology is precisely to 'insure a dynamic equilibrium between eco-, bio- and socio-logics in the perspective of insuring the greatest possible autonomy for a population in the long run' (Raffestin 1985: 148–9). Rather than trying to answer these objections and discussing the reality of this environmental bent of human ecology on a strictly theoretical level, where one can endlessly discuss different definitions of human ecology, this paper intends briefly to analyse to this effect the approach of a specific field of application of this theory, that being the one in which the author has some competence: the urban field. What will thus be scrutinized here is a human ecology at work.

No attempt will be made here to trace a genealogy of the approach to the city through human ecology, starting from the School of Chicago, passing through the works of Amos Hawley and ending at the present day. Instead, we will simply consider recent studies explicitly set by their authors in the domain of human ecology. It is worth mentioning that these studies have been chosen because of their founding and programmatic ambitions.

3 URBAN FIELD, CRISIS OF URBANITY AND HUMAN ECOLOGY

The approach to the city through human ecology is based in principle on the analysis of the interactions between the three angles of a triangle defined by Boyden as: natural environment, human society and human experience (Boyden 1984: 8). The three ecologies evoked by Guattari are thus at the origin of this approach. However, the passage to the case study is characterized by a shift towards environmental ecology. In effect, the ecological interrelations within the human ecosystems are considered as 'patterns of flow of energy and materials' (Boyden 1984: 24), a definition that excludes crucial forms of relations between individuals whether or not co-present in time and space: notably communication and symbolic exchange (one might say: 'information' in its broader sense).[2] The phenomena analysed through human ecology are therefore the production and consumption of energy, of water and of materials, in a city where humans are just a variable like any other, accounted for in terms of number and density of population. The biological metaphor organizes the empirical study since the city is considered as a 'metabolism' (Newcombe et al. 1978). Consequently the city is not the result of individual and collective projects of actors but is a big organism which feeds on its environment and modifies it in turn. Thus, in an article on the ecological study of Rome, Bonnes (1984: 54) supports the idea that 'the city has to be treated primarily as a biological system'.

It is obvious that this way of approaching the urban phenomenon produces in turn a particular representation of the city. The balance of flows of energy and material makes it appear predominantly as a source of waste and environmental nuisances. As an important consumer of material and energy, the city is the centre of an unregulated society and economy and, consequently, one of the most evident manifestations of the crisis of environment. The activities that it contains are thus considered as the principal elements responsible for this crisis (Boyden 1984: 54).

In addition to this study of the urban field as a biological system, certain works imply, however, an approach to the nature and the transformation of social relations in the cities. Thus, a part of the important study by Boyden *et al.* (1981) on Hong Kong attempts an assessment of this subject and tries to define among other things the 'sense of belonging' that the inhabitants feel for their neighbourhood and 'co-operative small group interaction'. However, this secondary part of the study can only remain very schematic because it refuses to got into any critical reflection concerning the questions studied (what is the sense of belonging to a place?) and because the only theoretical framework with which it is concerned is that of the four phases of the ecological development of humanity.[3] Assigning the observations made in Hong Kong to such a vast framework of interpretation obviously does not help in the understanding of, for example, the evolution of the patterns of social solidarity or the relations between private and public spaces, phenomena that are central for an appreciation of the dynamics of urban social organization.

Finally, at the level of the conclusions drawn from such studies, the diagnosis of the role played by the city leads to its characterization primarily as a source of environmental nuisance and waste, as we have already stated. Consequently, the practical solutions that are proposed follow from this acknowledgement and seek an ecological rationalization of the urban phenomenon. Nevertheless, a solution of the 'ecological' crisis of the city (through measures such as waste recovery, orderly traffic and common transport, rehabilitation, pedestrian precincts) would not avoid the degradation of social relations within the urban milieu; dimension of the urban phenomenon that human ecology, such as it is often put into practice, does not give itself the means to understand. These 'pathologies', linked both to the transformation of the physical setting and to social evolution, are fundamental aspects of urban contemporary life that demand specific theoretical instruments.[4] They can, in fact, be understood as the urban translations of an evolution of the forms of social life towards an increasingly important domination of technique and of impersonal systems of mediation of social exchange. The consequence of this evolution is what will be termed here the 'crisis of urbanity' and will be given a more precise content in the remainder of this text.

The outcome of these studies is therefore an appreciation of urban evolution which essentially restricts itself to the field of environmental ecology, without respecting the globalizing ambitions of the theoretical approach. This statement is based on an obviously partial analysis of the literature in this domain, but it could be supported by the analysis of numerous other works.[5] The literature in this domain would have to be exhaustively reviewed if a synthetic image of the approach to the urban phenomenon through human ecology were to be consistently established. That is not the purpose of this paper, which is, more modestly, to point out a 'natural' bent of human ecology and its inconveniences if it is set up as an integrative theory.

With regard to this ambition, it therefore seems, in short, that in the existing studies the 'ecological' approach to the city tends, on the one hand, to atrophy the domains of mental and social ecology and, on the other, if it takes them into consideration, to ignore the instruments of analysis of either social theory or psychology. It is therefore necessary to remedy this deficiency of human ecology if any credit is to be given for its merits as an integrative theory, because the study of urban reality today should not subsume the modalities of being together in this milieu under any kind of imperative based on an imminent environmental disaster and imposing as a priority the analysis of the ecological assessments.

This necessity of a global ecology, which implies the inclusion of a social theory capable of explaining the forms of social life in an urban milieu, is the basis on which we intend to continue, through an exploratory approach of two theories relevant to this task, the present discourse. The discussion will thus focus mainly on aspects of the social ecology of the city, leaving here its mental ecology in parentheses.

4 SOCIAL THEORY AND HUMAN GEOGRAPHY

The term 'urbanity' was introduced in the eighteenth century to replace the words 'courtesy' and 'civility' which had become obsolete because they were connected to the conduct at court. Its introduction is thus linked to a movement which characterizes the development of bourgeois society: the passage from the court to the city as the space of reference. It is therefore a polysemic term which refers both to the particular sociability which was in the process of establishing itself in the eighteenth century – based on anonymity and theatricality – and to the space in which it unfolded: the city. In the same way, the decline of public life beginning in the nineteenth century – and linked to a series of economic and cultural transformations – leads to talk of a 'crisis of urbanity' both as the crisis of a certain type of sociability and a crisis of the city. This diagnosis of urban evolution in the modern Western societies is developed notably by Richard Sennett (1974) and identifies a major change in the social ecology of the city. However, in order to avoid a certain idealization of public life during the enlightenment, which characterizes his work as a historian, and to extend this reflection, it is necessary to base this approach in a more comprehensive social theory (Quéré 1982).

Consequently, two important 'building-sites' in social theory will now be considered here because of both their globalizing intents and the active debate that they fuel within the geographic milieux: Anthony Giddens' theory of structuration, the object of an already well nourished dialogue within mainly Anglo-Saxon geography, and Jürgen Habermas' theory of communicative action, which, against the intellectual trend, insists obstinately on extending enlightenment's call for a founding reason. It is impossible to give here a detailed account of these two attempts to rethink a social theory and their critical reception, because of both their complexity and the extent of their sources. Therefore they will be confronted only as two different potential illuminations on the social pathologies of the contemporary cities. Such a mediation between contrasting theoretical frameworks also constitutes a means of questioning the specificity of the social ecology of the city. First, and in both cases, a brief summary of their arguments will be necessary.

Giddens' work, which he collects under the generic name of structuration theory, has the ambition of reconstructing social theory on a descriptive and comprehensive basis and presents itself explicitly as a rejection of the opposition between micro- and macro-social analysis. According to him, the interest of social analysis is rooted in the interaction between micro- and macro-processes. In order to theorize this interaction, an attempt which is at the very centre of his project as a whole, Giddens recovers and redefines the classical distinction between agency and structure. But instead of traditionally considering these concepts as expressions of dualisms, he proposes to read them as dualities: rather than opposing agency and structure as two incommensurable dimensions of social life – where agency would be exterior to structure and would command strictly distinct theories and empirical studies – Giddens conceives our everyday actions as contributing to the constitution and the reinforcing of social structures. The structures are no longer abstract motifs independent from action, but are the result of recursively reproduced practices, the structure being, so to say, inserted into the action itself. One of the fundamental propositions of his theory, of which the following is one of the numerous formulations, thus states that 'all structural properties of social systems ... are the medium and the outcome of the contingently accomplished activities of situated actors' (Giddens 1984: 191).

In order to apprehend further this link between agency and structure through which society constitutes and reproduces itself, Giddens introduces another conceptual distinction which is classic in social theory: that between social integration and system integration. These terms respectively define an interaction in a context where people are co-present in time and space, that is, 'what "society" meant before the eighteenth century: simply the company of others' (Gregory 1989: 188), and an interaction characterized by a great distance in time and space between actors or collectivities, that is, 'what "society" came to mean after the eighteenth century: the larger world stretching away from the human body and the human being' (Gregory 1989: 189). The nature of these definitions refers back to another feature of the structuration theory – the main reason for its meeting with the geographers – that is, the founding role of contextuality for the understanding of society. According to Giddens, the issue here is that of revoking the Kantian conceptions of space and time which reduce them to simple parameters, 'empty containers within which an activity occurs, or where material events are situated' (Giddens 1988: 170), and replacing them with a conception where space and time are considered as constituting the social system, because they express the very nature of the objects of which they are composed.

If we now consider these two conceptual pairs together, we can say that our everyday actions produce and reproduce the larger structures that regulate our life in society while the same structures canalize and orient our everyday activities. The reproduction of the social system is accomplished through the routines of everyday life or, in Giddens' words, 'the recursive nature of social activity'. Concretely, the spatio-temporal routines of everyday life thus constitute the link between human agency and social structures. At this point and notwithstanding certain critical reserves, Giddens refers to Hägerstrand's (1975) 'time-geography' and to his famous sketching out of the spatio-temporal paths, and in particular to the notion of 'bundles' of activities and that of meeting 'stations', as a means of materially grasping this link (Giddens 1984: 110–44). This time-geography has the triple advantage of being founded on the contextuality of action, of underlining the importance of its routine character and of proposing 'a mode of charting and of analysing the patterns of social reproduction' (ibid., p. 365).

However, Giddens fails here, for the time being, to produce a theory which is really equal to his ambition: we are left, in effect, with a missing link between everyday action and social structure when he finally observes that this link cannot be theorized in detail since agency is not simply determined by structure, social integration by system integration or vice versa. Their relation varies from one society to another. This relation can only be grasped, he concludes, through the analysis of concrete situations (Giddens 1984: 219). But the problem here is, as Gregson (1989) observes, that the empirical relevance of Giddens' theory is far from being straightforward. Therefore, more than a complete theory, his is a 'building site' as we have previously defined it.

In its present form, the structuration theory is primarily a theory of social reproduction through the central idea of the 'duality of structure'. In other words, Giddens' theory is an attempt to explain the composition of the 'social glue': it explores the logic which, from the micro- to the macro-social, gives society its coherence. What structuration theory fails – in part deliberately – to provide is a theory of social change that would allow us to define a 'locus' of the social crisis. In his critical attitude towards any evolutionary theory of human societies, Giddens indeed refutes all global theory of social change, although he proposes a conceptual apparatus allowing generalization in this domain (Giddens 1984: 244). As Edward Soja (1989: 146–7), perhaps unwillingly, emphasizes by stressing the importance of the glossary proposed by Giddens, structuration theory appears, in fact, primarily as the creation of a language which permits us solidly to introduce contextuality in the discourse on the constitution of society and not as a narration of its mutations. In this sense, it is a theory that provides useful tools for an integrative approach in human geography but does not respond to the critical ambition of such an approach (cf. Bernstein 1989 and Giddens

1989). This is the reason why the crisis of urbanity cannot be appropriately illuminated by this theory.

What Jürgen Habermas proposes in his theory of communicative action is essentially a critical account of social evolution through research on the processes of rationalization in modern Western societies. Leaving aside the details of his epistemological argumentation (see to this effect Habermas 1986: chap. 2), the way in which he structures his reading of social reality will be briefly and very schematically examined here. For Habermas, Western history is that of the increasing rationalization and institutionalization of the lifeworld. His theory of social reality is based on the methodological distinction between the two notions of lifeworld and system. These two spheres jointly form the social and they stage different activities: communicative action for the first and strategic action for the second. To these activities correspond different means of coordinating the action, which signifies (and this is what makes them fundamentally different) that they are regulated by different media: communication for the lifeworld, money and power primarily for the system.

Habermas' work, more than Giddens', presents itself as a critique of rationalization and modernity which – by basing it on a theory of communication which repudiates the validity of the philosophy of consciousness – refers to a theory of reification going from Marx and Weber to Horkheimer and Adorno. This critique concerns the 'colonization' of the lifeworld by the system; a formula which means that society has passed from a social interaction mediated by communication, based on mutual understanding and trust, to an interaction increasingly mediated by money and power which does not have the qualities of the first: a process of mechanization of the lifeworld that explains the appearance of today's main 'social pathologies' (cf. Habermas 1987, vol. 2: 365–410). However, in order not to simplify excessively the content of the theory of communicative action, it should be made clear that the progressive autonomy gained by the system with regard to the lifeworld, which characterizes social evolution in Habermas' work, is no 'paradise lost', since there is 'colonization of the lifeworld' only when forms of economic and administrative rationality force their way through into domains of the social which should remain regulated by communication, such as those specializing in cultural transmission, social integration and education.

To this distinction between lifeworld and system corresponds the pair, central also for Giddens, namely, social integration and system integration, the second steadily progressing at the expense of the first. Certain aspects of the relation between these two terms are similar to those described by the British sociologist, since his proposal is that societies be conceived of as 'connections of actions stabilized in systems' (Habermas 1987, vol. 2: 220). However, the difference with Giddens is in Habermas' more conflictual representation of the relation between social and system integration, since this is for him the place where the colonization of the lifeworld and the reification of the social relations characterizing the evolution of modern society come about,[6] whilst Giddens sees in it essentially a reproduction of the second on the part of the first.[7]

The nature of the connection between system and social integration for Habermas is thus hierarchical and has progressively reduced the lifeworld into a mere satellite of the system. This evolutionary theory of societal evolution, however, is not deterministic, since individuals constantly use defensive strategies in order to preserve the integrity of the lifeworld: there is a margin of freedom in everyday action which allows for the conception of such strategies. This permits him to reject the inevitability of reification and the 'pathos of alienation' which are so frequent in contemporary critical thought.[8]

The main problem with the theory of communicative action, as far as geography is concerned, lies of course in its neglect of the spatiality of social life; 'contextuality' is not as central a notion as it is with Giddens: space is for this reason not theorized as such. Because Habermas' theory does not specifically address the dimension of space, one might wonder what kind of interest it could have for geography. However, if one disregards this obvious absence, the theory of communicative action appears equally relevant to the understanding of the conflicts between technocratic space and the space of experience, which are so often perceived in a confused way by the citizens or the urban geographers. In light of this theory, the study of urban evolution leads indeed to the analysis of the kind of relations that structure social life – in terms of communicative or instrumental action – in different spaces and at various stages of their evolution, and to the attempt to understand the nature of urban pathologies and conflicts, the logic of their emergence, as phenomena linked to a colonization of the lifeworld. It is thus a reading of the urban reality that potentially allows for a critical evaluation of the social ecology of the city.

Concretely, the colonization of the lifeworld manifests itself in the evolution of urban space – which is but one of the fields in which it can be seen at work – in a de-realization of all that was traditionally covered by the notion of city, that is, in its evolution towards an abstract system mediated by power and money (Habermas 1985b: 11–29). For Habermas, the study of this phenomenon should be the task of social sciences today, all the more urgent in that the colonization of the lifeworld affects the 'grammar of the forms of life' (Habermas 1987, vol. 2: 432). From this point of view, discussing urban space results from a defensive strategy which is similar, for Habermas, to other social struggles precisely like ecology.[9]

A potentially productive way of elaborating a theoretical framework for comprehending the crisis of urbanity is by extending and further defining the concept of colonization of the lifeworld, through the articulation between social change and urban dynamics. In order not to go beyond the intended purpose here, we will only underline the fact that it is possible to isolate, on an empirical level, phenomena deriving from a colonization of the lifeworld, as in certain effects on the social system which are caused by a disaggregation of the spatial texture that of course plays an important role in the definition of identities and local communities within the cities. The analysis of gentrification, especially in North America, thus permits us to observe the disappearance of what one could call 'urban lifeworlds' caused by one population replacing another – obviously less affluent – and by the consequent change in social interaction involved. One can also note here that the 'urban lifeworld' has evolved into a town-planning issue, and this manifests itself in certain gentrified districts where architecture and urbanism are used – through signs and references of post-modernism in particular – by the economic and political power to create the simulacrum of an urban community. The language of architecture acts in this case as a 'symbolic refund', to use a well formulated expression by Jürgen Hasse (1988: 30).

The migrations of the 'Skid Row' in American cities and the pressure of the extension of the CBD that is exerted on these 'pockets' of social marginality form another field of analysis which can be apprehended through the point of view of the colonization of the lifeworld. A study by the author on the central districts of Vancouver (Söderström 1992), concerning the defensive strategies set up by a local association to fend against the transformation of the socio-economic contents of these districts, tries thus to demonstrate that the categories derived from the theory of communicative action have a heuristic value for grasping and organizing the mechanisms at work.

5 CONCLUSION

Habermas' theory, like any other social theory of 'globalizing' intents, poses just about as many problems as it solves. Various objections have been set, concerning firstly his idealization of a communicative activity which escapes conflicts and dissensions, but also his nostalgia for the public space of the enlightenment and the utopia of a transparency of social reality which underlies his emancipatory project, and Habermas has responded to these critiques.[10] The debate is therefore going on, and it is out of the question to subscribe blindly to the theory of communicative action, even less to apply its categories simplistically by mechanically assimilating institutions and actors to the concepts of system and lifeworld. This theory primarily sets a framework of analysis describing the logics of action, of which the relations have specific tendencies of evolution in Western society. Much more has to be done to make it an operational instrument of critical analysis in urban geography, in particular the creation of middle-range theories, or meso-level concepts. The question of Habermas' silence on space, which Derek Gregory (forthcoming) qualifies as 'strategic', is obviously one of the main questions left open.[11]

The main issue that we wanted to underline, by rapidly discussing this theory in connection with that of structuration, is the necessity to have a theory which is apt to seize the social relevance of urban evolution and, more generally, to illustrate the need for such an approach within the framework of an integrative theory in geography. There is, indeed, the necessity to operate theoretically and empirically in more than one domain: the three ecologies must be simultaneously apprehended through an articulation of a knowledge specific to each one of them (the city is not a biological system). In this case we shall be able to admit that the practical aims of such a theory are multiple, the crisis being multidimensional.

Once more, if we wish to give credit to the globalizing aims of human ecology, the latter has not only to set the postulate of an articulation of its three fundamental poles but also to articulate successfully and practically the three domains of mental, environmental and social ecology; a position that this text has wanted to develop at a moment when 'spectacle ecology' seems to impose its priorities.

6 NOTES

1 For example: *La Terre perd la boule* (The Earth loses its head) broadcast by TF1 on 22 June 1989 and *J'y crois dur comme terre* (My belief is as hard as rock), also by TF1 in September 1989.

2 Even though the implications of these flows of energy and material are considered at the level of society (on the social and political structure) and of the individual (on his or her mental and physiological health).

3 Boyden distinguishes a primordial phase of hunting and gathering, a phase of subsistence agriculture, a first urban phase and, finally, a modern and technological phase (cf. Boyden 1984: 9, and this volume).

4 Weichhart makes a similar comment when he states the necessity to turn to the methods of social and human sciences in human ecology (cf. Weichhart 1986: 70).

5 For an analogous but much more detailed and global critique of human ecology, see Ericksen (1980), who notably underlines the inadequacy of its social theory.

6 Habermas talks about a 'structural violence, exerted through a systematic reduction of communication' (Habermas 1987, vol. 2: 205).

7 This distinction derives from a noticeably different definition of these two terms. Giddens opposes them, as we have seen, on the basis of the spatio-temporal context of social interaction, whilst for Habermas social integration has its origin in the social connections uniting individuals who act and speak in view of a mutual understanding, and system integration goes back to impersonal relations between institutions and structures which lead to the material reproduction of society.

8 For further elements of comparative discussion of these two authors from the point of view of geography see Gregory (1988). For critical comments by Giddens on Habermas' work in relation to his own, see Giddens (1987: chap. 10). For other critical appreciations of recent works by Habermas, see Bernstein (1985) and Sfez (1988).

9 For a more detailed discussion of space in the theory of communicative action, see Mondada and Söderström (1991).

10 See in particular Habermas (1985a), and different articles in *Critique* 413 (1981), 464–5 (1986), 493–4 (1988).

11 It is necessary in particular to define to what extent it is possible to articulate lifeworld and system to spatial categories such as the local and the global.

7 REFERENCES

Bernstein, R.J. (1989) 'Social theory as critique', in D. Held and J.B. Thompson (eds) *Social Theory of Modern Societies: Anthony Giddens and His Critics*, Cambridge: Cambridge University Press.

Bernstein, R.J. (ed.) (1985) *Habermas and Modernity*, Cambridge, MA: MIT Press.

Bonnes, M. (1984) 'Mobilizing scientists, planners and the local community in a large-scale urban situation: the Rome case study', in F. DiCastri, F.W.G. Baker and M. Hadley (eds) *Ecology in Practice, Part II: The Social Response*, Dublin: Tycooly.

Boyden, S. (1984) 'Integrated studies of cities considered as ecological units', in F. DiCastri, F.W.G. Baker and M. Hadley (eds), *Ecology in Practice, Part II: The Social Response*, Dublin: Tycooly.

Boyden, S., Millar, S., Newcombe, K. and O'Neill, B. (1981) *The Ecology of a City and Its People*, Canberra: Australian National University Press.

Ericksen, E.G. (1980) *The Territorial Experience – Human Ecology as Symbolic Interaction*, Austin, Tex.: University of Texas Press.

Giddens, A. (1984) *The Constitution of Society*, Cambridge: Polity Press.

Giddens, A. (1987) *Reason without Revolution? Habermas' Theory of Communicative Action in Social Theory and Modern Sociology*, Cambridge: Polity Press.

Giddens, A. (1988) 'The role of space in the constitution of society', in D. Steiner, C. Jaeger and P. Walther (eds) 'Jenseits der mechanistischen Kosmologie – Neue Horizonte für die Geographie', *Berichte und Skripten* 36, Zürich: Department of Geography, ETH.

Giddens, A. (1989) 'A reply to my critics', in D. Held and J.B. Thompson (eds) *Social Theory of Modern Societies: Anthony Giddens and His Critics*, Cambridge: Cambridge University Press.

Gregory, D. (1988) 'La différenciation, la distance et la géographie post-moderne', in G. Benko (ed.) *Les nouveaux aspects de la théorie sociale: de la géographie à la sociologie*, Caen: Paradigme.

Gregory, D. (1989) 'Presences and absences: time–space relations and structuration theory', in D. Held and J.B. Thompson (eds) *Social Theory of Modern Societies: Anthony Giddens and His Critics*, Cambridge: Cambridge University Press.

Gregory, D. (forthcoming) *The Geographical Imagination*, London: Hutchinson.

Gregson, N. (1989) 'On the (ir)relevance of structuration theory to empirical research', in D. Held and J.B. Thompson (eds) *Social Theory of Modern Societies: Anthony Giddens and His Critics*, Cambridge: Cambridge University Press.

Guattari, F. (1989) *Les trois écologies*, Paris: Galilée.

Habermas, J. (1985a) 'Questions and counterquestions' in R.J. Bernstein (ed.) *Habermas and Modernity*, Cambridge, MA: MIT Press.

Habermas, J. (1985b) *Die Neue Unübersichtlichkeit*, Frankfurt a/M: Suhrkamp.

Habermas, J. (1986) *Morale et communication*, Paris: Editions du Cerf.

Habermas, J. (1987) *Théorie de l'agir communicationnel*, Paris: Fayard.

Hägerstrand, T. (1975) 'Space, time and human conditions', in A. Karlquist (ed.) *Dynamic Application of Urban Space*, Farnborough: Saxon House.

Hasse, J. (1988) 'Die räumliche Vergesellschaftung des Menschen in der Postmoderne', *Karlsruher Manuskripte zur Mathematischen und Theoretischen Wirtschafts- und Sozialgeographie* 91.

Mondada, L. and Söderström, O. (1991) 'Espace et communication, perspectives théoriques et enjeux sociaux', *Cahiers du Département des Langues et des Sciences du Langage* 11: 107–61.

Newcombe, K. , Kalma, J.D. and Aston, A.R. (1978) 'The metabolism of a city: The case of Hong Kong', *Ambio* 1: 3–15.

Quéré, L. (1982) *Des miroirs équivoques, aux origines de la communication moderne*, Paris: Aubier.

Raffestin, C. (1985) 'Ecologie générale et écologie humaine', *Cahiers Géographiques 1*, Genève: Département de Géographie, Université de Genève.

Sennett, R. (1974) *The Fall of Public Man*, New York: Knopf.

Sfez, L. (1988) *Critique de la communication*, Paris: Seuil.

Söderström, O. (1992) 'Les gestionnaires de la mémoire: la politique du patrimoine à Vancouver et Lausanne', PhD thesis, Lausanne: University of Lausanne.

Soja, E.W. (1989) *Postmodern Geographies*, London: Verso.

Weichhart, P. (1986) 'Diskussionsbemerkungen zur Konzeption und zum Programm einer "geographischen Humanökologie"', in D. Steiner and B. Wisner (eds) 'Humanökologie und Geographie – Human Ecology and Geography', *Zürcher Geographische Schriften* 28, Zürich: Department of Geography, ETH.

22

BOUNDARIES AND BARRIERS IN RELATION TO THE DIFFUSION OF AIDS

Peter Gould

For the sake of humane understanding, let us start by doing a thoroughly disparaged, increasingly shocking, and totally inhumane act. Let us reify the human family into a set of elements. Then, for the AIDS epidemic, there is only one boundary, one Peano curve separating the inside from the outside, one circle to the Venn diagram defining the set of the human family. We are all 'inside', all potentially 'at risk', because the human immunodeficiency virus (HIV) respects no one. Then, within that boundary, within that set of reified 'us', equivalence relations may partition the set into non-overlapping subsets. The elements of the human family divide; their intersections are null. As long as the partitions are maintained, the HIV is confined to infected elements capable of transmitting HIV only to others within the same subset. Thus, this abstract, reifying approach immediately centres our attention on barriers, those null sets of intersections, those breaks in the connective tissue, that tear apart the backcloth structure, and so produce the obstructions required to prevent the transmission of an HIV traffic. Bounds, those no-man's-lands of null intersections, are barriers, and barriers are bounds – from the human skin to international regulations, from the latex membrane of a condom to the walls of a quarantine hospital. Geographers acknowledge that processes appear at many spatial scales, and eventually we shall have to write the 'geographies' of the HIV and cell, the human body, and all of the human groups and institutions involved in a worldwide pandemic. All are made up of sets of elements ultimately influenced and structured by human relations – sexual, addictive, marital, medical, powerful, institutional, bureaucratic, governmental. . . many relations structure the human family, and so make and break structures upon which the HIV can travel.

For many reasons, not the least sheer ignorance, we cannot examine most of these in any detail, and our attention here must be confined to the diffusion of the AIDS epidemic at the mesoscale of the region. Nevertheless, it is important not to lose sight of what is going on at finer, more detailed scales, at scales of interpersonal and intergroup relations that ultimately form their own 'micro-geographies' and processes of spatial diffusion. These are controlled by barriers and 'breaks in the structure' appropriate to the spatial scales involved, but collectively they work their way up through a hierarchy of aggregating relations to the regional scale at which we are allowed, for reasons of medical confidentiality, to observe and analyse them.

1 MICRO-SPATIAL BARRIERS

The most powerful barrier to the transmission of the HIV at the micro-spatial scale is the human skin. As long as this can be maintained intact, the HIV in blood, semen and saliva appears unable to pass from one human body to another. Unfortunately, there are many possibilities of penetrating this barrier to produce a lesion through which the HIV can pass. The surest way is to share a hypodermic needle with an infected person. Needle sharing by addicts on intravenous drugs is well known, and the results are devastating in the United States. Out of a total of approximately 130,000 (July 1990) AIDS cases, nearly 31 per cent involved IV drugs at some point along the chain of transmission. For women alone, 71 per cent of the cases were IV drug related, either directly or through heterosexual transmission with a drug user. Athletes using steroid injections open up alternate pathways: a large national survey disclosed that about 30 per cent of the athletes in training use steroids, many subcutaneously.

Needle lesions also occur in legitimate uses. In many Third World hospitals, no facilities for sterilization are available, and transmission from infant to infant is common, similar to a chain of transmission through thirty children traced to a rural hospital in the Soviet Union. Four of these children then transmitted HIV to their mothers through lesions resulting from breast feeding. In parts of Africa, the World Health Child Immunization program has suffered severe setbacks. Not only is unsterilized needle sharing common, but attenuated vaccines, normally producing an immune response in a healthy child, may actually give the disease to a child whose immune system is already down from HIV. Surgeons operating on AIDS or HIV-infected patients are at particular risk: working quickly with surgical instruments in confined spaces means that 'sticks' are almost inevitable, producing a lesion surrounded by infected blood, and protective gloves thick enough to form barriers reduce the digital sensitivity that is necessary for operations. Needles are also used in blood transfusions and, while the needles breaking the skin barrier may be clean, the blood or blood product may be infected. In the early years of the epidemic, the US, for-profit blood industry, advised by the medical profession, noted that HIV could not be transmitted by blood transfusions, and anyway it was too expensive to test each pint of blood for HIV. It actually costs the US armed forces $ 4.31 to test a blood sample. Today, over 3,200 people have converted to AIDS as a result of blood transfusions. Similarly, 1,171 haemophiliacs receiving injections of Factor 8 blood clotting fraction have full-blown AIDS. It takes many pints of blood to make one clotting injection, so that, even with testing at 0.999 certainty, there is still a $(1-(0.999)^{200})$, or a 20 per cent, chance that the HIV will get through. With no testing, the probability of HIV transmission is mathematically indistinguishable from certainty.

Sexual transmission is also common. Of AIDS conversions in the United States, 67 per cent are homosexual or bisexual, but the homosexual community was only the first group to take the initial brunt of the epidemic. Anal tissue is extremely thin and easily subject to tearing, again producing lesions in the skin barrier for HIV to pass. This barrier is now being reinforced by latex condoms: for example, in 1984, new infections in the San Francisco homosexual community were at 19 per cent; by 1989, they were virtually zero. Unfortunately, heterosexual transmission is now rising relatively and absolutely, particularly among young people. Recent surveys in New York and Miami disclosed HIV infection rates of 1 per cent in 15 – 16 year olds, 2 – 3 per cent in 21 year olds, and 7 per cent in teenage runaways. Infection was roughly the same between men and women, suggesting mainly heterosexual transmission.

These rates are an order of magnitude greater than those previously seen in the general population of this age cohort, where rates of 2 – 3 per thousand were recorded from preinduction medical examinations for military service (now over two million). Perhaps these rates are not so surprising in view of the approximately 70 per cent rate of sexual intercourse during the high school years, over 50 per cent unprotected by any contraception, let alone a latex condom barrier. There are also very high rates of some form of sexually transmitted disease in the late teenage/young adult population, producing sores and lesions in the skin barrier. Drug use, particularly crack, tends to enhance transmission. The immediate 'high' of crack produces heightened sexual desire, combined with a high degree of impotence. Prolonged intercourse produces ideal conditions for heterosexual transmission of HIV.

2 MESO-SPATIAL BARRIERS

All barriers, whether interpersonal or institutional, have been generated on a voluntary basis in the United States, with the exception of physical restraint, or solitary confinement, placed upon a few prisoners in gaol who have made deliberate attempts to infect others. Some of these have been prosecuted for attempted murder. Mass, compulsory testing for HIV appears to be an anathema to most citizens, and a national blood screening of 20,000 families, gathering detailed and totally confidential sexual histories, appears to have been derailed politically. It was recommended in early 1988 and was tried at the pilot project stage in the summer at selected cities. These trials caused such an uproar, particularly among ethnic communities where HIV rates are high, that the project appears to have been abandoned. Thus, we have no estimate of HIV infection in the general population which could be used with confidence to estimate future conversions to AIDS. Meanwhile, 97 per cent of the more than 3,300 counties in the country report HIV infected people, and in certain poor communities, for example, the Bronx of New York City, rates are approaching those of East Africa. For example, 20 per cent of women coming in for prenatal care are seropositive, roughly the same rate as Kampala, Uganda, and rates among users of shooting galleries (heroin and cocaine), crack houses, and prostitutes are even higher. One epidemiologist noted, 'We are being eaten from the inside out by spatio-temporal processes', meaning that urban decay and the firing of buildings in the Bronx (itself a process of spatial diffusion), was only symptomatic of socio-economic decay. 'The AIDS epidemic is riding on the back of the wave of increasing poverty,' he added.

As for quarantine, it is presently unthinkable in the United States, and raises an important question of perception. Processes of spatial diffusion, like any self-limiting process, including epidemics, tend to follow a now-classic logistic curve in the absence of intervention. Diseases with high rates of mortality (AIDS is virtually 100 per cent) and with short periods of incubation, say seven days, result automatically in quarantine measures to protect the larger society. There is an ample legal basis at the Federal level for such measures. Conversely, diseases with long periods of incubation, say seven years, produce no quarantine measures, the only institutional mesolevel barrier available to a nation. Since the logistic curve is controlled by the ratio of two parameters, a and b, there is clearly a range over which a decision can be made. If a/b is small, the incubation period is short, the disease spreads rapidly, and the logistic curve rises quickly. If a/b is large, the curve rises slowly. Somewhere in that numerical range the decision to quarantine infected people must be taken.

Cuba, with tens of thousands of young men coming back from Angola, has chosen to set up the barrier of quarantine, testing three million of its citizens in 1988, and the remaining ten million in 1989. Those with HIV or AIDS are quarantined at Los Cocos, a former army hospital south of Havana, on the grounds that a society has the right to protect itself against a deadly disease that would have produced widespread HIV infection, and an AIDS epidemic in the next decade. China and certain eastern European and central Asian countries may be taking the same measures.

3 MACRO-SPATIAL BARRIERS

At the international level, barriers may be placed at borders to prevent the passage of infected people. Recent certificates of testing for HIV may be required before a visa is issued, or entry is allowed. In the Soviet Union, for example, all exchange students are required to produce recent certificates, or must submit to testing by Soviet medical authorities before taking up their studies. Those found to be infected are immediately returned to their home countries.

Such macro-spatial barriers could have a major impact in slowing the spread of the HIV virus, because the evidence is now overwhelming that it started in East (Type I) and West (Type II) Africa and was spread by infected international travellers. The earliest European cases, now traced by blood serum and medical histories to the late 1960s, both had African contacts through surgery and sexual transmission. The major pathway for international transmission today is the air network, the structural backcloth upon which people and the HIV travel. For example, it is no accident that Abidjan, in the Ivory Coast, has been called the 'sexual crossroads' of Africa, because even a cursory plotting of direct flights from other Francophone countries discloses the huge node in the network. The flights are used extensively by government ministers, and their large accompanying entourages, to travel to Abidjan for 'conferences', followed by periods of rest and relaxation from their arduous duties. Since HIV is mainly an urban phenomenon, infecting the modernized élite out of all proportion to their numbers, it is hardly surprising to find rates in the Ivory Coast rising to East African proportions, with HIV I rapidly gaining on the local HIV II variety.

4 A GLOBAL PLAGUE

We are now approaching the end of the first decade of this new global plague, and yet it is the first epidemic in modern history for which we have virtually no idea of its extent in geographic space and, as a result, it literally appears 'distant' and remote from most people. After all (many assume), it is essentially a disease of homosexuals and intravenous drug users: 'I'm not a homosexual, I'm not a drug user... why should I worry? AIDS is something that happens to other people at other places ... it is not going to happen to me.'

Why is there this huge void in understanding and perception? In the United States, and certainly some European countries, part of the ignorance results from the confidentiality issue, the sense that individuals should not be identified except to fully qualified medical personnel. Geographic location, specified by a very fine coordinate system or street address, could become an identifier; and, let me make it quite clear, I believe that this would be wrong. In the examples that we shall examine below, the spatio-temporal data sets that we have used for the mathematical modelling of the diffusion of AIDS have not identified anyone. In fact, we have no need to identify anyone, and we would not know what to do with such information if we had it. The loss of individual confidentiality is a genuine fear, but it has been taken to extreme and absurd lengths, so absurd that a vital component of our understanding is now in total disarray. In the United States, the Centers for Disease Control report only by state – Texas, California, etc. – a geographic scale so huge that it is analytically useless. It is also scientifically ridiculous: Rhode Island fits into Texas about 250 times. In Europe it is the same story: for example, Sweden reports Stockholm, Göteborg, Malmö and 'Other', in a country over

349

1500 kilometres in length. The location of the disease is disguised very nicely. Yet we now have large published maps in the United States showing the rates of AIDS in Los Angeles at the census tract level of three or four blocks, a spatial scale five orders of magnitude below the state level reported by the Centers for Disease Control. The maps were compiled as an undergraduate cartographic project at the University of California, and no one examining the green or yellow or pink colours could possibly identify an individual.

But the lack of geographic understanding is also symptomatic of a larger geographic blindness that runs through society, even to professional epidemiologists. I have spoken to PhDs in epidemiology who did not have even an introductory course to spatial modelling in their entire university careers and did not know of its existence. All thinking runs down that time line, and they do not even ask how many AIDS patients can dance on the head of a pin, let alone where they might be in geographic space. At the mathematical modelling group of the White House conference which I attended in July 1988, one mathematician, with distinguished contributions to modelling in quantum mechanics, thought that 'spatial modelling' meant running around from Alabama to Wyoming with a differential equation clutched in his hot little hand.

There are really two traditions of mathematical modelling in epidemiology, both concerned exclusively with the temporal dimension. First, a statistical tradition which tries to extrapolate forward, but finds that the bands of confidence widen so quickly into the future that not much can be said. As one British epidemiologist at the White House conference noted: by 1992 there will be between 750 and 25,000 AIDS cases in the UK. One wonders whether a model is really necessary to make such a statement. One doctor at the conference, a man who had been at the forefront of AIDS care in a San Francisco hospital from the beginning, said: 'I can extrapolate just as well with a plastic drawing curve!' Nevertheless, this statistical tradition has done valuable 'backcasting', retrospectively modelling the early years of the epidemic about which we know very little. After all, the class of retroviruses was discovered only in 1977, and the HIV itself in the early 1980s.

The second tradition is the mathematical – essentially differential equations. In the United States, this tradition is trying to divide the population into finer and finer units: young, old, male, female, black, white, homosexual, bisexual, heterosexual, haemophiliac, etc. Thus, the number and size of the equations grow, requiring more and more estimations of intra- and inter-transmission rates, rates that are changing even as they are measured. How does one estimate the transmission rate of HIV from black, intravenous drug-using, heterosexual teenagers to old, Hispanic, bisexual men? The mathematical modellers say that the social and behavioural scientists will have to give them the rates. In the meantime, they will take the first eigenvalue of the Jacobian, and if it is greater than one they know the epidemic will spread. Or did they know this already, without such massive mathematical apparatus? I think that it is plain to see how this tradition, which is undoubtedly powerful and dominant at the moment, can quickly deteriorate into silly computer games of nonlinear systems. We must do more than create just another sandbox for scholars to play in at the public expense. We have got to do better than this, which means breaking habits of thought that are tied exclusively to the historical or temporal domain. To understand the epidemic, we must think about it in both space and time. Everyone must be a geographer, because predicting the next number coming to us out of the future is not enough. We must predict the next map or maps.

5 SPATIAL MODELLING FOR PLANNING AND EDUCATIONAL INTERVENTION

A wide variety of spatial modelling methodologies are available to seek out the structure in spatio-temporal series, and to use this information to extrapolate and predict future geographic configurations. This is not the place to go into a detailed mathematical exposition, but briefly the approaches are expansion methodologies, transformational methodologies, and spatial adaptive filtering. Expansion methodologies generally make the numbers of AIDS cases cubic functions of time, and then express the parameters of the temporal function in terms of a variable such as population density whose geographic variation can be easily specified. Thus, the cumulative number of AIDS cases becomes a function of both time and space.

Transformational methods map (in the purely mathematical sense) the conventional cartographic representation of a region into a multidimensional 'AIDS space', usually under some gravity model type of transformation. In the multidimensional AIDS space, large cities are close together and at the centre of a configuration of points representing the centroids of subareas, perhaps counties or communes. The epidemic is starting by blowing a multidimensional bubble around the first epicentre, and as it expands in the transformed space it captures other areas. These can then start their own infectious bubbles in the local part of the AIDS or transformed space. If the order of infection is mapped inversely back on to the conventional carto- graphic map, the epidemic is seen to jump down the urban hierarchy, and then to spread by spatially contagious diffusion into the surrounding areas.

The third approach, that of spatial adaptive filtering, takes a succession of time slices, and each time calibrates a linear or quadratic function on the AIDS cases in the surrounding areas. Thus, it explicitly takes into account the geographic structure of the epidemic, and by feedback mechanisms tries to calibrate the function for each subarea based on contiguous values. In the early years of an epidemic, the so-called 'seeding stage', there is little spatio- temporal structure, and therefore predictions are relatively poor. However, as the epidemic unfolds on the map, like a photographic plate developing, the spatial adaptive filter takes the emerging structure into account and becomes the most powerful predictor (in the sense of minimizing error terms) of the next map or maps.

6 THE USES OF SPATIAL PREDICTIONS

Predicting the next map or maps is not an idle academic exercise. Conservative estimates in the United States predict roughly 400,000 AIDS victims by 1992, and these will require enlarged medical facilities, including hospitals and hospices for terminally ill people. For humane reasons, these should be placed as close to the home areas of the people involved as possible. This is the well-known geographic allocation and assignment problem, an area of intense research over the past quarter of a century, for which many algorithms are now available. For all practical purposes the spatial allocation problem has been solved, but it still requires estimates of the spatial distributions of AIDS patients before suggestions can be made about the effective assignment of new health facilities. Thus, the map predictions are at the core of any sensible, not to say sensitive, health planning in the future.

The second area involves generating materials for educational intervention campaigns. We have discovered that animated maps for television, sequences that include the predicted maps, have a very high visual and emotional impact. It would appear, from demonstrations that we have given to teenagers, university students, and medical personnel, that these animated cartographic sequences provide precisely those 'cues to action' that people in health education invoke to produce personal self-reflection about a health danger. It is quite clear that providing information is not enough: most people require a very immediate and personal sense of the danger before they reflect and change their own behaviour. Animated map sequences on television literally 'bring it home'.

7 THE DIFFUSION OF AIDS IN OHIO AND PENNSYLVANIA

Realizing that geographic processes can be examined at a variety of spatial scales, we are obliged to examine the AIDS epidemic in Ohio and Pennsylvania at the spatio–temporal scale of the county and quarter. Thus, we are allowed to observe the epidemic through this particular spatial–temporal filter or 'window', and all of the micro-spatial 'Brownian movement' disappears into this aggregate view. Nevertheless, and even at this great aggregated level, a high degree of spatio-temporal order appears. In fact, simply by observing carefully the first maps in the sequence, one could make an intuitive prediction of the next, which would probably be very close to the actual. It is this intuitive sense of emerging spatio–temporal structure, the developing 'photographic plate', that the mathematical modelling tries to capture in precise, well defined and computable ways.

In discussing the two sequences, it is important to bear in mind that the contour interval is geometric, with each value triple the succeeding one. It is unfortunate that we have to show the sequences in black and white, because they are much more dramatic in colour, particularly when they are animated on a television screen. As a stage aside, we are learning rapidly that we have entered a new, and virtually unexplored area of cartography. Colours have to be very carefully chosen for what is, essentially, a rhetorical effect. After all, these maps are meant to persuade young people that the possibility of HIV infection is not something remote, but close at hand. Moreover, much of conventional cartographic design can be forgotten, because what is the necessity of text and lettering when a voice is available on television to point out and describe?

The first AIDS case was recorded in Ohio in 1981 at Cleveland, and by 1982 it had intensified and jumped down the urban hierarchy to nearby Akron and Canton, and then to Columbus, the capital in the middle of the state (Figure 22.1). By 1983, the spatially contagious 'wine stain on a table-cloth' diffusion has continued around the original epicentre of Cleveland, and we can now see two other regional epicentres emerging in Columbus and Cincinnati. One or two cases are recorded elsewhere, but we do not know if these transmissions are the result of sexual intercourse, IV drug use, surgical transfusion, or a haemophiliac injection. They could be a result of in-migration, a person crossing the boundary of Ohio, perhaps coming home to die.

By 1984, there is an intensification in the original regional epicentres, and we begin to get the first hint of a strong northeast–southwest alignment that follows a major corridor of high speed road, rail and air linkages. This is hardly surprising, because the HIV is carried by one person to another, so basic geographic patterns of human spatial interaction should appear as major corridors of infection.

By 1985, we see the diffusion moving outward from major epicentres, and the beginning of another alignment across the northern part of the state, again a major alignment of interstate highways linking secondary urban centres. By 1986, the three regional 'amoebas' have linked together, and the uninfected rural areas are being squeezed smaller and smaller. The process continues in 1987, the last year for which we had reliable data at the time that the sequence was constructed. There are inevitable delays in reporting, but the 1988 prediction appears to very close to the mark.

Pennsylvania is a rather different case: whereas Ohio is relatively flat, with multiple urban centres, Pennsylvania is cut in a northeast–southwest direction by the folded ranges of the ridge-and-valley section, an alignment that still strongly controls the road system. The urban hierarchy is strongly influenced by the two major centres of Philadelphia (in the southeastern corner), and Pittsburgh (in the west), two epicentres that show up immediately on the first map (Figure 22.2). As noted previously, the maps are

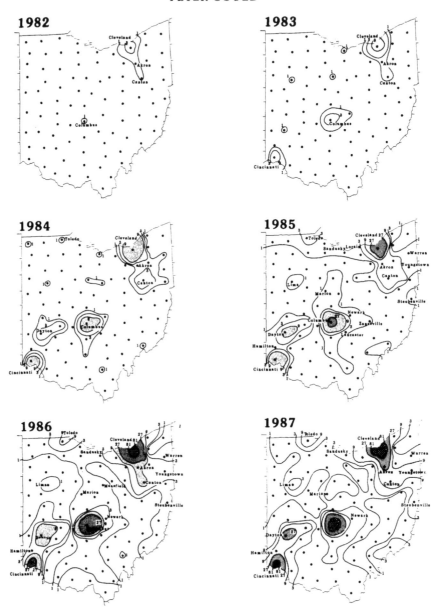

Figure 22.1   The diffusion of AIDS in Ohio, 1982–7

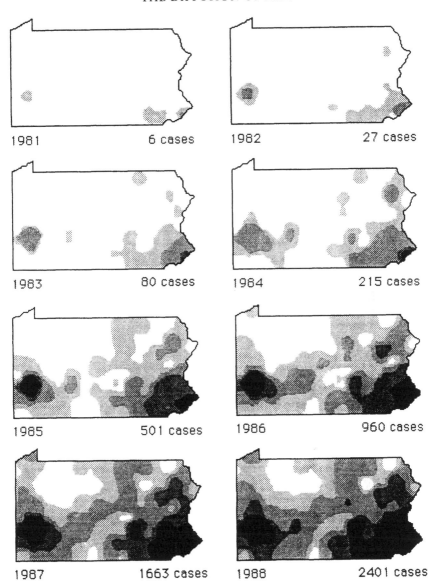

1981   6 cases

1982   27 cases

1983   80 cases

1984   215 cases

1985   501 cases

1986   960 cases

1987   1663 cases

1988   2401 cases

Figure 22.2   The diffusion of AIDS in Pennsylvania, 1981–8

very dramatic when animated in colour for television, and once again the contour interval is geometric.

By the end of 1981, six cases had been recorded, and notice that a few of these are just to the west of Philadelphia, directly in alignment with the Pennsylvania Turnpike. Once again the structuring of geographic space is crucial for understanding the spread of HIV and AIDS. For example, by 1982 the westward movement from Philadelphia has increased, and we can see the first effects of the northeast turnpike extension. Meanwhile, the epidemic is increasing in the major epicentres of Philadelphia and Pittsburgh.

In the following year we continue the 'developing photographic plate' effect, and by 1984 we can see hierarchical effects as the epidemic jumps to Scranton and Harrisburg, north and west of Philadelphia. Meanwhile spatially contagious effects are equally clear around the major and smaller epicentres.

By 1985, another slight 'channelling' effect in geographic space appears, this time east and northeast from Pittsburgh, following Route 220 to Altoona, State College, and Williamsport. The alignment is very clear by 1986, and by 1987 it is a distinct diagonal slash following the northeast–southwest alignment of the ridge-and-valley section of central Pennsylvania. By 1988, there are only a few areas left untouched, most of these in the north, with national and state forests, and very low population densities.

8 BREAKS AS BOUNDARIES AND BARRIERS

Boundaries and barriers are actually breaks in underlying structures, those backcloths necessary for the existence and transmission of traffic. The human immunodeficiency virus, considered as a traffic, needs human beings to exist, and it needs various sorts of connecting relations to form structures over which the HIV can be transmitted. We bound and bar transmission by creating breaks. In this way, by employing an abstract language of sets and relations, and the way in which these create algebraic hierarchies of structures, we see that the latex condom and the international health certificate form exactly the same structural obstructions, only at different geographical scales. Equally, the contaminated needle, and the infected airline traveller, form part of the many relations that connect people into transmitting structures. We know how to stop AIDS by creating boundaries and barriers at all scales, and we have had this knowledge since at least 1984. But there are also 'breaks', bounds and barriers in human thinking, reasoning and will. The question is: do we have the will to break, or the break in will? But, of course. Nietzsche was telling geographers that all along.

9 FURTHER READING

Publications on the geographic aspects of the AIDS epidemic are scarce. The following may prove useful.

AIDS, journal published monthly by Current Science, Cleveland Street, London, W1P 5FB, United Kingdom.

Bowen, W. (1989, 1990) 'AIDS in LA', two maps showing rates and cases by census districts, Northridge, CA: Department of Geography, California State University.

Gould, P. (1980) 'Q-analysis, or a language of structure: an introduction for social scientists, geographers and planners', *International Journal of Man-Machine Studies* 12: 169–99.

Gould, P. (1991) 'Modelling the AIDS epidemic for educational intervention', in R. Ulack and W. Skinner (eds) *Common Threads: AIDS and the Social Sciences*, Lexington, KY: University of Kentucky Press.

Gould, P., DiBiase, D. and Kabel, J. (1990) 'Le SIDA: la carte animée comme rhétorique cartographique appliquée', *MappeMonde* 1: 21–6.

Gould, P., Gorr, W., Golub, A. and Kabel, J. (1991) 'AIDS: predicting the next map', *Interfaces* 21: 80–92.

Shilts, R. (1987) *And the Band Played On*, New York: St Martin's Press.

Wallace, R. (1988) 'A synergism of plagues: "planned shrinkage", contagious housing destruction and AIDS in the Bronx', *Environmental Research* 47: 1–33.

INDEX

Printed and bound by CPI Group (UK) Ltd, Croydon, CR0 4YY

18/10/2024

01776243-0014